Stable Isotope Ecology

T0178060

Stable Isotope Ecology

Brian Fry

Springer

Brian Fry
Department of Oceanography and
    Coastal Sciences
Coastal Ecology Institute
School of the Coast and Environment
LSU
Baton Rouge, LA 70803
bfry@lsu.eu

Additional material to this book can be doawnloaded from http:// extras.springer .com

ISBN-10: 1-4899-9359-2   ISBN 0-387-33745-8 (eBook)
ISBN-13: 978-1-4899-9359-5

Printed on acid-free paper.

9 8 7 6 5 4 3 (corrected as of 3rd printing, 2008)

springer.com

I dedicate this book to my parents, Lois and Arthur Fry, who passed down to me their love of science.

Acknowledgments

Colleagues, friends, and family reviewed portions of this book. But any mistakes you find are due to the author, not to the careful vigilance of the reviewers. I thank all the reviewers: Valerie Allain, Edward Castaneda, Robert S. Carney, Matt Chumchal, Nicole Cormier, Kim de Mutsert, Amanda Demopoulos, Lawrence Febo, Marilyn Fogel, Arthur Fry, Marian Fry, Kari Galvan, Brittany Graham, John Hayes, Ayla Heinz-Fry, Jane Heinz-Fry, Caleb Izdepski, Natalie Loick, Skyler Neylon, Robert Olsen, Nathaniel Ostrom, Peggy Ostrom, Patrick Parker, Marie Perkins, Bruce Peterson, Brian Popp, Terri Rust, Alison Salmon, Joris Van der Ham, Maren Voss, Joseph Shannon, Michael Sierszen, Christopher Swarth, Chad Thomas, Len Wassenaar, and Jonathan Willis. Jake Vander Zanden and Simon Costanzo provided graphs used in the text. Todd Dawson supplied up-to-date information about measuring isotope values with laser technology. Charlotte Cavell and Sara Green assisted with graphics. Terry Wimberly assisted with manuscript preparation. Janet Slobodien and Ann Avouris provided editorial guidance at Springer. I thank Louisiana SeaGrant the NOAA Coastal Ocean Program for financial support during preparation of this book.

Contents

Supplemental Electronic Materials on the Accompanying CD

A. Chapter 1
   Color Figures and Cartoon
   Problems

B. Chapter 2
   Color Figures and Cartoon
   Problems
   Technical Supplement 2A: Measuring Spiked Samples
   Technical Supplement 2B: Ion Corrections
   Technical Supplement 2C: The Ratio Notation and The Power of 1

C. Chapter 3
   Color Figures and Cartoons
   Problems

D. Chapter 4
   Color Figures and Cartoons
   Problems
   I Chi Spreadsheets

E. Chapters 5
   Color Figures and Cartoons
   Problems
   I Chi Spreadsheets

F. Chapter 6
   Color Figure and Cartoon
   Problems
   I Chi Spreadsheet
   Technical Supplement 6A: How Much Isotope Should I Add?
   Technical Supplement 6B: Noisy Data and Data Analysis with
      Enriched Samples

1
Introduction

1.1 Discovery

I remember the first time I discovered the power of stable isotopes. It was by accident. It was thirty years ago, when I was a beginning graduate student along the south Texas coast. That summer I helped a visiting professor collect rodents (mice, rats, and ground squirrels) in a coastal sand dune community. Yes, I worked with the rodent traps, but I also got bored and wandered off during the hot afternoon hours, collecting plants and grasshoppers from the dunes. One evening later in that summer of 1976 we were at the mass spectrometer, watching the chart recorder display the isotope results for our collections. It was fascinating. One sample was very enriched in the heavy carbon stable isotope, ^{13}C, and the next sample was depleted in ^{13}C. A great divide was evident in the isotopes of the sand dune community. We watched the chart recorder for hours as sample after sample showed the basic ^{13}C distinction, or variations on this ^{13}C isotope theme.

We had discovered something new, something unexpected. Plants fell into two categories, recognizable as the C_3 and C_4 plants that had been described recently as distinct plant types (Bender 1968; Smith and Epstein 1971). But grasshoppers also fell into these same categories, so that the different grasshopper species were specialists in their diets. Rodents were intermediate in their ^{13}C values, and therefore dietary generalists. That evening at the mass spectrometer a food web took shape. And I realized that as I had wandered the dunes, making my collections, I had missed a fundamental order in the natural world. The isotope measurements showed carbon connections and flows in the sand dune community, from plants to specialist grasshoppers and generalist rodents (Fry et al. 1978). I could not perceive the isotopes with my senses, nor could many months of observation have shown so powerfully the distinctive C_3–C_4 structure of that coastal food web. The isotopes illuminated an unknown ecology. My experience was and is not unique, for scientists worldwide recognize that there is an "isotopically ordered world" (Wada et al. 1995) within ecological systems. As you embark on your own ecological adventures, I wish you good discovery with the isotopes. They will help you see and test ecological interactions, powerfully tracing otherwise invisible connections.

1.2 General Introduction

Ecologists make many types of measurements to understand ecological systems, measurements such as length, timing, or pH. Isotope measurements are chemical measurements that allow detailed nuanced views of element cycling in all systems that interest ecologists. It takes some time and practice to learn how to use isotopes, just as it takes time and practice to learn how to use a taxonomic key or how to use statistics. Learning to work with isotopes is usually time well spent, for it provides a different way to view ecological connections, a distinct tracer-based perspective that often leads to new discoveries.

The goal of the book is to help you use isotope tracers in correct and creative ways to solve environmental problems. This book works on accomplishing this goal in several ways. First, it focuses on fundamental principles of mixing and fractionation that govern isotope circulation in the biosphere, and aims to help you understand and use these principles. Second, it presents a new modeling approach that is fairly simple to use with computer spreadsheets. This approach enables you to circulate isotopes for yourself as you sit at your desk, mimicking in virtual reality the ways that isotopes circulate in nature. Lastly, this book presents many stories and illustrations to help you learn, hear, and speak the rich isotope language that permeates the natural world. Although the book is factual in essence, it seeks to stimulate your creativity.

A complete course of study for a graduate or post-graduate student interested in learning stable isotope ecology might include three elements: reading this book, finding and reading current research articles as supplements, and planning and carrying out her or his own pilot project using stable isotopes (Box 1.1). You should also realize that no matter what you encounter here, it is only a small part of what there is to learn about isotopes, so many sections point you towards further readings.

This book also emphasizes "learning by doing" in the living examples portrayed in many spreadsheet models. You can read about these models in the book, then go to the accompanying CD and open up the computer models to interactively enter parameter values and watch the isotope action unfold before your eyes. This book also contains a more traditional type of learning by doing, problems posed for each chapter to increase your isotope expertise.

For better and worse, this book reflects the author's experience and bias garnered over thirty years, and so focuses primarily on isotopes of the three elements, carbon, nitrogen, and sulfur. The majority of examples concern aquatic ecology. Nonetheless the book expounds and emphasizes

Box 1.1. How to Start Your Own Isotope Project

The most meaningful way to get started in stable isotope ecology is usually to design and carry out your own pilot project. This is much easier than you might think. Analyzing 10 to 25 samples is often enough to see if your idea is worth pursuing, and costs less than $300. Here are some of the steps you might take for your project.

1. Think about what interests you. Remembering that isotopes are everywhere and in everything, think about how isotopes might be circulating in something that interests you.

2. Literature review. Go to the library or check the Internet and the Web of Science to see if anyone is doing related work. Type in search words such as "isotope," ^{15}N, ^{13}C and the like and see what you find.

3. Think about a field site. What kinds of samples are available at your site that would be interesting to sample?

4. Find an isotope lab. Look up isotope laboratories on the Internet for prices and how to prepare samples.

5. Contact the isotope lab and discuss your idea. People working at the isotope lab may run a few trial samples at no cost, especially if they know you are just trying to get started and you seem to have a good idea. They will have some good advice in any case. If the isotope laboratory is nearby, go visit and see what is really involved with sample preparation.

6. Collect and analyze samples. Plant, animal, and soil samples can be collected by hand without fear of contamination, and small samples <100 mg typically suffice for isotope work. Laboratory preparation is also simple, typically drying at 60°C in an oven, then grinding to a fine power. Your contacts at the isotope lab will help you with specialized sampling, and how much sample you need to weigh out for the actual analysis.

7. Data interpretation. When results come back from the isotope lab, examine the data to see if your ideas worked out. Make this project an activity for one of your classes, so that you share the results and get some feedback discussion from your fellow students about your interpretations.

Following steps 1 to 7 is the fast track to learning about isotopes, and will probably teach you more than you will learn from reading several books (including this one). Louis Agassiz, one of the great scientists of the 19th century put it this way: "Study Nature, Not Books." (You can see this framed handwritten motto in Woods Hole, Massachusetts when you walk into the main library at the Marine Biological Laboratory. The motto provides a somewhat ironical introduction to a great collection of books and journals.) Overall, an isotope project will put you in touch with the world's greatest teacher, the natural world. And if you work through these seven project steps, you deserve promotion to the ranks of the working isotope scientists.

a centralized view of isotope circulation in the biosphere that underlies all branches of isotope research, so that the book is intended to be useful to all scientists working with stable isotope tracers.

This book generally has a very informal, conversational style, and represents my attempt to make a technical subject accessible in an uncomplicated and often fun way. You can read about my own scientific evolution in Section 1.3, and how I earned the nickname "Mr. Polychaete." But Mr. Polychaete has his own persona in this book where he offers wise isotope advice with a bit of humor and a wiggle. He is an isotope enthusiast who likes to laugh, and might have called this book *Isotope Sorcery for Ecologists* or *A Traveler's Guide to the Mysterious and Wondrous Realm of Isotopia*. Mr. Polychaete appears in this book to help explain isotopes to ecologists. Mr. Polychaete will also let you know when detailed technical sections are present that can be skipped during the first reading of this book.

A Little Encouragement for the Novice

Starting out as an amateur in the science of isotopes may seem daunting, but in reality, being an amateur is often a good thing in science. Here is something for you to think about, quoted from C.H. Hapgood's book, *Maps of the Ancient Sea Kings*.

Every scientist is an amateur to start with. Copernicus, Newton, [and] Darwin were all amateurs when they made their principal discoveries. Through the course of long years of work they became specialists in the fields which they created. However, the specialist who starts out by learning what everybody else has done before him is not likely to initiate anything very new. An expert is a man who knows everything, or nearly everything, and usually thinks he knows everything important, in his field. If he doesn't think he knows everything, at least he knows that other people know less, and thinks that amateurs know nothing. And so he has an unwise contempt for amateurs, despite the fact that it is to amateurs that innumerable important discoveries in all fields of science have been due ... when a difficult problem was being discussed, Thomas A. Edison said it was too difficult for any specialist. It would be necessary, he said, to wait for some amateur to solve it.

In other words, amateurs are great for their new thinking and initiatives. Your new start in this technical world of isotopes is welcome.

Isotopes and Their Elements

Isotopes are forms of the same element that differ in the number of neutrons in the nucleus. Extra neutrons in the nucleus of an element generally impart only subtle chemical differences, small differences that keep almost identical isotopes from being truly identical. In the world of chemicals, the real differences among elements lie in the numbers of protons and electrons (Figure 1.1). The negatively charged electrons react to form the bonds between atoms. The electrons also balance the number of positively charged

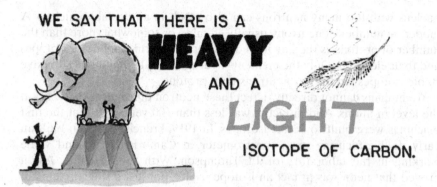

WE SAY THAT THERE IS A

HEAVY

AND A

LIGHT

ISOTOPE OF CARBON.

^{13}CARBON HAS ONE
MORE NEUTRON THAN
^{12}CARBON IN ITS NUCLEUS.

IN MOST CASES ^{12}CARBON AND ^{13}CARBON
BEHAVE THE SAME BECAUSE EXTRA NEUTRONS
DON'T CHANGE THE REACTIVE SPHERE OF
ELECTRONS AROUND THE NUCLEUS.

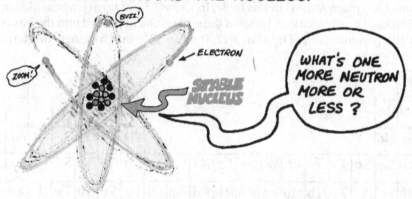

FIGURE 1.1. An extra neutron in the ^{13}C isotope makes the nucleus more massive or "heavier" than the ^{12}C isotope, but does not affect most chemistry that is related to reactions in the electron shell.

protons in the nucleus. So with this positive–negative balance in good working order, you might wonder who needs neutrons? The nontechnical answer is that having one or more neutrons is important because neutrons are the peacekeepers of the nucleus, keeping the highly charged, mutually repulsive protons from getting too close together. But overloading the nucleus with too many neutrons can also result in an unstable nucleus. A moderate number of neutrons, usually equal to or somewhat more than the number of protons, is the key ingredient for long-term stability of isotopes and their elements. Only the elements hydrogen (H) and helium (He) have stable isotopes with fewer neutrons than protons.

We humans cannot directly detect these neutron differences that exist at the level of atoms. And in fact, it was less than 100 years ago that the first machines were built to detect isotopes. In 1919, Francis W. Aston built an early double-focusing mass spectrometer in Cambridge, England while working in the laboratory of J.J. Thompson. With this machine, Aston proved that neon was in fact an isotope triplet, not just a singlet element. All three members of the neon triplet had the same number of electrons and protons, but extra neutrons accounted for the neon triplet at masses 20, 21, and 22. Each neon isotope had 8 electrons and 8 protons, but had different numbers of neutrons. There were 12, 13, or 14 neutrons in the neon isotopes, making the atomic number (protons + neutrons) totals come up to 20 (= 8 + 12), 21 (= 8 + 13), and 22 (= 8 + 14). Aston went on to discover that many elements come in more than one flavor or isotope form, and quickly received the 1922 Nobel Prize in Chemistry for his efforts (see: http://nobelprize.org/chemistry/laureates/1922). In the same year, he also published the seminal book in this area, *Isotopes*.

The word "isotope" comes from consideration of the periodic table of the elements (Figure 1.2), and means that isotopes of an element all occupy the same (*iso*) place (*topos*) in this table. In a sense, isotopes are whole hidden families of nearly identical forms of the same element packed into the boxes of the periodic table. Dr. Margaret Todd of Edinburgh coined the term

H																	He
Li	Be											B	C	N	O	F	Ne
Na	Mg											Al	Si	P	S	Cl	Ar
K	Ca	Sc	Ti	V	Cr	Mn	Fe	Co	Ni	Cu	Zn	Ga	Ge	As	Se	Br	Kr
Rb	Sr	Y	Zr	Nb	Mo	Tc	Ru	Rh	Pd	Ag	Cd	In	Sn	Sb	Te	I	Xe
Cs	Ba	La	Hf	Ta	W	Re	Os	Ir	Pt	Au	Hg	Tl	Pb	Bi	Po	At	Rn
Fr	Ra	Ac															

FIGURE 1.2. An abbreviated periodic table of the elements. Many elements have more than one isotope variety and differ in the number of neutrons. Stable isotopes of the circled elements HCNOS (hydrogen, carbon, nitrogen, oxygen, and sulfur) are emphasized in this book. Details about isotopes for many of these elements are available at the Web site http://wwwrcamnl.wr.usgs.gov/isoig/period/.

"isotope" (see footnote 15–90 in Dahl's *Flash of the Cathode Rays*), and it came to be preferred over a rival word, "pleiad," used for a group of stars recognized as sisters by the ancient Greeks. Frederick Soddy first introduced the term "isotope" in a formal way during a speech to the British Royal Society on February 27, 1913. Soddy was one of the first to suspect the presence of isotopes, and went on to win the 1921 Nobel Prize in Chemisty for "his investigations into the origin and nature of isotopes" (see http://nobelprize.org/chemistry/laureates/1921).

Today we know that all elements have multiple forms or isotopes. The ultimate source of elements and isotopes lies in the Big Bang that started our universe and in stars where nuclear reactions of fission and fusion produce new atoms (Penzias 1979, 1980; Broecker 1985; Clayton 2003). Given these sources, it is perhaps not surprising to learn that the fundamental proportions of heavy and light isotopes here on earth derive from ancient syntheses billions of years ago. Further reactions in the cold emptiness of interstellar space somewhat modify these proportions, especially for hydrogen isotopes (Clayton 2003). And radioactive decay on earth is also active, slowly but surely producing a mildly distinctive local mixture of elements and isotopes on this planet. At the time of this writing, scientists on our planet Earth recognize or suspect the existence of approximately 120 elements and 3100 accompanying isotopic forms or nuclides. However, most of the 3100 nuclides are very short-lived radionuclides. Among the nuclides, there are only 283 stable isotopes that do not undergo radioactive decay, so that stable isotopes comprise <10% of all known isotopes.

The term nuclide comes from a physicist's appreciation of fission and fusion processes in stars and Earthside reactors. For these nuclear reactions, the total number of neutrons plus protons (the nuclide number) is the important aspect of an element, not the chemistry deriving from electron interactions. Physicists keep track of the nuclides in a "chart of the nuclides." This chart is a fundamental starting point for understanding the physics of elements and isotopes, just as the periodic table is a fundamental starting point for chemistry. In the chart of the nuclides, all the isotopes are listed in plain view in separate boxes. There is quite a bit of chemical and physical "biographical" detail listed in each box, summarizing basic facts for each isotope. So in the chart of the nuclides, the isotopes are liberated from their crowded, hidden existence in the same place or *isotopos* element compartments of the periodic table. The chart of the nuclides is the reference to consult when you want to find out details about an isotope. Clayton (2003) provides excellent isotope-by-isotope information in a reader-friendly commentary on the early part of the chart of the nuclides that includes the elements of most interest in this book.

A current listing of the periodic table of the elements is available on the Web at http://pearl1.lanl.gov/periodic/default.htm. Current versions of chart of the nuclides are given at http://wwwndc.tokai.jaeri.go.jp/CN03/index.html and http://t2.lanl.gov/data/nuclides/map8.html. The number of stable isotopes given above as 283 comes from counting the stable isotope nuclides listed in those charts. Ecological applications for isotopes of many of these elements are described on numerous Web sites, for example, at http://wwwrcamnl.wr.usgs.gov/isoig/period/ and http://ecophys.biology.utah.edu/sirfer.html.

The most famous isotopes are undoubtedly the uranium isotopes used in nuclear reactors and in atomic bombs. The common form of uranium (^{238}U) has three more neutrons than the rarer, more reactive form (^{235}U), and much of the secret of dealing with uranium is figuring out how to separate these two isotopes which generally have very similar chemical qualities. You may hear in the news that nuclear proliferation worries often focus on centrifuge technology, because centrifugation is one of the ways that you can separate a more massive isotope twin (^{238}U) from its lightweight uranium counterpart (^{235}U). Radioisotopes such as these two uranium isotopes emit various kinds of particles and decay or change into other elements, liberating energy that can be used for both destructive and beneficial purposes.

But this book focuses on a different set of isotopes, the stable isotopes that persist in the same form for eons after they are formed. The stable isotopes have survived over billions of years of geological time here on Earth. They provide some of the few surviving records about early life on Earth and the early ecology of our planet. Stable isotopes are safe isotopes that do not decay and unlike the radioactive isotopes, are not at all hazardous to human health. In fact, stable isotopes are quite abundant and natural parts of each one of us (Figure 1.3; Wada et al. 1995).

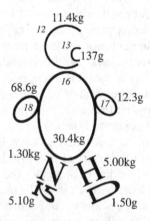

FIGURE 1.3. You are what you eat: stable isotopes in a 50 kg human who is composed mostly of light isotopes with a small amount of heavy isotopes. People are mostly water, so hydrogen and oxygen isotopes dominate at >35 kg. Next come C isotopes at >11 kg, then N isotopes. S isotopes are missing; they should be here at about 220 g for the light isotope ^{32}S and 10 g for the heavy isotope ^{34}S. Have you had your isotopes today? (From Wada and Hattori, 1990; reproduced with permission of CRC Press LLC.)

TABLE 1.1. Isotopes for the Light Elements HCNOS (Hydrogen, Carbon, Nitrogen, Oxygen and Sulfur).[a]

Element	Isotope Abundance				Mass Difference[b] (Relative)	Range in δ[c] (‰)
	Low Mass		High Mass			
Hydrogen[d]	^1H	99.984	^2H	0.016	2.00	700
Carbon	^{12}C	98.89	^{13}C	1.11	1.08	110
Nitrogen	^{14}N	99.64	^{15}N	0.36	1.07	90
Oxygen	^{16}O	99.76	^{18}O	0.20	1.13	100
Sulfur	^{32}S	95.02	^{34}S	4.21	1.06	150

[a] For each of these elements, the low-mass or "light" isotope is by far the most abundant of the isotopes, >95%. These fundamental isotope abundances prevailing on our planet Earth were determined long ago during element synthesis at the start of our universe, in interstellar space and in stars (Penzias 1979, 1980; Clayton 2003).
[b] Mass difference = high mass/low mass, e.g., 2/1 = 2 for the hydrogen isotopes.
[c] The listed range in δ values is representative for most natural samples that have not been artificially enriched with heavy isotopes (data from Anderson and Arthur, 1983). δ values are the common way to express isotope abundances (see Section 2.1).
[d] Hydrogen isotopes especially are in a different class in the isotope world, with large fractionations associated with the large 2× mass difference between protium (^1H) and deuterium (^2H, or also "D").

Stable isotopes often have skewed distributions on Earth, mostly reflecting details of their synthesis long ago in stars. For example, the lightest stable isotope accounts for more than 95% of all the isotopes for elements such as hydrogen (H), carbon (C), nitrogen (N), oxygen (O), and sulfur (S) (Table 1.1). But the reverse is true for some elements such as boron (B) and lithium (Li) where the heavy stable isotopes are the abundant isotopes, >80% of the total. Only a few elements such as bromine (Br), silver (Ag), and europium (Eu) show a roughly equal, 50–50, distribution between light and heavy stable isotopes. The element tin (Sn) has the most stable isotopes (10 isotopes), and there are elements such as fluorine (F) and phosphorus (P) that are endowed with only a single stable isotope form. Ecologists can only regret that stars did not make a second stable P isotope so that we could use differences between the isotope pair to track natural P dynamics in the biosphere.

But what about radioactive phosphorus, ^{32}P? In fact, ecologists added ^{32}P to terrestrial and aquatic ecosystems in the 1950s and 1960s to study P dynamics (reviewed by Odum, 1971). But today ecologists generally refrain from introducing radioactive isotopes into outdoor field settings. Instead, they study natural radiotracer distributions, or increasingly add stable isotopes instead to field experiments. There are also many medical uses for the nontoxic, safe stable isotope tracers (Boutton 1991; Fischer and Wetzel 2002).

For ecologists, stable isotopes provide a natural way to directly follow and trace details of element cycling. The isotopes function as natural dyes, colors, or tracers and their use can resolve many environmental problems. We use special machinery, especially an improved version of Aston's early mass spectrometer, to follow these isotope colors, like using special 3-D glasses at the movies to follow the action. Chapter 2 considers details of isotope measurements and the δ notation used to express measured isotope values. The special measurements let us follow the origins and fates of the many elements as they circulate in the biosphere. Elements of particular interest in today's environment are those that cycle tightly with organic matter (Table 1.1), especially H, C, N, O, and S. All these elements are blessed with two or more stable isotopes, that is, isotope twins, triplets, and quadruplets (Figure 1.4). The HCNOS elements are generally lightweights among the elements, but comprise most of the mass present in organic materials (see, e.g., Figure 1.3). Stable isotope studies of the sources and cycling of the HCNOS elements are sometimes supplemented by natural

FIGURE 1.4. This book focuses on the five elements (hydrogen, carbon, nitrogen, oxygen, and sulfur) and their 13 stable isotopes.

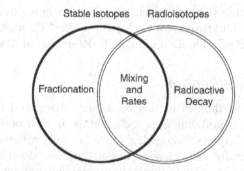

FIGURE 1.5. Stable isotopes are especially valuable for studying the origins and cycling of organic matter in the biosphere. Ecologists also use radioisotopes (especially ^{3}H, ^{14}C, and ^{32}P) to study cycling rates and to determine ages. Stable isotopes that pose no health risk are increasingly substituted for the radioisotopes in ecological and medical research.

radioisotope (especially ^{14}C) studies about rates of element cycling (Figure 1.5). Studies of natural radioisotope distributions are extremely valuable (e.g., Broecker and Peng 1982; Schell 1983) but the analytical work involved is currently too costly for most routine ecological applications. Fortunately, advances in measurement technologies have made HCNOS stable isotope analyses affordable for current ecological work, with costs in the $5–15 range for most samples.

This book deals with the HCNOS stable isotopes in the context of organic matter dynamics, and Chapter 3 gives five introductory reviews of how ecologists use isotopes in ecological research. The ecological focus of this book supplements other books that deal with isotope applications in geology (Hoefs 2004; Faure and Mensing 2004), H and O isotope applications in hydrology (Clark and Fritz 1997; Kendall and McDonnell 1998; Criss 1999), and six books of collected papers that deal with aspects of stable isotope ecology (Rundel et al. 1988; Ehleringer et al. 1993; Lajtha and Michener 1994; Wada et al. 1995; Griffiths et al. 1997; Ehleringer et al. 2005).

Isotopes are something like a mysterious hidden language written everywhere in the common chemicals and compounds circulating in the biosphere. We live surrounded by isotopes, in a sea of isotope information, with isotopes appearing even in the alphabetical letters of this text you are reading. The fundamental isotope information actually exists at an extremely fine and detailed level, on an atom-by-atom or "position-specific" basis within molecules (Rossman et al. 1991; Brenna 2001). But this book aims to help you work with a more aggregated isotope language used for ecological work with plants, animals, soils, and gases, so that you can start to write your own isotope ecology stories. Besides the HCNOS elements that are the focus of this book, isotope languages for other elements such as silicon (Si), calcium (Ca), and iron (Fe) are also being deciphered and read, yielding important results for ecological studies (Clementz et al. 2003; Varele et al. 2004; Basile-Doelsch et al. 2005; Rouxel et al. 2005).

Mixing and Fractionation

The two overarching themes of this book are mixing and fractionation. Chapter 4 outlines a modeling approach that deals with both mixing and fractionation, and Chapters 5 and 6 deal more carefully with mixing dynamics. Mixing is generally very easy to understand: we do this in cooking recipes every day, so it is easy to think about mixing isotope flavors. Mixing combines substances into a homogeneous whole.

But we also turn to the more difficult subject of fractionation, where the effects of an extra neutron are considered more closely. As the chemists and physicists studied isotopes, they found whole ranges of subtle behavior where an extra neutron or two made a detectable difference. For example, it turned out that right at the heart of the chemical universe, in the making and breaking of bonds, isotopes were not behaving exactly alike. Here the emphasis is on *exactly*, because it was very, very close, but still not *exactly*. Where bonds are forged or broken in forward-moving reactions, these slight rate or kinetic differences are important, leading to the important rule that

In kinetic reactions, the light isotopes usually react faster

(Figure 1.6). This differential isotope behavior known as fractionation means isotopes do not react exactly alike, as considered in detail in Chapter 7. In addition to this kinetic rule for reactions that move forward at slower and faster rates, there is also an important equilibrium rule for exchange reactions in which reactions proceed both forwards and backwards, eventually coming to a balanced equilibrium. This second rule is that

In exchange reactions, heavy isotopes concentrate
where bonds are strongest.

This differential concentration during exchange reactions is also a type of fractionation. Fractionation is the hidden power controlling isotope distributions on this planet, and the fundamentals of fractionation are in the chemical details.

Today, chemists can calculate maximum potential fractionations for many reactions, but unfortunately these detailed calculations are often of little help in understanding isotope distributions in most biological systems. The detailed chemical understanding is difficult to scale up for complex biological systems, so ecologists have been using isotopes in more empirical ways. Chapter 7 deals with these chemical and ecological approaches to understanding fractionation, why isotopes don't react exactly the same even though they are nearly identical in chemical structure. Most modern isotope studies focus on whole organisms or molecules, but with future technolog-

SOMETIMES THE EXTRA NEUTRON MAKES A DIFFERENCE. IT'S HARDER TO PUSH THE HEAVY MOLECULES UP AN ENERGY HILL...

... SO THAT PRODUCTS HAVE MORE OF THE LIGHT ISOTOPE AND LESS OF THE HEAVY ISOTOPE.

FIGURE 1.6. The extra neutron does make a very slight difference in some reactions; having an extra neutron usually results in slower reactions. This reaction difference is fractionation.

ical advance, the focus may shift downwards to the fundamental atomic level, where fractionation routinely alters isotope compositions of the atoms within molecules.

In reality, fractionation and mixing are just two sides of a coin, with fractionation acting to separate isotopes whereas mixing reunites them. These two processes of fractionation and mixing oppose and complement each other. You can think that these two processes operate separately and independently (Figure 1.7), but in reality they are linked in most contexts (Figure 1.8). They combine to continually recycle isotopes in the natural world.

Many ecologists do not like the complexities of fractionation. But fractionation is perhaps a necessary evil that sets the stage for mixing and tracing. Without fractionation, there would be only a uniform boring distribution of isotopes. Fractionation creates the artist's palette, the isotope colors that are later mixed and arrayed to form the grand isotope masterpieces of nature. Although many scientists think of two source materials uniting to create mixtures, Figure 1.9 also shows that fractionation of the mixture is necessary in the longer term to regenerate the source materials.

Chapter 6 considers how the individual scientist can create her or his own isotope tapestry and experiment. One important aim of these isotope

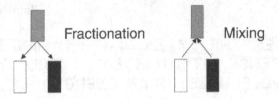

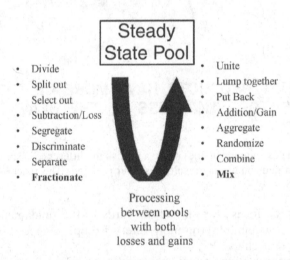

FIGURE 1.7. The two main themes of the book are fractionation and mixing. Fractionation splits apart mixtures to form source materials. These sources recombine via mixing. There is a strong general analogy between isotopes and colors, so that isotopes can be thought of as dyes or tracers. In this example, fractionation separates grey into black and white components, and conversely, black and white mix to form grey. (A color version of this figure on the accompanying CD gives this mixing in terms of blue and yellow sources mixing to form the intermediate green color, and conversely, fractionation regenerates blue and yellow from green.)

Steady
State Pool

- Divide
- Split out
- Select out
- Subtraction/Loss
- Segregate
- Discriminate
- Separate
- **Fractionate**

- Unite
- Lump together
- Put Back
- Addition/Gain
- Aggregate
- Randomize
- Combine
- **Mix**

Processing
between pools
with both
losses and gains

FIGURE 1.8. Fractionation and mixing together control isotope cycling and circulation. There are many words to use when thinking about isotope "fractionation" or "mixing," and as long as you remember that these words do not imply human intervention, control, or intent, most of these words can help you understand isotope cycling.

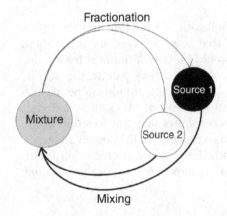

Fractionation

Mixture

Source 1

Source 2

Mixing

FIGURE 1.9. Isotopes cycle via fractionation and mixing, with fractionation splitting apart mixtures to form source materials. These sources recombine via mixing to complete the cycle.

addition experiments is to create very strong mixing signals so that fractionation becomes unimportant. To do this, scientists purchase isotope-labeled materials from a normal chemical catalogue, then add the label to a field or laboratory experiment to follow the downstream flow and fates of the tracer and element. In effect, these experiments try to outdo Mother Nature, putting the isotope signals where we humans think they will do most good so that we can see the mixing dynamics. In Chapter 6, we show that these isotope addition experiments are elegant, but not always simple.

Structure of This Book

Chapter 1 introduces stable isotopes and Chapter 2 presents the δ notation used to express isotope values. Chapter 3 reviews how ecologists use stable isotope tracers in modern research. Chapter 4 presents a new modeling approach for combining mixing and fractionation, circulating stable isotopes in any virtual ecological system you care to simulate or create. Chapters 5 and 6 give examples where mixing is the dominant force controlling isotope distributions, with Chapter 5 showing natural examples and Chapter 6 focusing on examples where experimenters add their own tracers. Chapter 7 shows how fractionation works in model systems and in real-world examples. Chapter 8 concludes the main part of the book by illustrating that isotopes are just part of the toolkit ecologists should use when investigating the natural world. The appendix gives the equations used throughout this book. The CD that accompanies this book contains a reading list, cartoons, and figures (many in color), interactive spreadsheet models, problems for the interested student, and answers to these problems. The CD also contains several detailed technical supplements for reference; the interested reader can print these out for use with this book.

Because every aspect of human endeavor has its own drama, you will find quite a few isotope stories throughout the book. These stories are definitely fiction, but build on the principles of mixing and fractionation to illustrate how isotopes move and act in the real world. These story examples are also extended problems, showing how scientific thinking can progress using isotope-based calculations.

Many of these story examples come equipped with practice spreadsheets on the attached CD. As you read these stories, open the accompanying spreadsheet models on the CD and start to manipulate isotopes for yourself to see what isotopes do in these living illustrations. The combination of stories, spreadsheet illustrations, and problems at the end of each chapter provides an underlying mathematical structure to this book to help readers develop a sound working understanding of stable isotope ecology.

Overall, about half the value of the book lies in reading the text for comprehension, and half in working through problems. If you neglect the simple algebra of the spreadsheets and problems, you will miss much of the point of this book. The point is that the spreadsheet math enables you to easily use and manipulate isotopes to explore the future, what isotopes might be doing in a system that interests you.

In many ways, this book has a how-to-do-it philosophy, teaching concepts and practice with stable isotopes. You can acquire the important concepts by reading, but also you have to work through problems to gain real mastery of isotope ecology. So I wish you both good reading and good problem-solving as you proceed through this book.

For those scientists who have little time and want the condensed version, reading the following sections will give a good overview: all chapter overviews and summaries, then Sections 1.1, 1.2 (this section), 2.1, 3.1–3.5, 4.1–4.4, 4.7, 5.1–5.5, 5.7, 5.9, 6.1, 7.1, 7.2, 7.5–7.7, and 8.2. These sections comprise about half the book in total. The appendix recapitulates the book in condensed mathematical form, summarizing equations used throughout the book. All mathematics in this book are simple algebra, so that advanced mathematical abilities are not needed for a good understanding of stable isotope ecology.

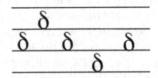

1.3 Just for Fun—An Isotope Biography of Mr. Polychaete

Sometimes it is good to know more about the author of a book you are reading. Here is a short sketch of my time in Isotopia, looking back through thirty years of isotope grime and fun.

Once upon a time those many years ago, I was an unsuspecting fellow sunnily and blissfully ignorant of all things isotope, just like you are now or like you were once upon a different time. But things began to change one summer on the Texas coast when I started graduate studies. It was a time spent out on the mudflats sifting through the sands looking for wiggly marine worms (yes, now you know what kind of person is writing this). I wasn't looking for worms for fishing, but searching for those polychaete worms because they were so unexpected, so full of color, and so cool to find. That summer, they called me Mr. Polychaete at the student awards festival, a distinction I wore with honor (and looking back, perhaps a true pinnacle of achievement in my life—something else that should warn you about this writing). Anyway, there I was and somehow, through family connections it was, I entered the school chemist's (Pat "The Chief" Parker's) lab, never to truly emerge again. There were these things, isotopes, that no one had ever heard of but were everywhere, all around, like secret writing. Definitely cool. Pick up something and put it in the measurement gizmo and get the secret decoded message, a new way to see the invisible connections out there on the mudflats. Sunburn and isotopes—who could ask for more!

It was a good summer, 1975. I decided to work towards a degree with isotopes in mind. I was driving down the island road that Fall and there was a dead coyote on the road. Road kill you might think and drive right on by. Not me, no sir. I jammed on the brakes and stalked back and darted between the cars to get a sample of coyote hair for isotopes. Yep, nothing could escape the young and budding Isotopeteer: isotopes everywhere, excitement at every turn, every twist in the road, every hair follicle. Glorious! Those singing youngsters on the Mickey Mouse program of my youth were proud "Mouseketeers," and now I had entered a similar exalted realm, that of the flushed "Isotopeteer."

But with every peak there comes a valley. And so it was. Not everything worked out all the time, and there were days of depression. One of my friends took pity on me, and asked what was wrong. I replied that as scientists we were supposed to test things, but sometimes the tests weren't working out. The isotopes were not the eternal key to life. In fact, sometimes they were worthless for finding out important stuff! I had morphed from an "Isotopeteer" into an "Isotopist," pissed off and angry at the isotope betrayal. My friend looked at me thoughtfully, then observed that for many months I had been so enthusiastic there must have been something to all this isotope business, even if it wasn't perfect. He told me to take the long-term view; remember the big picture. You will know that kind of free advice, I'm sure. So I thought and thought and thought some more. And I decided that my friend was right, that I was going to go where the isotopes would lead me, where the isotopes were useful in their own right, and quit trying to stuff them into my preconceived ways of thinking. That way I could sit back and observe the way isotopes were distributed over time and space in a topological manner, and could make grand pronouncements about the space–time continuum as the isotope Einstein. The grand view spread out before me as an "Isotopologist" and I entered another happy time.

This was a long and fun chapter in my career that took me to topics such as the origins of life, acid rain, and the cycling of organic matter in the sea. I also ran a big isotope lab, with samples sent in from all over the world. I helped scientists interpret their isotope results, and spent many hours imparting isotopological wisdom.

I found that the isotopes gave a curious combination of source and fractionation information, with most scientists wanting to use isotopes as source markers, dyes, or tracers. The idea is that Nature or the experimenter adds a few coded colors somewhere in the system, then tracks the colors, like releasing yellow and blue dyes in two upstream branches of a river, then seeing which stream contributes more color and water (the source information) and how fast the colors combine as the streams come together (the rate, process, or fractionation information). The demand was to help figure out the source information from the isotopes. With practice, I learned to part the curtains of isotope fractionation and extract the much-sought source information, becoming an "Isotope Sourcerer" magician. (Note: the true English word for a magician is sorcerer, not sourcerer.)

As I grew older, I decided to work on telling the many great isotope stories, imparting sage advice and wisdom about Isotopology and Isotope Sourcery from an ecological perspective. Mr. Polychaete still retains a sense of humor and fun about isotopes. He participates in this book as a wise, wiggly counsel, to help develop your skills with Isotopology and Isotope Sourcery.

Now that you have read this far, take some wise and wiggly advice from Mr. Polychaete and open up the CD that accompanies this book. There you should check out the several color cartoons and figures developed for this introductory Chapter 1. Also try out your new-found isotope knowledge with a set of ten true/false problems. Test yourself and see if you are indeed learning something with all this reading. And while you linger there on the CD, browse around in the various folders, to get an idea of what Mr. Polychaete has planned for you in the rest of this voyage into Isotopia. After your electronic explorations, which should leave you wiser and wigglier, return to the printed text and Chapter 2.

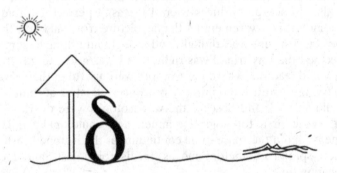

Further Reading

Section 1.1

Bender, M. 1968. Mass spectrometric studies of carbon 13 variations in corn and other grasses. *Radiocarbon* 10:468–472.

Fry, B., W.L. Jeng, R.S. Scalan, P.L. Parker, and J. Baccus. 1978. $\delta^{13}C$ food web analysis of a Texas sand dune community. *Geochimica et Cosmochimica Acta* 42:1299–1302.

Smith, B.N. and S. Epstein. 1971. Two categories of $^{13}C/^{12}C$ ratios for higher plants. *Plant Physiology* 47:380–384.

Wada, E., T. Ando, and K. Kumazawa. 1995. Biodiversity of stable isotope ratios. In E. Wada, T. Yoneyama, M. Minagawa, T. Ando, and B.D. Fry (eds.), *Stable Isotopes in the Biosphere*. Kyoto University Press, Japan, pp. 7–14.

Section 1.2

Anderson, T.F. and M.A. Arthur. 1983. Stable isotopes of oxygen and carbon and their application to sedimentologic and paleaoenvironmental problems. In M.A. Arthur, T.F. Anderson, I.R. Kaplan, J. Veizer, and L.S. Land (eds.), *Stable Isotopes in Sedimentary Geology*. SEPM Short Course #10, Society of Economic Paleontologists and Mineralogists, Dallas, TX, pp. 1–1 to 1–151.
Aston, F.W. 1922. *Isotopes*. E. Arnold & Co., London.
Basile-Doelsch, I., J.D. Meunier, and C. Parron. 2005. Another contintental pool in the terrestrial silicon cycle. *Nature* 433:399–402.
Bigeleisen, J. 1965. Chemistry of isotopes. *Science* 147:463–471.
Boutton, T.W. 1991. Tracer studies with ^{13}C-enriched substrates: Humans and large animals. In D.C. Coleman and B. Fry (eds.), *Carbon Isotope Techniques*. Academic Press, New York, pp. 219–242.
Brenna, J.T. 2001. Natural intramolecular isotope measurements in physiology: Elements of the case for an effort toward high-precision position-specific isotope analysis. *Rapid Communications in Mass Spectrometry* 15:1252–1262.
Broecker, W.S. 1985. *How to Build a Habitable Planet*. Eldigio. Palisades NY.
Broecker, W.S. and T.-H. Peng. 1982. *Tracers in the Sea*. Lamont-Doherty Geological Observatory. Palisades, NY.
Clark, I.D. and P. Fritz. 1997. *Environmental Isotopes in Hydrogeology*. Lewis. Boca Raton, FL.
Clayton, D. 2003. *Handbook of Isotopes in the Cosmos*. Cambridge University Press, New York.
Clementz, M.T., P. Holden, and P.L. Koch. 2003. Are calcium isotopes a reliable monitor of trophic level in marine settings? *International Journal of Osteoarchaeology* 13:29–36.
Criss, R.E. 1999. *Principles of Stable Isotope Distribution*. Oxford University Press, Oxford, UK.
Dahl, P.F. 1997. Flash of the Cathode Rays. Institute of Physics Publishing, Bristol.
Ehleringer, J.R., T.E. Cerling and M.D. Dearing 2005. *A History of Atmospheric CO_2 and Its Effects on Plants, Animals and Ecosystems*. Springer, New York.
Ehleringer, J.R., A.E. Hall, and G.D. Farquhar. 1993. *Stable Isotopes and Plant Carbon–Water Relations* (Physiological Ecology Series of Monographs, Texts, and Treatises). Academic Press, New York.
Faure, G. and T.M. Mensing. 2004. *Isotopes: Principles and Applications*. John Wiley and Sons, New York.
Fischer, H. and K. Wetzel. 2002. The future of ^{13}C-breath tests. *Food and Nutrition Bulletin* 23:53–56.
Griffiths, H. 1997. *Stable Isotopes: Integration of Biological, Ecological and Geochemical Processes (Environmental Plant Biology Series)*. BIOS Scientific. Oxford, UK.
Griffiths, I.W. 1997. J.J. Thompson—The centenary of his discovery of the electron and of his invention of mass spectrometry. *Rapid Communications in Mass Spectrometry* 11:1–16.
Hapgood, C.H. 1996. *Maps of the Ancient Sea Kings*. Adventures Unlimited, Kempton IL.
Hoefs, J. 2004. *Stable Isotope Geochemistry*, 5th ed Springer-Verlag, New York.
Kendall, C. and J.J. McDonnell. 1998. *Isotope Tracers in Catchment Hydrology*. Elsevier Health Sciences, New York.
Lajtha, K. and R.H. Michener. 1994. *Stable Isotopes in Ecology and Environmental Science*. Blackwell Scientific Publications, Oxford, UK.
Odum, E.P. 1971. *Fundamentals of Ecology*, 3rd ed, especially pp. 459–464 and references cited there. W.B. Saunders, New York.
Penzias, A.A. 1979. The origin of the elements. *Science* 205:549–554.

Penzias, A.A. 1980. Nuclear processing and isotopes in the galaxy. *Science* 208:663–669.

Rossman, A., M. Butzenlechner, and H.-L. Schmidt. 1991. Evidence for a nonstatistical carbon isotope distribution in natural glucose. *Plant Physiology* 96:609–614.

Rouxel, O.J., A. Bekker, and K.J. Edwards. 2005. Iron isotope constraints on the Archean and Paleoproterozoic ocean redox state. *Science* 307:1088–1091.

Rundel, P.W., J.R. Ehleringer, and K.A. Nagy. 1988. *Stable Isotopes in Ecological Research.* Springer Verlag, New York.

Schell, D.M. 1983. Carbon-13 and carbon-14 abundances in Alaskan aquatic organisms: delayed production from peat in Arctic food webs. *Science* 219:1068–1071.

Varele, D.E., C.J. Pride, and M.A. Brzezinski. 2004. Biological fractionation of silicon isotopes in Southern Ocean surface waters. *Global Biogeochemical Cycles* 18, GB1047, doi:10.1029/2003GB002140.

Wada et al. 1995. Listed above; see Section 1.1 readings.

Wada. E. and A. Hattori. 1990. *Nitrogen in the Sea: Forms, Abundance and Rate Processes.* CRC, Boca, Raton, FL.

2
Isotope Notation and Measurement

Overview

This chapter gives an introduction to isotope notation, calculations, and measurement. The beginner should probably read only the first section, 2.1, then skip on to Chapter 3 which reviews ecological applications of these isotope tracers. Reading Section 2.1 should allow you to understand the rest of the book, and reading the remaining sections 2.2 to 2.4 of this chapter should deepen your understanding as you read the wider isotope literature. There are also three detailed technical supplements for this chapter on the accompanying CD.

2.1. *The Necessary Minimum for Ecologists.* This section introduces the basics of isotope notation and how isotope measurements are made. If you are a beginner, read only this section, then skip the rest of Chapter 2 and go on to Chapter 3. After you finish the book, you can come back and read the rest of Chapter 2, as you have time and interest.

2.2. *Why Use the δ Notation?* Isotope values are expressed in the δ notation which turns out to be slightly inexact but convenient notation. This book uses the convenient δ notation, but this section also shows how to convert from the δ notation to exact alternative notations, the ratio (R) notation and the atom percent (AP) notation.

2.3. *Why Is δ a Good Substitute for % Heavy Isotope?* For natural samples, δ values are linearly related to % heavy isotope, and this section gives the algebra that explains why this is so.

2.4. *δ and the Ratio-of-Ratios.* If you examine the δ definition carefully, you find a ratio-of-ratios which is often hard to understand when first encountered. This section explains the advantages of a ratio-of-ratios definition.

Technical Supplement 2A. The Atom Percent Notation and Measuring Spiked Samples (see accompanying CD, Chapter 2 folder). Stable isotopes can be separated commercially and added to ecological systems, enriching natural abundances. This section shows that measuring enriched samples usually requires calculations made with the atom percent notation rather than the δ notation, and can also require some modifications of normal measurement systems.

Technical Supplement 2B. Ion Corrections (see accompanying CD, Chapter 2 folder). Most isotope measurements are made with mass spectrometers that produce a variety of ions from gases such as H_2, N_2, O_2, CO_2, and SO_2. The main ion beams provide the desired isotope information, but are also contaminated by a variety of minor ion products. This section shows some details of how calculations routinely assess and correct for these ion problems.

Technical Supplement 2C. The Ratio Notation and the Power of 1 (see accompanying CD, Chapter 2 folder). Because the light and heavy isotopes of elements react nearly identically, reaction ratios are very close to 1. As it turns out, many special mathematical properties apply to calculations made near the value of 1, making algebraic shortcuts possible and convenient in isotope ratio calculations. The technical supplement also shows the easy algebra for writing exact fractionation equations in ratio notation.

Main Points to Learn. Isotope notation is confusing in many ways, yet simple at its core. The δ values are difference measurements made with respect to recognized standards, and are related in a straightforward, essentially linear way to % heavy isotope. This leads to the rule that the higher the δ value, the greater the amount of heavy isotope, and the lower the δ value, the lower the amount of heavy isotope, or "higher heavier, lower lighter." Most isotope ecology applications use simple addition and subtraction with δ values to understand isotope circulation, but in special cases, using the alternative ratio and atom percent notations becomes important. Currently, mass spectrometer machines developed over the last 85 years routinely measure most isotope values with great precision and accuracy. But fast real-time lasers may eventually replace these expensive machines for isotope measurements as laser technology improves.

2.1 The Necessary Minimum About Isotope Notation and Measurement

Isotope values have their own special notation, the δ notation that signifies difference. The δ values denote a difference measurement made relative to standards during the actual analysis. The isotope compositions of standards are given in Table 2.1, and are used routinely in the calculation of δ values where they appear as the $R_{STANDARD}$ term:

$$\delta^H X = [(R_{SAMPLE}/R_{STANDARD} - 1)] * 1000.$$

TABLE 2.1. Isotope Compositions of International Reference Standards.

	Ratio, H/L[a]	Value, H/L[a]	% H	% L
Standard Mean Ocean Water	$^2H/^1H$	0.00015576	0.015574	99.984426
(SMOW)	$^{17}O/^{16}O$	0.0003799	0.03790	99.76206
	$^{18}O/^{16}O$	0.0020052	0.20004	99.76206
PeeDee Belemnite (PDB)	$^{13}C/^{12}C$	0.011180	1.1056	98.8944
and Vienna-PDB (VPDB)	$^{17}O/^{16}O$	0.0003859	0.0385	99.7553
	$^{18}O/^{16}O$	0.0020672	0.2062	99.7553
Air (AIR)	$^{15}N/^{14}N$	0.0036765	0.36630	99.63370
Canyon Diablo Troilite (CDT)	$^{33}S/^{32}S$	0.0078772	0.74865	95.03957
and Vienna-Canyon Diablo	$^{34}S/^{32}S$	0.0441626	4.19719	95.03957
Troilite (VCDT)	$^{36}S/^{32}S$	0.0001533	0.01459	95.03957

[a] H and L indicate heavy and light isotope components, respectively.
Source: Ratio values are taken from Hayes (2002) for H, C, N, and O isotopes and from Ding et al. (2001) for S isotopes. Historical values for PDB and newer values for VPDB are considered equivalent (based on data in Coplen 1983, 1996), and similarly, historical values for CDT (Coplen and Krouse 1998) and newer values for VCDT also are considered equivalent (Ding et al. 2001). See Ding et al. (2001) and Hayes (2002) for errors associated with the ratio measurements.

In this definition, the δ notation is specified for a particular element ($X =$ H, C, N, O or S), the superscript H gives the heavy isotope mass of that element (^2H, ^{13}C, ^{15}N, ^{18}O, or ^{34}S), and R is the ratio of the heavy isotope to the light isotope for the element, $^2H/^1H$, $^{13}C/^{12}C$, $^{15}N/^{14}N$, $^{18}O/^{16}O$, or $^{34}S/^{32}S$. The δ^2H measurements are also known as δD, where D stands for deuterium, the heavy stable isotope of hydrogen.

The δ definition involves a final multiplication by 1000, and this multiplication amplifies very small differences measured between samples and standards. Small differences of 1 percent become 10 permil δ units, because of the final multiplication by 1000. Thus, the δ definition makes the small neutron-related isotope differences seem large. The units of δ are "‰" or "permil" (also per mill), from Latin roots for parts per thousand, just as "percent" or "%" is derived from Latin roots for parts per hundred. A sample that measures 10‰ (ten permil) is only 1% (one percent) different than the standard, and even a seemingly large 100‰ difference is still only a 10% difference. Most of us should say "permil" out loud a few times to get familiar with this term, until it begins to sound different than "percent."

Most δ values range between −100 and +50‰ for natural samples, the so-called "natural abundance" range, with the exception that δ measurements for hydrogen span a broader range. Many δ values are negative values, and these negative δ values are usually quite confusing when we first encounter them. It often takes some time before these negative numbers start to seem natural and familiar. But negative δ values just indicate relatively less heavy isotope than is present in the standard.

Standards have a δ value of 0‰, which makes sense from the δ definition because when a standard is measured versus itself, the difference will be zero. Standards contain appreciable, nonzero amounts of heavy and light isotopes (Table 2.1), so that 0‰ means no difference from the standard, not "0% isotope," not "no isotope," and not "no heavy isotope."

Samples with higher δ values are relatively enriched in the heavy isotope and are "heavier." Samples with lower δ values are relatively enriched in the light isotope and are "lighter." This leads to the convenient mnemonic for δ values, "higher heavier, lower lighter." Remembering this mnemonic will help when we think about fractionation and how light isotopes react slightly differently than heavy isotopes.

Viewed in a more detailed, technical way, the δ definition actually contains two separate ratios (R_{SAMPLE} and $R_{STANDARD}$) and a ratio-of-ratios, $R_{SAMPLE}/R_{STANDARD}$. This leads many scientists to write about isotope variations in terms of ratios. Although use of the ratios has its advantages, practical use of the δ values does not involve a focus on "ratios." Instead, δ values are straightforward indicators of "% heavy isotope" because there is a simple, essentially linear relationship between δ values and isotope content (Figure 2.1). Thus in this book, the terms "heavier" and "enriched" refer to

---

FIGURE 2.1. Linear relationships of H, C, N, O, and S heavy isotope contents to δ values. Large natural abundance δ variations from −100 to +100‰ correspond to only slight variations in percent heavy isotope, so that the effect of using δ values is to greatly magnify the small natural differences found in nature. Also, the strong relationships shown between δ values and "% heavy isotope" means that δ values can be used to track heavy isotope dynamics in accounting and budget equations used later in this book for I Chi modeling. Lastly, the range shown here is for natural samples. Isotope can be purchased and added to natural systems, raising values to 1000‰ and above. Outside the natural abundance range, the depicted linear relationships do not hold, and become increasingly curvilinear. Data used for the lines in these graphs were calculated from the definition of δ and using standards listed in Table 2.1, with SMOW used as the standard for oxygen isotopes. The basic equation used for the calculations derives from the δ definition as $^{H}AP = 100*(\delta + 1000)/[(\delta + 1000 + (1000/R_{STANDARD})]$ where % heavy isotope is atom % of the heavy isotope, or ^{H}AP. Calculations with this equation were modified for oxygen and sulfur that have more than two stable isotopes, assuming that the minor O and S isotopes were fractionated according to mass-dependent rules (Hulston and Thode 1965; Hoefs 2004; $\delta^{17}O = 0.515*\delta^{18}O$, $\delta^{33}S = 0.515*\delta^{34}S$, and $\delta^{36}S = 1.9*\delta^{34}S$). Depicted best-fit lines reflect natural conditions and have r^2 values of 1.0000. Equations for the lines are as follows: hydrogen (% $^{2}H = 0.0000156*\delta^{2}H + 0.0155726$), carbon (% $^{13}C = 0.00109*\delta^{13}C + 1.10559$), nitrogen (% $^{15}N = 0.000365*\delta^{15}N + 0.366295$), oxygen (% $^{18}O = 0.000200*\delta^{18}O + 0.200041$), and sulfur (% $^{34}S = 0.00400*\delta^{34}S + 4.19652$).

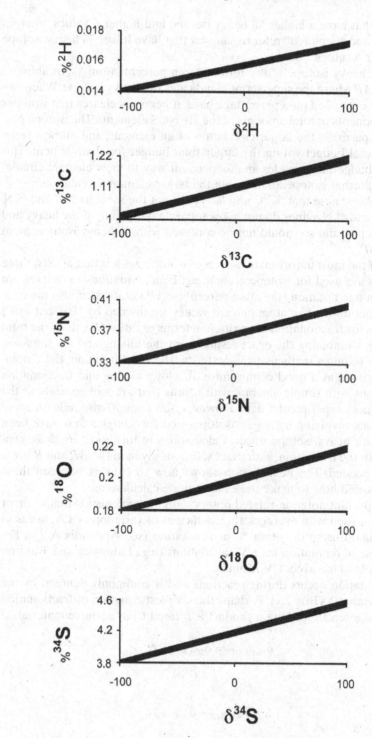

samples that have a higher % heavy isotope and higher δ values, whereas "lighter" and "depleted" refer to samples that have lower % heavy isotope and lower δ values.

The % heavy isotope is also termed atom percent, atom %, or abbreviated as ^{H}AP where the superscript H indicates the heavy isotope. When isotopes are expressed on a percentage basis, it becomes clearer that isotopes are components of total amounts of the HCNOS elements. The isotope percentages partition the larger circulation of an element, and allow a separate subtotal budget within the larger total budget for that element. This subtotal budgeting provides an independent way to view element circulation, an internal isotope audit within the larger element circulation.

Also please note that %^{13}C and %^{15}N are not the same as %C and %N. The %C and %N values do not allow separate budgeting of the heavy and light isotopes, and so should not be confused with % heavy isotope, atom %, or ^{H}AP.

One of the most unfortunate aspects of isotope work is that at least three notations are used for isotope accounting. Exact equations for mixing are written in one notation (the atom percent or AP notation, or also the fractional F notation when atom percent values are divided by 100), but exact equations for fractionation are written in terms of ratios (R values, the ratio notation). Combining the exact results from the mixing and fractionation notations requires mathematical dexterity. The third notation, the δ notation, functions as a good compromise. It allows mixing and fractionation calculations with simple algebra, with results that are still accurate at the level of most experimental data. However, this compromise fails for some calculations involving hydrogen isotopes and for samples that have been spiked with heavy isotope tracer. Calculations in this book are done generally with the δ notation, with exact solutions given in the AP and R notations as needed. The next section shows how to convert between these notations, and how to make the exact isotope calculations.

An important notation-related observation for ecologists is that an error term often used with averages, the coefficient of variation or CV, needs to be calculated using the atom % or F notation (see Appendix A.2 in Fry 2003). Use of δ notation for CV calculations (e.g., Lancaster and Waldron 2001) leads to incorrect CV results.

Fractionation occurs during reactions and is commonly denoted by the Greek symbol Δ (Box 2.1). Perhaps the simplest equation of fractionation applies to a reaction where a product is formed from a source material,

$$\delta_{PRODUCT} = \delta_{SOURCE} - \Delta,$$

or

$$\Delta = \delta_{SOURCE} - \delta_{PRODUCT}.$$

For example, when a plant fixes carbon dioxide during photosynthesis, a fractionation of 20‰ occurs between the source atmospheric CO_2 at −8‰ δ ^{13}C and the −28‰ plant sugar product that is formed from atmospheric CO_2. The Δ fractionation values are expressed in positive permil units (e.g., Δ = 20‰), and are usually quite similar to the simple difference between two δ values. Box 2.1 considers fractionation definitions and terminology in more detail, as do other authors (Farquhar et al. 1989; Mook 2000; Hayes 2002, 2004).

Box 2.1. What Is a Fractionation Factor?

There are several terms scientists use to denote fractionation factors, and several derivations of these fractionation terms. The following gives a derivation of fractionation factors for common biological reactions that are one-way kinetic reactions, following the derivations presented by Farquhar et al. (1989). For these kinetic reactions, one can think of two rates: one for molecules with the heavy isotope substitution, and one with the more usual light isotopes. These reaction rates or kinetic "k" constants can be designated for the light (L) and heavy (H) isotope molecules as ^{L}k and ^{H}k. The ratio of rate constants gives the fractionation "alpha" or "$\alpha_{L/H}$" or most simply "α":

$$\alpha = \frac{^{L}k}{^{H}k}.$$

If there is no effect of the isotope substitution of light for heavy isotopes, then the reaction rates would be equal and α would have a value of 1. But because molecules with a light isotope substitution usually react slightly faster, this ratio is normally slightly greater than 1. A 1% faster reaction of ^{L}k versus ^{H}k translates to an α value of 1.01. You might note that if you look at the decimal places following the 1, you can see that a value shows this 1% difference in reaction rates as the fraction .01. To make this fractionation difference easier to see and work with, Δ values are derived from α values:

$$\Delta = (\alpha - 1)*1000,$$

where Δ gives the fractionation in positive permil (‰) units. Thus, if there is a 1% faster reaction rate for the light-isotope molecules, this is also a 10‰ faster reaction. The α and Δ terms express these differences: α = 1.01, and Δ = 10‰. When scientists talk about isotope fractionation, they commonly use the positive permil units of the Δ values, such as 10‰. Because common parlance favors expressing fractionation in this way, positive Δ values are used in this book.

You should also know that many scientists use alternative definitions of α and Δ. Hydrologists and geochemists especially favor a definition of $\alpha = {}^{H}k/{}^{L}k$ for the common case where the lighter isotope reacts faster so that α is typically less than 1 by this definition, and $(\alpha - 1)*1000$ is redefined as ε, a negative permil number (Mook 2000). This usage is based on inversions of the terms given above, H/L instead of L/H. Choosing between H/L and L/H resembles the disputes of the Lilliputians who fought over which end of the egg is better, the top or bottom. This book adopts the L/H formulations that perhaps are less conventional but favored by chemists and some biologists. In case you need them, here are the formulas to convert the hydrological and geochemical H/L fractionations to the L/H fractionations used in this book:

$$\alpha_{L/H} = \left(1/\alpha_{H/L}\right) \quad and \quad \Delta_{L/H} = \left(-1000*\varepsilon_{H/L}\right)/\left(1000+\varepsilon_{H/L}\right)$$

or approximately $\Delta_{L/H} = -\varepsilon_{H/L}$ for values in the 0 to 20‰ range. For completeness, and not forgetting hydrologist and geochemist readers of this book, it is also possible to calculate the H/L fractionation values from the L/H values of this book, using the formulas:

$$\alpha_{H/L} = \left(1/\alpha_{L/H}\right) \quad and \quad \varepsilon_{H/L} = \left(-1000*\Delta_{L/H}\right)/\left(1000+\Delta_{L/H}\right).$$

Lastly, no matter how you define α, Δ, and ε there are occasional "inverse" isotope effects that have the opposite sense of your chosen definition, when the heavy isotope molecules react faster than their light isotope counterparts. We meet one of these inverse cases in Section 7.8.

Here we also consider how the isotope measurements are currently made, usually with an isotope ratio mass spectrometer. But first you collect a plant, animal, soil, or gas sample from nature or the laboratory, then grind, pulverize, and combust your precious sample until it emerges as a simple gas that the isotope machine can conveniently analyze. For this preparation work, you can use an elemental analyzer, a gas chromatograph, or a laser. These different devices plus various combustion interfaces are the front-end engines that convert samples to the common denominator gases. For the HCNOS isotope measurements, these common denominator gases are generally hydrogen (H_2), carbon dioxide (CO_2), nitrogen (N_2), oxygen (O_2), and sulfur dioxide (SO_2). Most ecologists currently use dual CN isotope measurements made with elemental analyzers coupled to mass spectrometers. Technical advances now are allowing triple CNS isotope measurements from the same sample, and future advances will likely allow quadruple HCNS isotope measurements all from the same sample at affordable costs of <$10 U.S. per sample (Sieper et al. 2006).

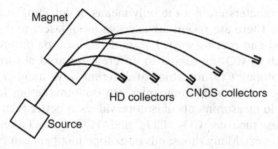

Magnet

HD collectors   CNOS collectors

Source

FIGURE 2.2. Schematic of an isotope ratio mass spectrometer used to make isotope determinations. In the source region, gas molecules are ionized as they encounter electrons boiling off a hot filament. The charged ions are accelerated via electric fields through a stainless steel flight tube (not shown) maintained under vacuum. In the central magnetic field, charged ions are separated according to inertia, and dispersed towards collectors for automated counting by computers. Due to their small masses and consequent low inertia, the hydrogen ion beams are sharply bent by magnet focusing. Magnet focusing results in much more gradual bends in flight paths of the ion beam for gases with higher masses, especially N_2, O_2, CO_2, and SO_2.

The mass spectrometer (Figure 2.2) uses carefully calculated physics to make the actual measurement, with the following main steps. Gases first enter a source region where a white-hot filament is boiling off electrons. Close encounters with these electrons are violent, and the sample gas molecules are ionized, losing electrons of their own, and often even fragmenting to simpler molecules. Ionized molecules lacking an electron have a positive charge, and via electric fields, the ions are accelerated out to the flight tube. The positively charged ions pass through a magnetic field that separates them according to their atomic mass and isotopes, with the resulting ion beams focused into collectors for counting (Figure 2.2).

The main principle of the mass spectrometer separation is inertia, simple inertia. The gas molecules with the extra neutrons require more force to displace them from their flight paths. Thus their flight paths are straighter than those of their lighter-isotope counterparts (Figure 2.2).

In the end, computers tally up the counts from the multiple collectors and calculate the final isotope values. Simple algebra outlined in the appendix gives common ways to recalculate these laboratory results relative to international standards for final use in talks and publication. Technical Supplements 2A and 2B on the accompanying CD give details about how computers actually count the ion beam currents and calculate "raw" or laboratory δ values. These calculations are done routinely by computer software in most modern laboratories, so that most users don't have to bother with these calculations. But these calculations are included in the Technical Supplements for the interested reader.

Mass spectrometers are not the only means of detecting and measuring isotope values. There are recent advances in using lasers to detect isotope values in rapid and precise ways. Lasers detect isotope differences in gas molecules such as CO_2 by tuning to different infrared absorption bands, with heavy-isotope $^{13}CO_2$ absorbing at different wave numbers than light-isotope $^{12}CO_2$. Commercial companies are now marketing laser devices that permit field measurements of isotope values at better than 0.5‰ precision for gases such as CO_2, CH_4, and H_2O (Los Gatos Research, www.LGRinc.com). Many thousands of isotope numbers can be generated each day by lasers to track real-time field experiments, and this new technology contrasts with the slower mass spectrometers that typically produce 50 to 500 values per day in laboratory conditions.

Here are some final notes on terminology. (1) Many ecologists currently write about isotopes in terms of "isotope signatures" or "isotope fingerprints." Although isotopes are often good "descriptors" of processes, and multiple isotope measurements can indicate a fairly distinctive isotope "profile," isotope values rarely provide a truly unique fingerprint. Also, there is usually substantial natural time and space variation for isotope compositions of plants, animals, and soils. A more neutral terminology that recognizes this variability is "isotope values" rather than "isotope signatures." And common parlance favors use of "isotope values" rather than "isotopic values." For these reasons of common parlance and respect for natural variation in isotope compositions, this book consistently uses "isotope values" or "δ values" when discussing stable isotopes. (2) Future isotope scientists may adopt the recent recommendations that ϕ values be substituted for δ values (Brenna et al. 1997; Corso and Brenna 1997). ϕ values are defined in a parallel manner to δ, but with fractional abundance F values substituted for ratio R values:

$$\phi^H X = [(F_{SAMPLE}/F_{STANDARD} - 1)] * 1000.$$

For most natural samples, the ϕ notation gives nearly identical values to the commonly used δ values, and algebraic calculations are generally both precise and easier with ϕ. However, the ϕ notation is not widely used at present by isotope scientists, and for this reason and for continuity with the published literature, this book uses the traditional δ values.

2.2 Why Use the δ Notation?

There are actually four important notations used in isotope work. These notations are: δ, R, F, and AP or atom %. This section helps you navigate among these notations, with the δ notation providing the starting point. The definition of δ already involves 3 of these notations, δ, R, and F:

$$\delta = [(R_{SAMPLE}/R_{STANDARD} - 1)]*1000,$$

where

$$R = {}^{H}F/{}^{L}F \quad \text{and}$$

F = fractional abundance of the heavy $({}^{H}F)$ or light $({}^{L}F)$ isotope.

The fourth notation is AP = atom percent = $F*100$. Although it seems a little complex at first, it is actually easy to calculate R, F, and AP from δ using the formulas above and a computer spreadsheet.

Usually, δ suffices because of the close correlations between δ and the other notations (Figures 2.1, 2.3). Because of these correlations, we usually use δ instead of R or F values. But using δ does generate very small errors in calculations of fractionation and mixing and when considering enriched samples. Most of the time we ignore all these small problems, but occasionally we have to fall back from δ to the R, F, and AP values. For mixing, it is exact to use the F values. For fractionation, it is exact to use the R values (which are actually ratios of F values, so using F values is also exactly correct for fractionation calculations). And for some algebra, you have to learn to navigate among all three notations. Here are several examples where the F notation is needed: (1) for some statistics, especially the coefficient of variation, (2) when truly exact values are needed for mixing, (3) when working with hydrogen isotopes that have δ values outside the −100 to +100‰ range,

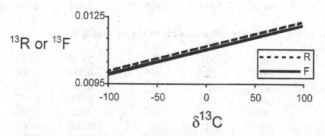

FIGURE 2.3. There is a linear relationship between the three types of isotope notation (δ, R, and F) for natural samples in the −100 to +100‰ δ range. This example shows how R and F are related to δ for carbon isotopes, $\delta^{13}C = [(R_{SAMPLE}/R_{STANDARD}) - 1]*1000$ where $R = {}^{H}F/{}^{L}F = {}^{13}C/{}^{12}C$, ${}^{H}F = {}^{13}C$, ${}^{L}F = {}^{12}C$, the standard is VPDB (see Table 2.1), and $R_{STANDARD} = {}^{H}F/{}^{L}F_{STANDARD} = 0.011056/0.988944 = 0.011180$.

and (4) when working with artificially enriched samples that have very high enrichments, >10% heavy isotope. For all these cases, one converts between the notations using the δ definition above. An important final rule is that the most fundamental quantity in these calculations is F or AP.

Calculations based on F or AP are always the correct ones, whereas calculations based on δ and R are sometimes misleading.

Table 2.2 illustrates some of these problems. When the δ notation is used, small errors arise for both mixing and fractionation, especially when δ values differ greatly from the natural abundance range near 0‰. For example, in an added tracer experiment, you might consider a 50/50 mixture from two sources whose δ values were 0 and 601‰. Using the δ notation, the answer for this mixing problem is 300.5‰, but the correct value is actually 299.5‰, obtainable only by basing the calculations on the F notation. This is still a small error of 1‰, but an error that grows larger at higher enrichments.

Errors can also arise during fractionation calculations with δ. For example, in an enrichment experiment where the starting substrate pool is labeled at 1000‰, the actual correct δ value for a product made with a 10‰

TABLE 2.2. Examples of Errors Encountered in Working with the Three Different Isotope Notations, δ, R, and F.[a]

	Results of Mixing or Fractionation			Difference (‰) vs. Correct, F-Based Solution		
	δ	R	F	δ	R	F
A. Mixing, average of two samples, one has δ = 0‰, δ of second sample as follows. 2nd Sample:						
−100	−50	−50	−50.028	0.028	0.028	0.000
20	10	10	9.999	0.001	0.001	0.000
60	30	30	29.990	0.010	0.010	0.000
190	95	95	94.900	0.100	0.100	0.000
601	300.5	300.5	299.5	1.000	1.000	0.000
B. Fractionation, Δ = 10‰ or α = 1.01 versus the following starting δ value. Starting δ value:						
−100	−110	−109	−109	−1.000	0.000	0.000
−50	−60	−59.5	−59.5	−0.500	0.000	0.000
0	−10	−10	−10	0.000	0.000	0.000
50	40	39.5	39.5	0.500	0.000	0.000
100	90	89	89	1.000	0.000	0.000
1000	990	980	980	10.000	0.000	0.000
10000	9990	9890	9890	100.000	0.000	0.000

[a] Results are given as δ values (left four columns), or $\Delta\delta$ values (difference by simple subtraction) versus correct, F-based calculations (right three columns).
Source: Used with permission from Fry, B. 2003. Steady-state models of stable isotope distributions. *Isotopes in Environmental and Health Studies* 39:219–232, Published by Taylor & Francis Ltd, http://www.tandf.co.uk.

fractionation is 980‰, not 990‰. Oddly, the overall effect of working with the δ notation is that fractionation is underestimated at high enrichments, but at enrichments with light isotope that drive values towards no heavy isotope (δ = –1000‰), the opposite effect occurs, and fractionation effects are overestimated with the δ notation. The consequence of all of this is that when working outside the natural abundance range, it is good to check calculations based on δ against calculations based on F. The right answers are the F-based calculations.

Lastly, it is instructive to consider the relative importance of errors in mixing and fractionation, using the –100‰ examples listed at the top of the mixing and fractionation sections in Table 2.2. In these examples, the error for mixing is about 0.0278‰, and the error for fractionation is 1‰, a 36-fold larger error. This large difference in errors is representative for many situations, so that with the δ notation, fractionation calculations have much larger errors than do mixing calculations. For this reason, first efforts to reduce errors should be focused on fractionation terms, and using R notation is often adequate for eliminating major errors. Technical Supplement 2C in the Chapter 2 folder on the accompanying CD shows ways to routinely use the R notation instead of δ notation in detailed isotope fractionation calculations.

You may well ask, given all these uncertainties, why would anyone actually use δ? The answer is this: because it is more convenient, and because all these uncertainties are normally very, very small and safely ignored compared to other normal sources of sampling and measurement error for most HCNOS natural abundance samples. Here is an example that actually converts among all these notations, and looking at such examples has convinced most scientists that it is simpler to use δ.

The problem is this. If the δ ^{13}C value of the atmospheric CO_2 is currently –8‰ versus the VPDB standard, that is, δ ^{13}C = –8.0‰, what are the corresponding values for R_{SAMPLE}, ^{H}F, and AP (atom %)? Here is the solution.

To calculate R_{SAMPLE} for carbon isotopes, remember first that the $R_{STANDARD}$ value for VPDB is 0.01118 (from Table 2.1). Rearranging the definition of δ,

$$R_{SAMPLE} = [(\delta/1000) + 1] * R_{STANDARD} = [(\delta/1000) + 1] * 0.01118 = 0.0110906.$$

The next step is to calculate ^{13}F from R_{SAMPLE} from the δ definition

$$^{H}F = (\delta + 1000)/[(\delta + 1000 + (1000/R_{STANDARD}))]$$

so that

$$^{13}F = (\delta + 1000)/[(\delta + 1000 + (1000/0.01118))] = 0.0109689.$$

Calculating atom % ^{13}C from ^{13}F is easy, ^{13}AP = Atom % = 100 * F = 1.09689.

When you look at these numbers, $\delta = -8‰$, $R_{SAMPLE} = 0.0110906$, $^{13}F = 0.0109689$, and $^{13}AP = 1.09689$, you might agree that to prevent headaches from looking at too many decimal places, it is better to use the δ notation that gives $-8‰$. But when exactness is truly needed, you should use the R, F, and AP notations.

2.3 Why Is δ a Good Substitute for % Heavy Isotope?

Here we start with the definition of δ and show how it contains a very nearly linear equation for the relationship between δ and the fractional abundance of heavy isotopes. The definition for δ is:

$$\delta = [(R_{SA}/R_{ST}) - 1] * 1000.$$

$R = H/L$ where H is the fraction of heavy isotope and L is the fraction of light isotope. H and L range from 0 to 1. The subscripts SA and ST denote sample and standard, respectively.

The next step is to realize that R_{ST} is a constant. For carbon isotopes, for example, the PDB standard has an R_{ST} value of 0.01118 (Table 2.1), and dividing R_{SA} by this amount yields:

$$\delta = [(R_{SA}/0.01118) - 1] * 1000 = (89.44544 * R_{SA} - 1) * 1000$$
$$= 89{,}445.44 * R_{SA} - 1000.$$

The third step is to realize that R_{SA} can be rewritten as $H/(1 - H)$, so that $R_{SA} = H/(1 - H)$, and

$$\delta = (89{,}445.44/(1 - H)) * H - 1000.$$

This is still not the equation for a line, and now comes the most nonintuitive part of this derivation. When H varies over a narrow range corresponding to 100‰ or less (for example, H varies from about 0.010 to 0.011 for most natural carbon isotope measurements, a range of 0.001), the denominator term is nearly constant and can be approximated as a constant c. This approximation yields the equation for a linear relationship between δ and H:

$$\delta = (89{,}445.44/c) * H - 1000.$$

Overall, it is the small variation in the $(1 - H)$ term, the term used in the denominator of the ratio $H/(1 - H)$ in the above equations, that accounts for the linear relationships between H and δ. This small variation is relative, and is small only across the relatively "narrow" isotope ranges of 100‰.

Also, although the denominator minimizes impacts of H variations, the numerator vastly multiplies any slight variations in H. The combination of these two factors determines the almost perfect straight-line relationships between δ and H at natural abundance levels.

However, as isotope ranges increase from 100‰ to 1000‰ and beyond, when both natural and isotope-enriched samples are being studied together, the variation in H in the denominator of the term $H/(1-H)$ cannot be ignored, and the relationship between H and δ is increasingly nonlinear. Technical Supplement 2A on the accompanying CD illustrates this problem that occurs when combining work with two types of samples, natural abundance samples and highly enriched samples. In these cases, it is necessary to convert the δ values to fractional abundances (H values) or $100*H$ = atom percent so that natural and enriched samples can be compared directly. But for almost any "narrow" range of 100‰ variation, be it at natural abundance or 99% enrichment, the linear relationship between δ and H applies.

2.4 δ and the Ratio-of-Ratios

The δ definition used throughout the isotope world is an odd parameter, because it is calculated from a ratio-of-ratios:

$$\delta = (R_{SAMPLE}/R_{STANDARD} - 1)*1000,$$

where the ratio-of-ratios is

$$R_{SAMPLE}/R_{STANDARD} = (H_{SAMPLE}/L_{SAMPLE})/(H_{STANDARD}/L_{STANDARD}),$$

where H and L represent the fractions of heavy and light isotope, respectively. Although odd in several ways, this ratio-of-ratios definition has four useful functions.

Foremost, the δ definition expands very small absolute differences in isotope compositions into much larger numbers that typically fall in the −100 to +100‰ range, numbers that are easier to use in everyday communication. And, as it turns out, for narrow ranges such as −100 to +100‰, the relationship between δ and % heavy isotope is almost exactly a straight line (see Figure 2.1). So, in spite of the confusing ratio-of-ratios calculation, δ values give very good approximations of the % heavy isotope and, by difference, also % light isotope in a sample. The conclusion is that δ values can be understood rather simply as percentages of heavy isotope, or atom %.

A second use has to do with the ratio-of-ratios itself, for this "double ratio" is a way to normalize for initial conditions when the standard is the starting point. To see this more clearly, we rearrange

$(H_{SAMPLE}/L_{SAMPLE})/(H_{STANDARD}/L_{STANDARD})$ by multiplying by

$(L_{STANDARD}*L_{SAMPLE})/(L_{STANDARD}*L_{SAMPLE})$ to obtain

$(H_{SAMPLE}*L_{STANDARD})/(H_{SAMPLE}*L_{SAMPLE})$ then restate

$L_{STANDARD}/L_{SAMPLE}$ as $1/(L_{SAMPLE}/L_{STANDARD})$ to obtain

$(H_{SAMPLE}/H_{STANDARD})/(L_{SAMPLE}/L_{STANDARD})$.

In this formulation, one sees that the heavy isotopes and the light isotopes in the sample are both normalized to contents of the standard material $(H_{SAMPLE}/H_{STANDARD})$ and $(L_{SAMPLE}/L_{STANDARD})$, respectively. Using this normalizing ratio-of-ratios, it becomes apparent that deviations from the standard are what count, regardless of the starting composition of the standard. In practice, this gives a common scale to isotope variations for elements such as carbon and hydrogen that differ a great deal in their normal H/L compositions. For example, a 1% deviation from the standard yields the same 10‰ δ value for carbon or hydrogen, even though the heavy isotope contents of standard materials differ more than 70-fold, about 1.1% heavy isotope for carbon and 0.015% heavy isotope for hydrogen. Although it seems odd, the double normalization of heavies and lights via the ratio-of-ratios calculation makes the δ notation flexible and comparable across elements. That is, a 10‰ difference means the same 1% difference for H isotopes as for C isotopes.

A third advantage to using a ratio-based definition of δ can be found in laboratory measurements. Mass spectrometer measurements are typically more precise when a ratio is monitored, rather than just monitoring the light isotope ion beam by itself or the heavy isotope ion beam by itself. Machine noise and fluctuations in the source and magnet focusing affect all ion beams and largely cancel out when multiple ion beams are monitored and used to calculate a ratio.

A last advantage of the ratio-of-ratios lies in consideration of equilibrium reactions, where one is comparing the isotope fluxes in an exchange between two substances. The forward fluxes derive from the reaction rates for the light and heavy isotopes, $L_{FORWARD}$ and $H_{FORWARD}$, and from the reverse fluxes, $L_{REVERSE}$ and $H_{REVERSE}$. The fractionation in the forward direction is $L_{FORWARD}/H_{FORWARD}$ and the fractionation in the reverse reaction is $L_{REVERSE}/H_{REVERSE}$, with the overall fractionation α between two substances in equilibrium given as

$$\alpha = (L_{FORWARD}/H_{FORWARD})/(L_{REVERSE}/H_{REVERSE})$$

or

$$\alpha = (H_{REVERSE}/L_{REVERSE})/(H_{FORWARD}/L_{FORWARD}).$$

This last quantity is again a ratio-of-ratios that can be calculated from δ values of substances participating in the exchange reaction. In effect, the δ values are convenient expressions of the overall isotope fractionation occurring between the two equilibrated substances (see also Section 7.6 in Chapter 7).

In conclusion, using ratios allows tracking of two quantities at once, in this case the differential behavior of the light versus heavy isotopes. Although ratios can be confusing, they are invaluable at many levels for tracking the many isotope twins, triplets, and "multiplets" of the HCNOS elements.

2.5 Chapter Summary

Most ecologists currently do not make isotope measurements themselves, and instead send off samples to specialized laboratories for the analytical work. But it is helpful to have some idea of how the measurements are made in those analytical laboratories, and to understand the notation used in the reports that come back from the isotope labs. This chapter aims to help the novice understand the isotope notations and how isotope measurements are made.

Mass spectrometers currently make most isotope measurements, comparing samples to standards for a difference measurement, or δ value. Most standard materials currently used in isotope work are distributed from the International Atomic Energy Agency in Vienna, Austria. But some standards are free, especially nitrogen and oxygen gas in the atmosphere that provide reference points for $\delta^{15}N$ and $\delta^{18}O$ measurements.

Measurement shows that natural materials can be enriched or depleted in heavy isotopes relative to the standard materials, with positive δ values reflecting enrichment and negative δ values reflecting depletion. It is often confusing for beginners that the isotope values can be positive or negative, but this confusion starts to clear once one remembers that δ values are a difference measurement, not an absolute concentration measurement. With this in mind, it is also good to reiterate that a δ value of zero means no difference from the standard, or the same as the standard, not zero amounts or zero concentrations. All natural samples and standards have appreciable, nonzero isotope concentrations. Negative δ values mean a smaller percentage of heavy isotope than is present in the standard, not negative amounts of isotopes.

The δ difference measurements also have unfamiliar units, units given as ‰, parts per thousand or permil. Note that permil is similar to percent, the difference being that permil denotes parts per thousand whereas percent denotes parts per hundred. Samples that are ten permil different than a standard are one percent different, so that 10‰ = 1%. Most ecologists should say "permil" out loud a few times to get used to these ‰ units used for the δ measurements.

These are basic points for beginners to read about in Section 2.1 of this chapter. More advanced readers can consult the remaining sections of this chapter for details of notation and measurement. Unfortunately, many isotope terms are used differently throughout the literature, but consulting these more advanced sections can help clarify alternative usages and terminologies. Sections 2.2 to 2.4 present some of the philosophy and algebra that underlies the δ scale, and show how to convert to the main alternate notations, the ratio (R) notation and the atom percent (AP) notation. The atom percent notation is generally recommended when dealing with enriched samples, and Technical Supplement 2A on the accompanying CD shows that using the δ notation for enriched samples often will result in errors that can be easily avoided by converting the δ values to atom percent values. Technical Supplement 2B on the accompanying CD shows how mass spectrometer data is used to calculate the δ isotope values. Technical Supplement 2C on the accompanying CD shows some elegant mathematics of the ratio notation, and how to write exact fractionation equations with this notation.

Most of the important information about notation and measurement is contained in Section 2.1, and beginning readers should probably focus on that section. The advice is thus to read Section 2.1 then skip on to Chapter 3, until more advanced information about notation and measurement is needed. The more advanced material is in Sections 2.2 to 2.4 and in Technical Supplements 2A, 2B, and 2C in the Chapter 2 folder on the accompanying CD.

Further Reading

Section 2.1

Avak, H., A. Hilkert A, and R. Pesch. 1996. Forensic studies by EA-IRMS. *Isotopes in Environmental and Health Studies* 32:285–288.

Barrie, A. and S.J. Prosser. 1996. Automated analysis of light-element stable isotopes by isotope ratio mass spectrometry. In T.W. Boutton and S. Yamasaki (eds.), *Mass Spectrometry of Soils.* Marcel Dekker, New York, pp. 1–46.

Brand, W.A. 1996. High precision isotope ratio monitoring techniques in mass spectrometry. *Journal of Mass Spectrometry* 31:225–235.

Boutton, S.W. 1991. Stable carbon isotope ratios of natural materials: I. Sample preparation and mass spectrometric analysis. In D.C. Coleman and B. Fry (eds.), *Carbon Isotope Techniques.* Academic Press, San Diego, CA, pp. 155–172.

Bowling, D.R., S.D. Sargent, B.D. Tanner, and J.R. Ehleringer. 2003. Tunable diode laser absorption spectroscopy for stable isotope studies of ecosystem-atmosphere CO_2 exchange. *Agricultural and Forest Meteorology* 188:1–19.

Brenna, J.T., T.N. Corso, H.J. Tobias, and R.J. Caimi. 1997. High-precision continuous-flow isotope ratio mass spectrometry. *Mass Spectrometry Reviews* 16:227–258.

Coplen, T.B. 1983. Comparison of stable isotope reference standards. *Nature* 302:236–238.

Coplen, T.B. 1996. Ratios for light-element isotopes standardized for better interlaboratory comparion. *EOS* 77:255–256.

Coplen, T.B. and H.R. Krouse. 1998. Sulphur isotope data consistency improved. *Nature* 392:32.

Corso, T.N. and J.T. Brenna. 1997. High-precision position-specific isotope analysis. In *Proceedings of the National Academy of Sciences* 94:1049–1053.

Ding, T., S. Valkiers, H. Kipphardt, P. De Bievre, P.D.P. Taylor, R. Gonfiantini, and R. Krouse. 2001. Calibrated sulfur isotope abundance ratios of three IAEA sulfur isotope reference materials and V-CDT with a reassessment of the atomic weight of sulfur. *Geochimica et Cosmochimica Acta* 65:2433–2437.

Farquhar, G.D., J.R. Ehleringer, and K.T. Hubick. 1989. Carbon isotope discrimination and photosynthesis. *Annual Review of Plant Physiology and Plant Molecular Biology* 40:503–537.

Fry, B. 2003. Steady state models of stable isotope distributions. *Isotopes in Environmental and Health Studies* 39:219–232.

Ghosh P. and W.A. Brand. 2003. Stable isotope ratio mass spectrometry in global climate change research. *International Journal of Mass Spectrometry* 228:1–33.

Griffiths, I.W. 1977. J.J. Thompson—the centenary of his discovery of the electron and his invention of mass spectrometry. *Rapid Communications in Mass Spectrometry* 11:1–16.

Hayes, J.M. 2002. Practice and principles of isotopic measurements in organic geochemistry. http://www.nosams.whoi.edu/docs/IsoNotesAug02.pdf

Hayes, J.M. 2004. An introduction to isotopic calculations. http://www.nosams.whoi.edu/docs/IsoCalcs.pdf

Hoefs, J. 2004. *Stable Isotope Geochemistry*, 5th Edition. Springer-Verlag, New York.

Hulston, J.R. and H.G. Thode. 1965. Variations in the S^{33}, S^{34} and S^{36} contents of meteorites and their relation to chemical and nuclear effects. *Journal of Geophysical Research* 70:3475–3484.

Lancaster, J. and S. Waldron. 2001. Stable isotope values of lotic invertebrates: Sources of variation, experimental design, and statistical interpretation. *Limnology and Oceanography* 46:723–730.

Los Gatos Research. www.LGRinc.com.

Mook, W.G. 2000. *Environmental Isotopes in the Hydrological Cycle, Principles and Applications*. Available on-line from http://www.iaea.org/programmes/ripc/ih/volumes/volumes.htm

Mulvaney, R.L. 1993. Mass spectrometry. In R. Knowles and T.H. Blackburn (eds.), *Nitrogen Isotope Techniques*. Academic Press, San Diego, CA, pp. 11–58.

Sieper, H-P., H-J. Kupka, T. Williams, A. Rossmann, S. Rummel, N. Tanz, and H-L. Schmidt. 2006. A measuring system for the fast simultaneous isotope ratio and elemental analysis of carbon, hydrogen, nitrogen and sulfur in food commodities and other biological material. *Rapid Communications in Mass Spectrometry* 20:2521–2527.

Section 2.2

Fry, 2003. Listed above; see Section 2.1 readings.

3
Using Stable Isotope Tracers

Overview

Isotopes are forms of an element that differ in the number of neutrons. Isotopes function as natural dyes or colors, generally tracking the circulation of elements. Isotopes trace ecological connections at many levels, from individual microbes to whole landscapes. Isotope colors mix when source materials combine, and in a cyclic process that ecologists can appreciate, the process of isotope fractionation takes the mixed material and regenerates the sources by splitting or fractionating the mixtures. Elements and their isotopes circulate in the biosphere at large, but also in all smaller ecological plant, animal, or soil systems. Chapter 3 reviews this circulation for each of the HCNOS elements, then gives four short reviews that may stimulate you to think about how you could use isotopes in your own ecological research.

3.1. *Isotope Circulation in the Biosphere.* This section shows that elements and their isotopes circulate in characteristic ways in the biosphere. This circulation leads to regular isotope patterns and distributions that are described in this section. Ecologists should learn these isotope distributions as starting points for using isotopes.

3.2. *Landscape Ecology and Isotope Maps.* This is the first of four mini-reviews about how ecologists currently use stable isotope tracers in their ecological research. Isotopes often have strong spatial signals that can be summarized in maps. The isotope maps are useful for understanding the ecology of changing landscapes.

3.3. *Community Ecology and Invasive Species in Food Webs.* Many ecologists use isotopes to help study feeding habits of species and communities, including human communities. The isotope measurements track effects of species invasions on ecosystem biogeochemistry, food webs, and diets.

3.4. *Life History Ecology and Animal Migrations.* Many species move and migrate, and isotopes are powerful tools for studying the migrations of individual animals.

3.5. *Plants, Microbes, and Scaling Up.* There is a large-scale planetary circulation of elements that we humans increasingly perturb. Ecologists interested in plants and microbes are using isotopes to help track changes in this cycling at both small and large scales.

Main Points to Learn. Reading this chapter may leave you with the impression that there is no limit to the ecological uses of the chemical isotope tracer measurements. This is an accurate impression. The isotopes provide a second opinion in amazingly diverse studies, helping to test ecological ideas from a different, tracer-based perspective. Ecologists do well to learn the general outlines of isotope circulation in the biosphere (Section 3.1), then apply this knowledge in more detail in specific studies such as those reviewed in Sections 3.2 to 3.5. The reviews highlight some of the classical strengths of isotope studies. Section 3.2 shows that isotopes measured along transects and in depth profiles record shifts in ecosystem organization related to space and time, information valuable for landscape ecology and historical ecology. Isotopes are also valuable chemical descriptors of niche space (Section 3.3), with shifts in nutrition, food web organization, and community ecology readily observed and measured with isotopes. Studies of individual species and animal behavior are also feasible with isotopes, so that it is possible to follow animal movements and bird migrations (Section 3.4). Lastly, isotopes are useful at the largest ecological scales, at the level of global ecology. Isotope measurements are currently part of a global effort to budget sources and sinks of atmospheric CO_2, a gas important in climate change that affects all ecology on this planet (Section 3.5).

3.1 Isotope Circulation in the Biosphere

Elements and isotopes circulate in the biosphere, and fractionation and mixing combine to produce regular, characteristic isotope distributions on this planet. The amounts of elements and isotopes involved in the circulation are important, with large pools providing points of stability and poise in the overall isotope circulation. Several of these large stable pools are in the ocean, including ocean water itself for hydrogen and oxygen isotopes, the inorganic carbon pool in the ocean for carbon isotopes, and sulfate in the sea for sulfur isotopes. The atmospheric reservoir of N_2 provides an important large pool for the N cycle. All these biosphere pools are part of even larger geochemical reservoirs circulating through the planet on geological time scales. Of course, humans such as you and I are very tiny droplets in these big pools. It is good to maintain a sense of perspective, isn't it?

Against this slowly changing background of the large, well-buffered pools of elements and isotopes, fractionation is an agent of change. During uptake of nutrients and CO_2 by plants and microbes, fractionation typically results in the strongest CNS isotope signals seen in the biosphere. For H and O isotopes, fractionation during reactions involving water (H_2O) are important controls. These fractionation reactions for all five HCNOS elements produce labeled substances that then mix and recombine, offsetting and ultimately erasing the effects of fractionation. That is the overall circulation, a process of generating isotope signals from large reservoirs via fractionation, then erasing these signals via mixing. This is the give and take of isotope ecology in action.

You might imagine that anything is possible in this world of circulating isotopes, especially if you are thinking of some of the charts of complex reactions in metabolism or material flows in ecosystems. But in fact there are characteristic patterns of isotope values in the biosphere. For the CNS isotopes, these characteristic isotope patterns (Figures 3.1 to 3.3) result from several factors, including large reservoirs, a few key enzymes involved in resource uptake, and common ecological stoichiometries that link together the organic matter cycles. Probably, it would have been difficult to predict in advance what these isotope patterns would be, but the advice now is just to accept them as they are. As you progress through this book you will understand more about how these patterns develop. Spend some time

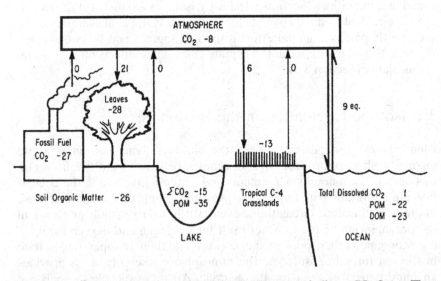

FIGURE 3.1. $\delta^{13}C$ distribution in ecosystems. Single arrows indicate CO_2 fluxes. The double arrow signifies an equilibrium isotope fractionation. Numbers for pools indicate $\delta^{13}C$ values (‰) and numbers of arrows indicate the fractionation (Δ, ‰) occurring during transfers. Negative $\delta^{13}C$ values indicate that less heavy isotope is present than in the standard (which has a 1.1% ^{13}C content; Table 1.2), not that isotope concentrations are less than zero. (From Peterson and Fry, 1987. Reprinted, with permission, from the *Annual Review of Ecology and Systematics*, Volume 18, copyright 1987 by Annual Reviews www.annualreviews.org.)

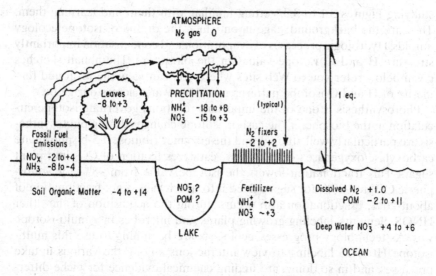

FIGURE 3.2. Representative $\delta^{15}N$ values in natural systems. See Figure 3.1 for explanation of symbols. (From Peterson and Fry (1987). Reprinted, with permission, from the *Annual Review of Ecology and Systematics*, Volume 18, copyright 1987 by Annual Reviews www.annualreviews.org.)

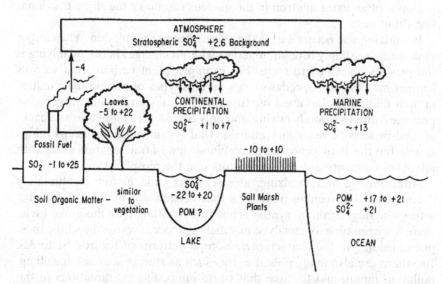

FIGURE 3.3. Representative $\delta^{34}S$ values in natural systems. See Figure 3.1 for explanation of symbols. (From Peterson and Fry (1987). Reprinted, with permission, from the *Annual Review of Ecology and Systematics*, Volume 18, copyright 1987 by Annual Reviews www.annualreviews.org.)

studying Figures 3.1 to 3.3, getting familiar with them and learning them. They are the background stage upon which the drama of isotope ecology unfolds. Hydrologic processes of evaporation and condensation importantly structure H and O isotope values in the biosphere. Thumbnail sketches given below refer you to Web sites where you can view the profound regularity of H and O isotope patterns found for water across our globe.

Photosynthesis is one of the important reactions governing isotope circulation in the biosphere. The carbon isotope changes during photosynthesis are particularly well-studied, and the enzyme "ribulose-1,5-bisphosphate carboxylase/oxygenase" or "Rubisco" catalyzes fixation of CO_2 into plant sugars. This fractionation lowers the isotope value from $-8‰$ for atmospheric CO_2 to $-28‰$ for sugars fixed into tree leaves, a net fractionation of about $20‰$. Fractionation also occurs during the acquisition of the other HNOS elements, labeling growing plants and microbes in a multi-isotope way. As technology progresses, ecologists are beginning to use this multi-isotope HCNOS labeling to view interactions among the various uptake processes, and in so doing, are finding chemical evidence for niche differentiation among different species of plants and microbes. Fractionation is thus useful for understanding how plants and microbes function and differentiate in the biosphere. When bonds are formed or broken at the atomic level, this is the site of fractionation and label introduction. Said another way, when chemical bonds change during uptake and loss reactions, the presence of an extra neutron in the nucleus can make the slight fractionation difference.

But mixing also occurs during the overall isotope circulation. When larger molecules are simply brought together without changes in bonds, mixing is the process on which to focus. For example, plant carbon from various sources mixes in soils, predators mix various types of prey organic matter in their diets, microbes use a mixture of substrates, and so on. Most ecologists use isotopes to study mixing, and only consider fractionation as a background process of secondary interest. That is fair and helps keeps things simple, but this book promotes a combined view of fractionation along with mixing as elements and isotopes circulate in the biosphere.

After mixing and remixing, almost all organic matter is ultimately decomposed into simple molecules that accumulate in large reservoirs, setting the stage again for synthesis and fractionation. With this grand cycle, there has been time for steady-state labeling to occur across the whole biosphere, leading to the characteristic isotope patterns of Figures 3.1 to 3.3. But there are also many pulsed events such as storms, seasons, upwelling, pollution inputs, and the like that cause interesting perturbations in the isotope patterns. Ecologists do well to take advantage of both the average isotope distributions and deviations from these averages, and to remember that there is a general biogeochemical framework for isotope circulation. This framework includes a variety of geological and chemical reactions as well as biological reactions, and developing a wider perspective that includes chemical and geological considerations will add breadth to eco-logical interpretations.

This section concludes with short thumbnail descriptions of the isotope distributions for each of the HCNOS elements, then the chapter moves on to four short essay reviews in Sections 3.2 to 3.5. The reviews illustrate how ecologists currently take advantage of the following isotope patterns in the biosphere.

Overview of HCNOS Distributions in Nature

The next paragraphs focus on the CNS elements that are linked in organic matter cycling and then on the HO elements that are more linked in the hydrological cycle. Figures 3.1 to 3.3 give a comparative overview of CNS stable isotopic distributions in the biosphere, and are taken from Peterson and Fry (1987). The interested reader should consult the references in that study for origin of many statements made in the following. Also, the reader should note that the δ values shown in Figures 3.1 to 3.3 are representative and do not encompass the full spectrum of observed values. For more indi-vidual element-by-element isotope information, Hoefs (2004) gives many useful summaries for the HCNOS elements in the book *Stable Isotope Geochemistry*.

The Carbon Cycle

The carbon cycle involves active exchanges of CO_2 among the atmosphere, terrestrial ecosystems and the surface ocean (Figure 3.1). The $\delta^{13}C$ value of atmospheric CO_2 is decreasing in response to inputs of ^{13}C depleted CO_2 from fossil fuel plus biomass burning and decomposition. Over the past 100 years the decrease may have been almost 1‰, from about −7‰ to −8‰. Carbon uptake by the dominant C_3 plants on land involves a net frac-tionation of about 20‰ between the atmospheric CO_2 and plant biomass (−28‰). Carbon uptake by C_4 plants, mainly tropical and salt grasses, involves a small net fractionation of about 5‰. Soil organic matter globally contains severalfold more carbon than either the atmosphere or living plant biomass and in general is similar or slightly enriched in ^{13}C in comparison with the dominant vegetation. Although either differential preservation or mineralization of soil components with different $\delta^{13}C$ values does lead to gradual shifts in soil ^{13}C content, on average there is little fractionation of respired CO_2.

The exchange of CO_2 between the atmosphere and the surface of the ocean involves an equilibrium chemical fractionation between atmospheric CO_2 (−8‰) and the total CO_2 (ΣCO_2, mostly bicarbonate) in surface ocean water (about 1‰). The withdrawal of carbon to form carbonates involves small isotope fractionations whereas uptake of dissolved inorganic carbon in planktonic photosynthesis involves larger kinetic fractionation that results in algal values of about −19 to −24‰. Both the dissolved and the particulate organic matter in the oceans predominantly have a marine planktonic origin.

The ^{13}C contents of components of the carbon cycle of fresh waters vary widely depending on the source of dissolved CO_2 in the waters. These sources include carbonate rock weathering, mineral springs, atmospheric CO_2, and organic matter respiration. Where respiration inputs are strong, $\delta^{13}C$ values for dissolved inorganic carbon may approach −20‰, and algae that further fractionate during carbon uptake can measure −45‰.

Animal carbon isotopes usually reflect and index the time-integrated average diet, with little-known French research performed in the early 1970s first firmly establishing this now time-honored relationship, "You are what you eat." Section 3.3 below considers revising this isotope maxim to a slightly more general formulation, "You are what you eat less excrete."

The Nitrogen Cycle

Most nitrogen in the biosphere is present as N_2 gas in the atmosphere. This massive reservoir is well mixed with an isotope composition that is essentially constant at 0‰. Nitrogen in most other parts of the biosphere also has an isotope composition near the 0‰ value, from −10 to +10‰ (Figure 3.2), primarily because the rate of nitrogen supply often limits reactions such as plant growth and bacterial mineralization. Under these conditions all available nitrogen can be consumed, without regard to isotope content and with no overall isotope fractionation. Thus, slow rates of N supply and limiting amounts of substrate N are often important for understanding nitrogen isotope distributions.

Some cumulative and large fractionations do occur in the nitrogen cycle. A cumulative faster loss of ^{14}N than ^{15}N during particulate N decomposition results in ^{15}N increases of 5 to 10‰ with increasing depth both in soils and in the ocean. Nitrification and denitrification in the sea both proceed with substantial isotope effects (Δ = 10 to 40‰), and where nitrate is abundant, assimilation by phytoplankton proceeds with a smaller effect (Δ = 4 to 8‰).

Lakes appear more variable in isotope composition than the large world ocean. Large isotope contrasts might be expected between lakes in which primary production is limited by N (little fractionation by phytoplankton) versus P (abundant N → large possible fractionations during N uptake by phytoplankton). Where phytoplankton have different $\delta^{15}N$ values than terrestrial vegetation, the nitrogen isotopes may function as source markers for autochthonous and allochthonous organic matter. This approach has been successfully applied in marine environments.

There is a wide range reported for nitrogen isotope values for ammonium and nitrate in precipitation from about −20 to 10‰ (Figure 3.2). Some of the more negative values are related to soil and anthropogenic emissions in highly industrialized areas. Section 3.2 in this chapter presents an example using $\delta^{15}N$ measurements to trace human pollutant plumes. Further study may show that stable isotope studies are helpful in identifying the sources and fates of N that human activities are currently adding to many forests and lakes. $\delta^{15}N$ increases regularly in natural and human-affected food webs, and has been used widely to estimate trophic levels in natural systems (see Section 3.3).

The Sulfur Cycle

Sulfate in the ocean is a large well-mixed sulfur reservoir whose isotope composition is 21‰ heavier than primordial sulfur in the earth and solar system at large. This primordial sulfur is represented by the isotope standard, Cañyon Diablo Troilite, sulfur from a meteorite that crashed near Flagstaff, Arizona about 50,000 years ago. Fixation of sulfate by phytoplankton occurs with a small isotope effect ($\Delta = 1$ to 2‰), but dissimilatory sulfate reduction in marine sediment occurs with a large effect of 30 to 70‰. Over geological time, and partially in response to global-scale fluctuations in sulfate reduction activities, the $\delta^{34}S$ values of oceanic sulfate have varied from about +10 to +33‰. Uplift and preservation of marine sedimentary sulfides and sulfate-containing evaporites on land have produced a patchwork of sulfur in terrestrial environments, each with different $\delta^{34}S$ values for bedrock sulfur. Thus, large $\delta^{34}S$ ranges must be assigned in general sulfur cycle diagrams (Figure 3.3). In spite of this, continental vegetation seems to average near +2 to +6‰ over large areas and is quite distinct from the ±17 to +21‰ values of marine plankton and seaweeds.

The stable isotope composition of sulfur entering the atmosphere can also be quite variable. For instance, SO_2 emissions from a sour gas plant in Alberta, Canada vary between +8 and +25‰, but in eastern Canada and the northeastern United States initial studies show average ambient $\delta^{34}S$ values of 0 to 2‰ for SO_2. The oxidation of SO_2 to sulfate occurs with an overall inverse effect ($\Delta = -4$‰) that favors concentration of the heavy ^{34}S isotope in the product sulfate; this inverse effect arises from an equilibrium step between SO_2 and HSO_3^- prior to final oxidation to sulfate. Oxidation of other sulfur-containing molecules also occurs with small isotope effects, $\Delta < 5$‰.

Rainfall sulfate over the open oceans has a significantly lower $\delta^{34}S$ value than sea-spray sulfate (±13 versus +21‰), possibly because of slow oxidation of reduced sulfur gases. Continental sulfates usually have much lower $\delta^{34}S$ values of 0 to 10‰. The isotope compositions of gases such as H_2S, carbonyl sulfide, and dimethyl sulfide are poorly known, but further study may clarify the relative contributions of human versus natural sources of these atmospheric sulfur compounds.

H and O Isotope Varieties of Water, H_2O

$^1H^1H^{16}O$	>99%
$^1H^1H^{18}O$	0.2%
$^2H^1H^{16}O$	<0.01%
$^2H^2H^{16}O$	heavy water
$^2H^2H^{18}O$	heaviest water

FIGURE 3.4. There are several stable isotope varieties of water, some of which are shown here. Heavy water $^2H^2H^{16}O$, which is double-deuterated water or D_2O, is very rare in nature, but can be produced in quantity in specialized isotope-separation laboratories. D_2O is a common laboratory solvent for nuclear magnetic resonance (NMR) studies of chemical compounds.

The Hydrogen Cycle

Much of the hydrogen cycle involves water (Figure 3.4), with various processes in the water cycle leading to characteristic, large-scale geographic patterns of hydrogen isotopes in water (see, e.g., Figure 3.9 in Section 3.4). Ocean water is the main reservoir of hydrogen in the biosphere and the standard reference material (standard mean ocean water or "SMOW") for hydrogen isotope measurement. The isotope composition of ocean water represents a good starting point for following isotope dynamics in the hydrological cycle. The transitions between liquid water and water vapor during evaporation and condensation involve kinetic and equilibrium reactions with isotope fractionation. Water vapor evaporating from the sea has δ^2H (δD) values of −10 to −20‰, and as this process reverses during condensation and formation of rain and snow, this trend towards lower atmospheric δD values is amplified. As water vapor moves inland and up mountains, it progressively loses moisture and δD values decline further.

These processes can be amplified yet again in colder regions where low temperatures promote stronger fractionations between vapor and condensate. A combination of high elevation and low temperature can result in δD values of −200 to −400‰ for water in high-elevation glaciers and for snow in polar regions. In less dramatic examples of these same inland "rainout" effects, large rivers that often have continental origins and are fed by snowmelt can have much lower δD values than coastal marine waters. This makes δD source signals valuable tracers in coastal estuaries and floodplains linked to these rivers.

Isotope hydrology studies often consider the water isotopes (hydrogen and oxygen isotopes) as markers for water sources and water circulation. Global-scale maps and animations of hydrogen and oxygen isotope variations in water are conveniently available on the Web (see, e.g., http://isohis.iaea.org/userupdate/waterloo/index.html or www.waterisotopes.org). δD can also be used to track sources of urban water, and the increasing influence of humans on our planetary water cycle.

Analytical advances are making it easier to investigate the origins and cycling of hydrogen bound in organic matter. About 10 to 20% of hydrogen in organic materials is exchangeable with water vapor present in normal laboratory air, but this exchange effect is understood and can be corrected for during routine analysis. With these corrections, studies of plants indicate strong hydrogen isotope fractionations during photosynthesis, $\Delta = 170‰$ versus source water. Lipids appear to preserve much of this original fractionation because they have δD values of about $-200‰$ versus source water. However, most plant hydrogen is affected by additional fractionation occurring during cellulose formation, in processes that leave organic material enriched in δD and largely offset the original photosynthetic fractionation. In the end, fractionations in photosynthesis, cellulose formation, and (for terrestrial plants) transpiration largely cancel so that resulting δD values are strongly correlated with those of local water. With care, it is possible to use plant δD as an indicator of source water δD, a parameter widely useful in reconstructions of past climates, ecologies, and hydrological regimes.

Hydrogen in animal tissues can be divided into three main pools: hydrogen derived from dietary sources, hydrogen from drinking water, and exchangeable hydrogen. One investigator drank deuterium-enriched water and monitored the hydrogen isotopes in hair of his beard, finding that 10% of all hydrogen was exchangeable, 30% came from ingested water, and the remaining 60% came from food. A similar finding was observed for quail fed in captivity. The overall finding for animals is that δD values are primarily controlled by diet, which in turn is strongly correlated with δD values of local water. Section 3.4 considers an example in which hydrogen isotopes help track diets and movements of migrating animals.

The Oxygen Cycle

There are three oxygen isotopes that act as tracers when the many common oxygen-containing molecules circulate in the biosphere (Figure 3.5). The water cycle controls much of the oxygen dynamics and oxygen isotope dynamics. Evaporation and condensation result in predictable variations in

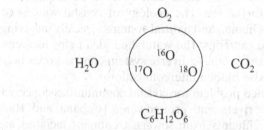

FIGURE 3.5. Three stable isotopes of oxygen (center) are present in common compounds (periphery) that circulate in the biosphere.

isotope compositions of water that are now routinely tracked at regional and global levels (see http://isohis.iaea.org/userupdate/waterloo/index.html, http://www.waterisotopes.org/ and http://ecophys.biology.utah.edu/labfolks/gbowen/pages/Isomaps.html#IAEA).

But dioxygen O_2 gas is also important, comprising about 21% of the gas in our atmosphere. Several sections of Chapter 4 consider O_2 dynamics in metabolic contexts, with photosynthesis producing oxygen and respiration consuming that oxygen. Isotopes are used increasingly to track the global balance in these two processes. Oxygen isotopes in CO_2 are briefly considered in Sections 3.5 and 7.6.

Oxygen in organic matter is partly exchangeable with environmental water, and most studies of organic oxygen remove or factor out this exchangeable oxygen before focusing on the bound nonexchangeable oxygen. The $\delta^{18}O$ values of nonexchangeable oxygen in cellulose is generally enriched in ^{18}O by about 27‰ versus source water, possibly reflecting fractionations involved in the equilibration of CO_2 with water. Oxygen isotope studies with animals have focused on determining which local sources of water are used. The degree to which food influences $\delta^{18}O$ variations in animals has not been determined fully.

3.2 Landscape Ecology and Isotope Maps

One important landscape for human management is the watershed. Along marine coasts, watersheds are lands that share a common hydrological connection, a drain to the sea. The ecology of coastal watersheds is an interlinked ecology of human and natural systems typically involving towns, forest, grasslands, and estuaries. The watershed idea helps us focus on the unified and linked nature of these diverse systems, and isotopes help clarify many of these otherwise hidden interconnections.

A common watershed problem for coastal communities concerns wastewater discharge into rivers and groundwater (Cabana and Rasmussen 1996). Nitrogen-rich effluents from sewage treatment facilities are especially potent fertilizers in the downstream coastal zone where N is the element typically limiting primary production. Watershed N additions lead to downstream algal blooms. Some of these blooms involve toxic algal species, and some can increase respiration so much that low oxygen conditions and fish kills develop. Generally algal blooms and eutrophication lead to unsightly conditions and public concerns to restore water purity.

But it is not always easy to pinpoint the sources of these polluting nutrients. There are often multiple sources involved, and algae rapidly take up the nutrients then disperse with the tides, in effect diluting the evidence. Against this background, one team of Australian researchers decided to deploy a set of bioindicator algae in fixed containers (Costanzo et al. 2001). The algae would take up the nutrients and acquire a high ^{15}N signal characteristic of pollution N from watersheds. The algae grew in incubation chambers deployed across the bay in a spatial grid, so that the results could be mapped and matched to potential sources along the coastal watershed. Algae were placed in clear, flow-through chambers and allowed to grow and absorb N nutrients for four days before harvest and analysis. The results clearly identified two ^{15}N hotspots with high δ^{15}N values along the populated western shore of Bramble Bay, Brisbane, Australia (Figure 3.6).

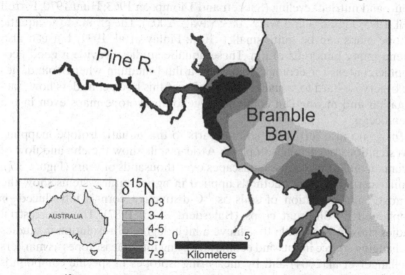

FIGURE 3.6. δ^{15}N values of algae in Bramble Bay, Australia where the city of Brisbane occupies the western shore. High δ^{15}N values along the western shore indicate N pollution inputs from watershed rivers and local sewage treatment facilities. The coastal pollution plumes are hard to identify by conventional measurements of ammonium and nitrate nutrients, because tides rapidly disperse nutrients and algae use up the nutrients during growth in algal blooms of the region. But the isotope values persist as nutrients are incorporated into the algae, tracing the nitrogen linkage to coastal inputs. Results are contoured for macroalgae that were incubated four days in situ at approximately 100 sites in September 1997, then analyzed for δ^{15}N (Costanzo et al. 2001). This δ^{15}N work continues now as a monitoring technique termed "sewage plume mapping" (Costanzo et al. 2005). (Reprinted from *Marine Pollution Bulletin* 42:149–156, S.D. Costanzo, M.J. O'Donohue, W.C. Dennison, N.R. Loneragan, and M. Thomas, A new approach for detecting and mapping sewage impacts. Copyright 2001, with permission from Elsevier.)

Sewage treatment facilities were at the center of these hotspots and thus implicated as sources of the high ^{15}N signals from ammonium and nitrate nutrient releases. This identification helps guide efforts to clean up the bay (Costanzo et al. 2001, 2005). Isotope maps in another marine study showed the time-course of cleanup and ecosystem recovery after closure of a coastal sewage plant (Rogers 2003). Overall, isotopes may be particularly well suited as indicators of ecosystem restoration because their distribution in plants and animals reflects the integrated biogeochemical cycling of elements plus the added effects of species-level interactions within food webs. Here the message might be, "If you want to restore the ecosystem, work on restoring the isotope distributions," or more simply, "Restore the isotopes, restore the system."

There are more and more examples of isotope maps used to investigate landscape ecology. In the sea, coastal nutrient inputs from humans and natural patterns of nutrient cycling combine to create landscape- (or "seascape-") level isotope maps that identify regions that differ in their inputs and nutrient cycling (Sackett and Thompson 1963; Hunt 1970; Farrell et al. 1995; Jennings and Warr 2003; Savage 2005). The grain of the aquatic isotope maps can be quite small (<10 m; Finlay et al. 1999) but can also extend many hundreds of km. These aquatic studies provide a good geographic context for ecological study in a fluid medium where boundaries are otherwise hard to visualize. Section 7.6 of this book considers how fractionation and mixing can combine to generate isotope maps even in the open ocean.

There are also terrestrial counterparts to this aquatic isotope mapping. Investigations of carbon isotopes in African soils show the ebb and flow of savannah and forest across landscapes over thousands of years (Figure 3.7). Similar carbon isotope methods applied in agricultural systems show the turnover and evolution of soils as ^{13}C-distinctive corn is introduced or supplanted by different crops (Balesdent et al. 1987). These terrestrial studies show it is possible to achieve a high degree of resolution in studies of changing source inputs and ecology by using multiple isotopes as markers (Bellanger et al. 2004) and by measuring isotopes in specific compounds (Wiesenberg et al. 2004).

Overall, the isotope maps should help us monitor and manage our changing landscapes in this time of great human influence (Vitousek et al. 1997), the new era some geologists term the Anthropocene. Chapter 6 considers deliberate isotope additions to natural systems to create isotope maps that change in time as well as in space. Isotope maps clearly identify and track large-scale ecological processes in today's world.

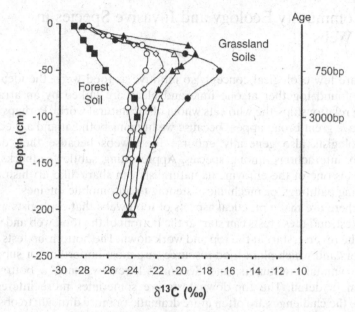

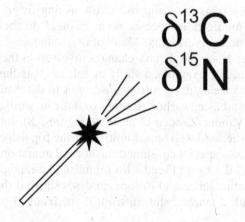

FIGURE 3.7. $\delta^{13}C$ values of soils from six sites in Gabon, Africa where C_4 savannah grasses (–12‰) and forest trees (–29‰) contribute to soil organic matter. Low values near –29‰ indicate landscapes dominated by forests, whereas high values approaching –12‰ indicate landscape-level shifts to open savannah. The square symbols give the isotope values for forest soils in a reference undisturbed system that has not been invaded by savannah. Considering the isotope profiles of the other nonreference soils as a history and reading from the bottom up, forests dominated the landscape until about 3000 years ago when the landscape shifted to open savannah, but this trend reversed about 750 years ago, with forests now dominating again. (From Delegue, M.-A., M. Fuhr, D. Schwartz, A. Mariotti, and R. Nasi. 2001. Recent origin of a large part of the forest cover in the Gabon coastal area based on stable carbon isotope data. *Oecologia* 129:106–113. This is reprint of Figure 2, p. 109 from the article and is used with permission from Springer.)

3.3 Community Ecology and Invasive Species in Food Webs

There are few ecological concepts so intuitive as food webs, the idea that species found together at one time and place are linked by an array of feeding relationships, the who eats whom of the natural world. Perhaps food webs have great broad appeal because we humans both eat and are edible, but ecologists also generally endorse food webs because they depict complex interactions among species. Appreciating subtle feedbacks and linkages is one of the enjoyments naturalists can share, like art historians discussing paintings, or mechanics listening to automobile engines.

But there are many practical aspects of food webs that ecologists would like to test, and these tests can start at the bottom of the food web and work up, or the reverse, start at the top and work down. The bottom-up tests seek to understand which plant food resources are most important for supporting the animal consumers, and Section 5.1 reviews such a bottom-up problem in detail. The top-down tests are sometimes more interesting because the challenges are often more dramatic or more difficult: to observe predation in action, or to infer its indirect effects. Today, ecologists recognize that both bottom-up and top-down effects are important in food webs, but there are also other influences that are not so easy to characterize, such as the effects of spatial pattern in structuring food webs or the effects of invasive species.

Invasive species introduced from other locations can have profound effects, with successful invaders often increasing to very large numbers and sometimes dominating ecosystem dynamics. In these cases, one would expect changes in element cycling that is part of the food web, and a corollary is that isotope distributions might be expected to change as well. Reversing this logic, isotope distributions might help us understand and document the effects of invasive species. Lake studies show this to be the case.

Many lake ecosystems are "improved" for fishing by introductions of nonnative and nonindigenous species. What changes do these new neighbors bring for the normal residents? Many of these changes are diet-linked, and isotope studies track the dietary changes involved in these ecosystem-level perturbations. For example, a study of fish in Canadian lakes tested effects of invasive species, using nitrogen isotopes to determine changes in fish trophic level, and using carbon isotopes to indicate shifts in the sources of fish nutrition (Vander Zanden et al. 1999; Figure 3.8). Introductions of bass fish changed the food web opportunities for the top fish carnivore, lake trout. Introduced bass species consumed the normal nearshore forage fishes important for the lake trout. Faced with diminished nearshore forage, the trout declined in abundance and isotope analyses showed that the decline was associated with a marked shift in trout diets. Trout in lakes with bass

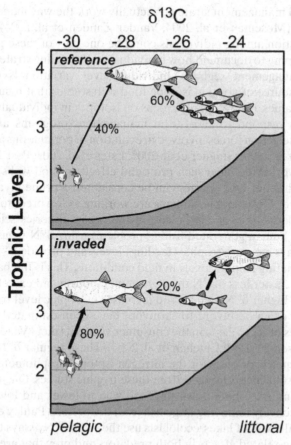

FIGURE 3.8. Effects of species introductions measured in lake ecosystems. Introduction of nearshore bass species forces the native top predator, lake trout, offshore. Reflecting this spatial displacement, lake trout diets shift towards feeding in a more pelagic food web (as measured by lower $\delta^{13}C$) and at a lower trophic level (as measured by lower $\delta^{15}N$; with $\delta^{15}N$ translated into the y-axis "trophic level" in this figure). Dietary shifts help explain the decline of lake trout in the invaded lakes. This figure summarizes results from comparative studies in different lakes and results for single lakes studied over time. (From Vander Zanden et al. 1999; used with the permission of the author and Nature Publishing Group. Copyright 1999.)

fed less on fish in general and also moved offshore to feed in a more planktonic food web. Trout were essentially deprived of many of the benefits of nearshore feeding by the competing bass. In this case, it would be interesting to try to reverse the effects by capturing and removing bass, and then use isotopes as part of the monitoring program to test whether trout quickly resumed feeding in nearshore areas.

Ecologists are wondering generally what measures are effective in restoring ecological relationships, and isotopes can help measure whether

implemented management strategies actually work the way managers hope
they should (Moseman et al. 2004; Vander Zanden et al. 2004). Studying
isotope variation among individuals could be one part of these restoration
attempts, helping to document how individuals shift feeding strategies under
different management regimes. Individual-level variation is commonly
treated as a source of noise in isotope food web studies that focus typically
on average values. But retaining a focus on isotopes in individuals can show
how animals respond to short-term management programs and also to
longer-term selection forces involved in evolution of populations and species
(Beaudoin et al. 1999; Guiguer et al. 2002; Estes et al. 2003; Post 2003).

Humans worldwide have such profound effects in food webs (Vitousek
et al. 1997) that our own species can be considered a prime invader at the
planetary level. For example, humans are working as top predators in many
food webs (Jackson et al. 2001), and in the ocean, fisheries scientists are
studying our human-generated fishing problems using $\delta^{15}N$ measurements.
The $\delta^{15}N$ assays measure trophic (feeding) levels and provide a "trophometer" for estimating trophic levels in field conditions. This is the basis for the
trophometer: faster loss of ^{14}N than ^{15}N in metabolism and excretion leaves
animals with higher $\delta^{15}N$ values. And increases in trophic level from a plant
to herbivore or an herbivore to carnivore have been estimated to involve
$\delta^{15}N$ increases of 2.2 to 3.4‰ in the consumer versus its diet (Vander Zanden
and Rasmussen 2001; McCutchan et al. 2003). These regular $\delta^{15}N$ increases
provide the metric or basis for the nitrogen isotope "trophometer".

Several results have emerged from these trophic studies. One conclusion
is that humans are fishing down the food web to lower and lower trophic
levels by selectively removing large fish from the oceans (Pauley et al. 2000).
At more local levels of lakes, ecologists use the same $\delta^{15}N$ assays to measure
subtle feeding-related effects in both predators and prey that are otherwise
difficult or impossible to measure (Branstrator et al. 2000; Kelly 2000;
Vander Zanden et al. 2003). Isotope studies show that even tourists visiting
wilderness preserves can have measurable incidental effects on lake food
webs, apparently via low-level nutrient inputs (Hadwen and Bunn 2004).
The ^{15}N-based estimates of trophic level are also very valuable for understanding how contaminants such as mercury, PCBs, and selenium circulate
in food webs (Yoshinaga et al. 1992; Kidd 1998; Stewart et al. 2004). Given
the profound, multilevel human effects on nutrients, contaminants, and
larger animals that all influence food webs, and that much of this cycles back
towards our own species, perhaps isotope ecology will evolve from the early
days of proclaiming, "You are what you eat" (DeNiro and Epstein 1976),
towards ecologically more balanced rules such as, "You are what you eat
less excrete in the planetary garden of the Anthropocene." Section 4.7 considers an eat–excrete example in detail, in part because it is increasingly
clear that metabolism and losses as well as dietary gains can influence
isotope values of animals (Tieszen and Fagre 1993; Ambrose 2000;
Sponheimer et al. 2003).

Effects of humans are also profound in terrestrial food webs, and isotopes are used to help characterize some of these effects. For example, isotope studies trace the chemical evolution of natural soils after clearing for crop production (Arrouays et al. 1995). For noncrop species such as N-fixing plants that can also invade landscapes with large effects (Rice et al. 2004), isotope studies should also provide a way to detect altered nitrogen dynamics for soils (Hobbie et al. 1998, 1999) and soil food webs (Ponsard and Arditi 2000). Nitrogen fixation provides a new source of N to ecosystems, a source with a well-characterized $\delta^{15}N$ value near −1‰, so that isotopes are useful for tracing N fixation inputs (Shearer and Kohl 1988). These nitrogen studies are increasingly important because new human N inputs to the biosphere from fertilizers and waste materials are increasing dramatically, with strong effects for this planet (Galloway et al. 1994, 2004).

Altered landscapes and altered plant productivities also affect animal ecologies and are of great concern for conservation. Stable isotope investigations currently target animal feeding relations in natural and human-altered landscapes. Studies focus on a very wide diversity of individual animal species, including, for example, elephants, bears, chimpanzees, and hummingbirds (Schoeninger et al. 1999; Felicetti et al. 2003; Carleton et al. 2004; Cerling et al. 2004). Terrestrial ecologists also use isotopes to estimate field metabolic rates of many types of animals including, for example, penguins and reindeer (Culik and Wilson 1992; Gotaas et al. 1997). Overall, using stable isotope tools to study animal feeding and metabolism will help conservation efforts.

Finally, we humans are the center of many food web and nutritional studies. Reading isotope studies about ancient humans will convince you rapidly that food has always been a centerpiece of human ecology and culture. Archaeological and anthropological studies have used isotope assays to illuminate topics such as the diet of hominids that lived millions of years ago, how food crops such as corn played an important organizing role in the nutrition and overall culture of agrarian societies, and when human mothers weaned their infants in these cultures (van der Merwe 1982; DeNiro 1987; Fogel et al. 1997; Sponheimer and Lee-Thorp 1999; Fogel and Tuross 2003; Lee-Thorp et al. 2003). These human-related concerns are moving forward into the current day with many medical studies using stable isotope tracers that are nonradioactive and safe for human consumption.

For example, there is a standard isotope assay now widely used by doctors for ulcer detection. The test involves patients ingesting ^{13}C-labeled urea that is not normally degraded in human stomachs. But when the ulcer-causing bacterium *Heliobacter pylori* is present, this bacterium readily degrades urea, liberating ^{13}C-enriched CO_2 that can be detected in the exhaled breath of the patient. Lasers as well as mass spectrometers are used to detect the ^{13}C-label in exhaled CO_2 in increasingly simple and inexpensive ways. Overall, this isotope test has become the reference "gold" standard for detecting the presence of the ulcer-causing bacterium (Fischer and Wetzel 2002).

Medical uses of the stable isotopes will likely accelerate as isotope measurements become easier, faster, and cheaper. Progress is being made routinely in this regard as laser technology gains sensitivity and accuracy, starting to replace mass spectrometric methods in some applications (Murnick and Peer 1994) with rapid recent advances in laser detection of isotope values for gases such as CO_2, CH_4, and H_2O (Los Gatos Research, www.LGRinc.com). It also seems likely that metabolic isotope studies will be made with increasing specificity at the biochemical level (Hayes 2001), targeting, for example, molecules such as cholesterol and specific atomic positions within that molecule to trace exact origins and fates of important compounds. Like DNA profiling, metabolic isotope profiling is likely to become increasingly sophisticated and useful, opening up new vistas on physiology and disease in the field of "isotopics" (Brenna 2001). Isotope profiling will be a powerful supplement to current food web studies that thus far focus almost exclusively on what is in the diet, rather than on the metabolic fates and performance achieved with the various diets. Individualized nutritional strategies are present in nature and important for evolutionary ecology (Mayntz et al. 2005). Combining metabolic isotope profiling with DNA fingerprinting is likely to give future scientists a very refined view of what diets are in the field, and what these diets mean for the metabolic performance of animals, including humans.

P.S. In case you were wondering, a diet free of heavy isotopes would not result in much weight loss. Once the heavy isotopes were replaced with light isotopes in an average 50 kg human, that human would have lost only about 0.03 kg (30 g). But if you were wondering in the other direction—would too much heavy isotope be a bad diet—there is this cautionary tale from the world of laboratory studies done in the 1940s and 1950s. There appears to be an upper limit of the amount of heavy isotope organisms can incorporate before normal metabolic enzymes start to fail. Katz (1960) summarizes research that metabolism shifts and pH values fall as low as 3 when algal cultures are transferred from normal H_2O to pure D_2O. Also, greatly enlarged "monster" cells appear for a time, cells that are apparently unable to divide. But after weeks or months of adaptation, cultures of algae in pure D_2O eventually produce cells that have metabolic capabilities very similar to those of cells grown in normal H_2O. The same applies for bacteria. But for mice, rats, and dogs, when D_2O levels are raised above 35%, death ensues, "death by isotopes." At lower D_2O levels near 20%, hyperirritability, anemia, convulsions, and sterility are sublethal symptoms that appear. The fortunate end of this story is that it is actually quite challenging to produce purified D_2O, although the nuclear industry does this routinely. Natural levels are always less than 0.02% D_2O on this planet. In summary, stable isotopes are nonradioactive safe tracers, and they won't help you or hurt you under any normal circumstance in your personal dietary planning.

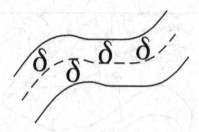

3.4 Life History Ecology and Animal Migrations

Some animal behaviorists tell students to sit quietly and watch for a year, until you begin to think like the animal that interests you. This requires patience and a peeling away of the human veneer to see things from another perspective not centered on ourselves. Most ecologists spend their careers gradually attuning to the natural world, and the study of individual species or autoecology is a centerpoint for many ecologists (Schaller 1966). There are a great many mysteries found at the species level, and adopting the viewpoint of another species gives provocative insight into the functioning of natural and human-altered systems (Sapolsky 2002).

One of the most spectacular aspects of species ecology is migration, especially bird migrations visible in the sky in the Northern Hemisphere during spring and fall months. Ecologists are interested in the costs and benefits of migration as a way to understand why some birds migrate short distances and others move thousands of kilometers. But before assessing costs and benefits, there is basic work to be done detecting where animals go and what they do once they get there. To work on this descriptive ecology, generations of bird enthusiasts have helped tag and band birds, but unfortunately tag returns are meager, and give a very fragmented view of average migration patterns. Isotope studies are providing a much increased resolution about bird migrations, and hydrogen isotopes are proving especially useful tools for tracking long-distance migrations.

The reasons why hydrogen isotope tracers work well for studies of long-distance migration rest on the fundamental chemistry of isotope fractionation. The water cycle of evaporation from oceans and precipitation inland involves isotope fractionations that leave behind the light isotopes. During chemical equilibrium fractionation, the vapor phase is enriched in the light isotopes, leaving the liquid phase heavier by difference or mass balance. The isotopically heavier liquid phase falls out as rain or snow, so that residual cloud-borne water moving inland or upwards is isotopically lighter and has lower δ^2H or δD values, where D stands for deuterium, ^2H. Larger fractionations accompanying cold conditions in polar regions magnify some of these patterns, creating low δD values nearer the poles. This sounds complex. However, it is straightforward in practice and leads to continen-

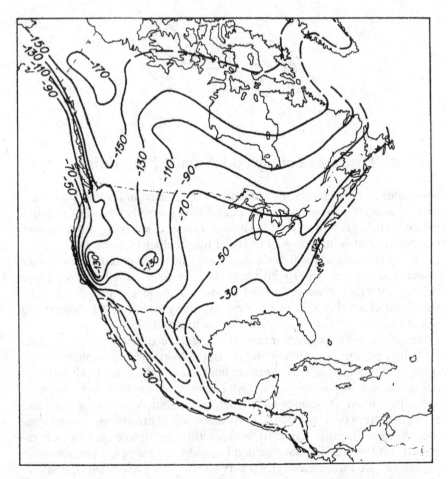

FIGURE 3.9. Isotope map of North America for precipitation δD values. Plant and animal δD values reflect this continental-level map. (From Taylor, Jr., H.P., 1974, *Economic Geology* 69(6), p. 850, Fig. 6.)

tal-level patterns in the isotope compositions of water, in effect a giant isotope map created by the water cycle (Figure 3.9).

   Bird migrations play out across this chemical landscape, with birds at high latitudes having low δD values and birds near the equator having high δD values. The isotopes in the water provide a bottom-up source signal that first labels plants during photosynthesis and carbohydrate metabolism, then leads to general labeling of the local food web. The end result is that organic matter in materials such as bird feathers will have δD values that reflect the local water δD values. Migrating birds typically molt and form new feathers at the end of the summer, and feathers retain that late-summer isotope chemistry until the next year's molt. Capturing birds by nonlethal netting, ecologists can weigh and measure a bird, extract a feather, then release the animal. Later analysis of the feather and matching feather δD values to the isotope map will show where the bird was at the time of the last molt. This matching strategy works best at continental levels for birds that live at lower altitudes. Birds that spend extended periods in high mountains may have low δD values because of mountain water values, not because of migrations to high latitudes.

A recent study of Wilson's Warblers used this natural tracer experiment to advantage (Kelly et al. 2002). These warbler birds summer in the northern part of North America, from 35 to 70°N, so birds from the farthest north should have the lowest δD values. When migrants were collected in the southern, winter part of their range, feathers showed an unexpected pattern, with the lowest δD values nearest the equator (Figure 3.10). These results suggest the following scenario. Birds that start from the farthest north in the late summer go farthest south in the fall and winter, bypassing or "leapfrogging" other warbler populations that move shorter distances. The advantages of this leapfrog migration are not entirely clear, but seem part of an overall cost–benefit strategy that has evolved over time. Energy costs associated with migration are large, and autoecological studies show much differentiation in migration patterns, probably reflecting differentiated strategies to minimize energy losses and maximize reproductive output.

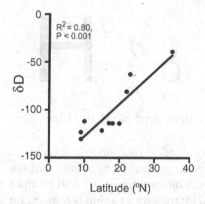

FIGURE 3.10. δD values of feathers collected from Wilson's Warblers that overwintered at sites from central America (10° N) to the southern United States (35° N). Animals collected farthest south at 10° N had the lowest δD values, so that their point of origin for the migration was in the far north (see Figure 3.9). These long-distance migrators moved past and leap-frogged over other populations that move much less during their fall and winter migrations (From Kelly, J.F., V. Atudorei, Z.D. Sharp, and D.M. Finch. 2002. Insights into Wilson's Warbler migration from analyses of hydrogen stable-isotope ratios. *Oecologia* 130:216–221. This is a reprint of Figure 6 on p. 219 of the article, used with permission from Springer.)

Migrations are common for many species besides birds, and natural isotope tags are useful generally for studying the ecological phenomenon of migration. The natural isotope tags require no prehandling or disturbance of animals, the tags are present in all members of a species, and they reflect the fundamental feeding and habitat use patterns of individuals (Wassenaar and Hobson 1998; Hobson 1999, 2002; Rubenstein and Hobson 2004). The isotopes also can be assayed from small amounts of material that are collected without killing an animal, for example from feathers, blood, feces, or respired CO_2 (Podlesak et al. 2004). The isotope assays can be made for historical specimens preserved in museums or in nature. Analysis of isotopes in migratory species has become important for conservation biology, not only for bird species (Rubenstein et al. 2002), but also for species such as endangered' whales (Killingley 1979; Schell et al. 1989) and turtles (Killingley and Lutcavage 1983). An important perspective from these studies is that sustaining the habitats and migration corridors is not only important for species, but also for the ecosystems hosting these species. Isotope studies of migrating salmon that return to spawn in small streams (Kline et al. 1990; Naiman et al. 2002) especially show that migrants provide numerous subsidies and feedbacks that link ecosystems across landscapes. Studies of commercial shrimp migrations also show that such linkages are important for sustainable fisheries (Fry 1981, 1983).

Problem 5 in Chapter 7 challenges you to follow animal migrations by building your own I Chi isotope model.

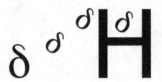

3.5 Plants, Microbes, and Scaling Up

Many of the examples cited in the previous sections occur against a background of isotope signals generated by plants and microbes. Animals are actually quite unimportant in terms of overall biomass and genetic diversity in our biosphere, interesting as animals may be for their close kinship to our human species. But working with plants and microbes is important for understanding circulation of elements and isotopes in the biosphere, partly because many of the fundamental isotope signals that mix through the biosphere are generated during fractionation as plants and microbes acquire nutrients and CO_2. These fundamental fractionations by plants and microbes give one the chance to scale up process measurements and regional observations to the global level. In fact, the isotope extrapolation doesn't stop there, but extends to other planets as well in the field of astrobiology where studies of isotope patterns generated by ancient earth ecologies and modern laboratory cultures become templates for what we might expect elsewhere in our solar system. Thus, one fun advantage of working with plants and microbes is that imaginative isotope extrapolations are possible to global levels and beyond.

But let's return to earth. One interesting discovery of the late 1960s was that terrestrial plants show a clear-cut distinction in $\delta^{13}C$ values (Bender 1968). The C_3 plants such as trees, shrubs, and many grasses have lower $\delta^{13}C$ values that average near −28‰ whereas C_4 plants such as corn, sugar cane, and dryland grasses have higher values that average near −13‰ (O'Leary 1988; Ehleringer and Cerling 2001). These isotope distinctions arise during photosynthesis as carbon atoms are incorporated into 3-carbon (C_3) or 4-carbon (C_4) sugars by formation of new chemical bonds. This basic C_3/C_4 distinction created by fractionation has been exploited in terms of source mixing in a very wide variety of ecological studies, some of which are cited above (e.g., Figure 3.7). The C_3/C_4 distinction has also led to some fun laboratory activities. For example, isotope ecologists have studied adulteration of supposedly pure food products such as beer, wine, honey, and maple sugar. These products are made from C_3 plants but can be diluted for monetary gain and resale by adding inexpensive high fructose corn syrup that is a C_4 product (e.g., Brooks et al. 2002). Some of these alcohol-related forensic studies can lead to general merriment, especially if sampling involves larger quantities of beer and wine.

The effects of plant photosynthesis are far-reaching for our biosphere, and extend even to the pool of atmospheric CO_2 that plants use in photosynthesis. You probably know that CO_2 concentrations are increasing generally in the atmosphere due to human consumption of fossil fuels, but you may not know that there are interesting seasonal variations in this record due to plants and microbes (Figure 3.11, top). For example, at a far northern site in Canada, CO_2 levels drop in the summer growing season when plant photosynthesis is active. Plants are carbon sinks in the summer causing the depletion in the source pool of atmospheric CO_2. Later in the year as fall and winter set in, plant photosynthesis wanes but respiration from plants and microbes in soils becomes dominant, regenerating CO_2 to the atmospheric pool.

This very regular seasonal dynamic of CO_2 concentrations also has an isotope dynamic associated with it (Figure 3.11, bottom). The carbon isotopes show an inverse pattern to the concentrations, with higher $\delta^{13}C$ values in the summer, and lower values in the winter. This inverse relationship is an expression of isotope fractionation at work, with summer photosynthesis withdrawing carbon with low $\delta^{13}C$ values near −28‰, so that the residual atmospheric pool suffers an isotope enrichment and obtains the observed higher $\delta^{13}C$ values. Fractionation in photosynthesis is splitting its effects between the plants and the atmosphere, creating a light product (plants) and leaving behind a heavy substrate (atmospheric CO_2). As you

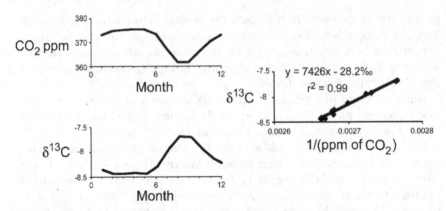

FIGURE 3.11. Atmospheric CO_2 records from 82.5° N at Alert, northerneastern Canada, part of a global monitoring network for CO_2 (http://cdiac.esd.ornl.gov/trends/co2/contents.htm; data shown are for the year 2000; you can access more data for other years from this Web site and make your own plots). The CO_2 concentrations decline during the summer growing season (top left panel) when isotope fractionation during photosynthetic withdrawal of CO_2 leaves the residual atmosphere enriched in ^{13}C with higher $\delta^{13}C$ values (bottom left panel). An inverse technique that plots $\delta^{13}C$ versus $1/CO_2$ concentration yields a y-intercept that is the isotope value of the source dominating the CO_2 dynamics, in this case −28.2‰ carbon from C_3 plants (middle right panel).

might infer, fall and winter conditions reverse this process and regeneration of plant and soil carbon via respiration adds back low $\delta^{13}C$ carbon to the atmospheric CO_2 pool, and atmospheric $\delta^{13}C$ values decrease in the fall and winter. This respiration is mostly from soil microbes.

It is possible to use a combination of the concentration and isotope data to extrapolate the isotope composition of CO_2 added back in the fall and winter, plotting the isotope data versus the inverse of the CO_2 concentration (Figure 3.11, right panel). In this plot, the y-intercept is the expected isotope value of the CO_2 added when the amount of this material is extrapolated to infinity, so that background contributions of other materials become insignificant. The extrapolated value of −28‰ agrees well with values expected for C_3 plants that dominate ecosystems and soils of the far north. This elegant extrapolation plot is known as a Keeling plot in honor of Charles Keeling who used this technique in the 1950s to investigate sources of CO_2 in air samples from the western United States (Keeling 1958; Pataki et al. 2003). This inverse isotope plot reappears in this book in Section 5.7 where it is again used to infer sources.

A global network of CO_2 monitoring stations is active today (see http://cdiac.esd.ornl.gov/trends/co2/contents.htm). Samples are collected monthly for concentration and isotope determinations. The concentration data are used to budget the amount of CO_2 circulating in the atmosphere, and the isotopes help budget which sources and sinks are active in the CO_2 cycle. Year-to-year CO_2 variations are evident (Trolier et al. 1996) and global models help identify regions and seasons contributing to these variations (Fung et al. 1997; Bousquet et al. 2000). The stable isotopes help by "sourcing," distinguishing the effects of –28‰ C_3 photosynthesis on a global scale from the effects of –13‰ C_4 photosynthesis and –21‰ marine photosynthesis (Mortazavi et al. 2005).

In this "sourcing" or "sourcery", the isotopes provide a second budget within the larger overall budget of total amounts, with the isotope mixture in each sample providing information that is analogous to a color, flavor, or variety. Another way to think about this is that if you had a handful of stones you might just count the total number of stones during a dark night, but in the day, you might also make a separate but related inventory of stones by their colors. Measured isotope values represent distinctive mixes of heavy and light isotopes, and each mix or isotope color can be entered in budgets along with the total amount. In the final accounting, both the total amounts and the isotopes have to balance, and the advantage of this dual accounting or dual budget approach is that it gives converging lines of evidence that point more quickly to answers. Increasingly, ^{14}C contents and $\delta^{18}O$ values of atmospheric CO_2 are measured in addition to $\delta^{13}C$ (see http://cdiac.esd.ornl.gov/trends/co2/contents.htm), so that atmospheric CO_2 budgets are actually triple and quadruple, helping to focus imaginations and models on correct answers. Detailed measurement of these multiple isotope tracers is a powerful way to budget circulation of carbon on this planet. This useful isotope budgeting applies to many other atmospheric gases such as methane, nitrous oxide, carbon monoxide, and oxygen that also are changing due to human activities (Tyler 1986; Ostrom et al. 2000; Snover et al. 2000; Hoffman et al. 2004; Nakagawa et al. 2004). The budgeting is a type of input–output or mass balance accounting that applies to all ecological systems and underlies the modeling approach adopted in the next chapters.

There are many other ways isotopes are used for plant and microbial studies. Plant ecologists use isotopes to study how plants compete, acquire nutrients, and balance these needs with water acquisition and water retention strategies (Boutton et al. 1999; Chambers et al. 2001; Dawson et al. 2002; Chimner and Cooper 2004; Oakes and Connolly 2004). Microbial isotope studies often focus on biogeochemical cycling, helping identify which microbial groups are active by labeling specific bacterial and algal groups or isolating biomarker compounds from these groups (Boschker et al. 1998; Boetius et al. 2000; Boschker and Middelburg 2002; Pel et al. 2003; Pearson et al. 2004). Under special circumstances, it is also possible to measure isotope values of individual microbial cells (Orphan et al. 2001).

Later sections of this book further consider plant and microbial examples. Section 7.7 gives a detailed analysis of carbon isotope fractionation during photosynthesis and how $\delta^{13}C$ values can indicate plant water use efficiency. Section 8.2 discusses how isotopes can help measure realized niches and competition among tree species. Section 7.8 discusses microbial dynamics of the sulfur cycle that may be relevant for planetary ecologies of not only our own home planet, but also for other planets and satellites rich in sulfur compounds (Carlson et al. 2000). Spectral measurement of isotope compositions of gases found in the atmospheres of other planets and moons is now possible (Lammer et al. 2000), and ultimately may provide a remote viewing tool to help identify where life is present and metabolically active in our wider universe.

3.6 Chapter Summary

Stable isotope ecology is a branch of chemical ecology, a tool-based or measurement-based approach that uses chemical tracers to follow ecological dynamics. Ecologists studying plants, animals, and microbes might first count the number and kinds of species present in a community, but then do more detailed studies that involve measurements such as DNA genomics, pH, and isotopes. These measurements all require a more sophisticated, chemical approach. Isotopes alone usually are not enough to solve ecological puzzles, but need to be considered together with other lines of evidence. This chapter provides diverse ecological examples to stimulate your interest and imagination about how isotopes might be useful in your own ecological research.

This book emphasizes stable isotopes of five elements, hydrogen, carbon, nitrogen, oxygen, and sulfur (HCNOS) that are naturally present everywhere in our biosphere. These elements are a fundamental part of all ecology, carried along through food webs, deposited in soils, and present in trace gases. For tracing the ecological action recorded in the HCNOS element cycles, there are no better markers than isotopes.

Isotope tracers are used in an enormous diversity of applications, from the origins of human diets to the sources and fates of gases involved in greenhouse warming. This chapter on using isotope tracers starts with an overview of isotope distributions in the biosphere, then gives four short review essays that enumerate some of the examples relevant to ecologists. These essays show how isotopes can trace the widespread effects of humans in this current era, an era that some geologists call the Anthropocene. In detail, the essays deal with isotope mapping of ecological gradients and regions (Section 3.2), tracking effects of invasive species (Section 3.3), tracing animal migrations (Section 3.4), and using isotopes at large scales to track planetary effects of plant and microbial metabolism on the atmospheric CO_2 pool (Section 3.5).

These are but a sampling of the topics where isotope measurements have proven helpful and useful. There is no simple answer to the beginner's question: Which isotopes will work best for me? But reading about past successes and doing some literature review of your own are important first steps in using stable isotope tracers.

Further Reading

Section 3.1

Hoefs, J. 2004. *Stable Isotope Geochemistry*, 5th Edition. Springer-Verlag, New York.
Peterson, B.J. and B. Fry. 1987. Stable isotopes in ecosystem studies. *Annual Review of Ecology and Systematics* 18:293–320.

Carbon

Balesdent, J., C. Girardin, and A. Mariotti. 1993. Site-related $\delta^{13}C$ of tree leaves and soil organic matter in a temperate forest. *Ecology* 74:1713–1721.
DeNiro, M.J. and S. Epstein. 1976. You are what you eat (plus a few ‰): The carbon isotope cycle in food chains. *Geological Society of America Abstracts Program* 8:834–835.
Ehleringer, J.R. and T.E. Cerling. 2001. C_3 and C_4 photosynthesis. In H.A. Mooney and J. Canadell (eds.), *Encyclopedia of Global Environmental Change*, Volume II, John Wiley and Sons, New York, pp. 186–190.
Friedli, H., H. Loetscher, H. Oeschger, U. Siegenthaler, and B. Stauffer. 1986. Ice core record of the $^{13}C/^{12}C$ ratio of atmospheric CO_2 in the past two centuries. *Nature* 324:237–328.
Fry, B. 2002. Conservative mixing of stable isotopes across estuarine salinity gradients: A conceptual framework for monitoring watershed influences on downstream fisheries production. *Estuaries* 25:264–271.
Lacroix, M. and F. Mosora. 1975. Variations du rapport isotopique $^{13}C/^{12}C$ dans le meatbolisme animal (Variations in the $^{13}C/^{12}C$ isotopic ratio in the animal metabolism). In *Isotope Ratios as Pollutant Source and Behaviour Indicators*. IAEA, Vienna, pp. 343–358.
Mosora, F., M. Lacroix, and J. Puchesne. 1971. Recherches sur les variations du rapport isotopique $^{13}C/^{12}C$, en function de la respiration et de la nature des tissues, chez les animaux superieurs. *Compte Rendus de l'Academie des Sciences, Serie D* 273:1752–1753.
O'Leary, M.H. 1988. Carbon isotopes in photosynthesis. *BioScience* 38:328–336.
Popp, B.N., E.A. Laws, R.R. Bidigare, J.E. Dore, K.L. Hanson, and S.G. Wakeham. 1998. Effect of phytoplankton cell geometry on carbon isotopic fractionation. *Geochimica et Cosmochimica Acta* 62:69–77.

Nitrogen

Altabet, M.A. and R. Francois. 1994. Sedimentary nitrogen isotopic ratio as a recorder for surface ocean nitrate utilization. *Global Biogeochemical Cycles* 8:103–116.
Casciotti, K.L., D.M. Sigman, M.G. Hastings, J.K. Bohlke, and A. Hilkert. 2002. Measurement of the oxygen isotopic composition of nitrate in seawater and freshwater using the denitrifier method. *Analytical Chemistry* 74:4905–4912.
Heaton, T.H.E. 1987. $^{15}N/^{14}N$ ratios of nitrate and ammonium in rain at Pretoria, South Africa. *Atmospheric Environment* 21:843–852.

Hobbie, E.A., S.A. Macko, and H.H. Shugart. 1998. Patterns in N dynamics and N isotopes during primary succession in Glacier Bay, Alaska. *Chemical Geology* 152:3–11.

Hobbie, E.A., S.A. Macko, and H.H. Shugart. 1999. Insights into nitrogen and carbon dynamics of ectomycorrhizal and saprotrophic fungi from isotopic evidence. *Oecologia* 118:353–360.

Liu, K.K. and I.R. Kaplan. 1989. The Eastern tropical Pacific as a source of ^{15}N-enriched nitrate in seawater off southern California. *Limnology and Oceanography* 34:820–830.

Mariotti, A. 1983. Atmospheric nitrogen is a reliable standard for natural ^{15}N abundance measurements. *Nature* 303:685–687.

Mariotti, A., J.C. Germon, P. Hubert, P. Kaiser, R. Letolle, A. Tardieux, and P. Tardieux. 1981. Experimental determination of nitrogen kinetic isotope fractions: Some principles; illustration for the denitrification and nitrification processes. *Plant and Soil* 62:413–430.

McClelland, J.W. and I. Valiela. 1998. Linking nitrogen in estuarine producers to land-derived sources. *Limnology and Oceanography* 43:577–585.

Minagawa, M. and E. Wada. 1984. Stepwise enrichment of ^{15}N along food chains. Further evidence and the relation between δ^{15}N and animal age. *Geochimica et Cosmochimica Acta* 48:1135–1140.

Peters, K.E., R.E. Sweeney, and I.R. Kaplan. 1978. Correlation of carbon and nitrogen stable isotope ratios in sedimentary organic matter. *Limnology and Oceanography* 23:598–604.

Saino, T. and A. Hattori. 1987. Geographical variation of the water column distribution of suspended particulate organic nitrogen and its ^{15}N natural abundance in the Pacific and its marginal seas. *Deep-Sea Research* 34:807–827.

Sulfur

Cameron, E.M., G.E.M. Hall, J. Veizer, and H.R. Krouse. 1995. Isotopic and elemental hydrogeochemistry of a major river system—Fraser-River, British Columbia, Canada. *Chemical Geology* 122:149–169.

Canfield, D.E. 2001. Isotope fractionation by natural populations of sulfate-reducing bacteria. *Geochimica et Cosmochimica* 65:1117–1124.

Fry, B., H. Gest, and J.M. Hayes. 1988. ^{34}S/^{32}S fractionation in sulfur cycles catalyzed by anaerobic bacteria. *Applied and Environmental Microbiology* 54:250–256.

Goldhaber, M.B. and I.R. Kaplan. 1975. Controls and consequences of sulfate reduction rates in recent marine sediments. *Soil Science* 119:42–55.

Jorgensen, B.B. 1990. A thiosulfate shunt in the sulfur cycle of marine sediments. *Science* 249:152–154.

Krouse, H.R. 1980. Sulphur isotopes in our environment. In P. Fritz and J. Ch. Fontes (eds.), *Handbook of Environmental Isotope Geochemistry, vol. 1, The Terrestrial Environment, A.* Elsevier, Amsterdam, pp. 435–471.

Mayer, B. and H.R. Krouse. 1996. Prospects and limitations of an isotope tracer technique for understanding sulfur cycling in forested and agro-ecosystems. *Isotopes in Environmental and Health Studies* 32:191–201.

Rees, C.E., W.J. Jenkins, and J. Monster. 1978. The sulphur isotopic composition of ocean water sulphate. *Geochimica et Cosmochimica Acta* 42:377–381.

Richards, M.P., B.T. Fuller, M. Sponheimer, T. Robinson, and L. Ayliffe. 2003. Sulphur isotopes in palaeodietary studies: A review and results from a controlled feeding experiment. *International Journal of Osteoarchaeology* 13:37–45.

Trust, B.A. and B. Fry. 1992. Stable sulphur isotopes in plants: A review. *Plant, Cell and Environment* 15:1105–1110.

Vanstempvoort, D.R. and H.R. Krouse. 1994. Controls of δ^{18}O in sulfate—Review of experimental data and application to specific environments. *Environmental Geochemistry of Sulfide*, ACS Symposium Series 550:446–480.

Hydrogen

Epstein, S., P. Thompson, and C.J. Yapp. 1977. Oxygen and hydrogen isotopic ratios in plant cellulose. *Science* 198:1209–1215.

Estep, M.F. and H. Dabrowski. 1980. Tracing food webs with stable hydrogen isotopes. *Science* 209:1537–1538.

Fogel, M.F. and T.C. Hoering. 1980. Biogeochemistry of the stable hydrogen isotopes. *Geochimica et Cosmochimica Acta* 44:1197–1206.

Friedman, I., A.C. Redfield, B. Schoen, and J. Harris. 1964. The variation of the deuterium content of natural waters in the hydrologic cycle. *Reviews of Geophysics* 2:177–224.

Fry, B. 2002. Listed above; see Section 3.1, Carbon readings.

Gleason, J.D. and I. Friedman. 1970. Deuterium natural variations used as a biological tracer. *Science* 169:1085–1086.

Hobson, K.A., L. Atwell, and L.I. Wassenaar. 1999. Influence of drinking water and diet on the stable-hydrogen isotope ratios of animal tissues. *Proceedings of the National Academy of Science* 96:8003–8006.

Kendall, C. and T.B. Coplen. 2001. Distribution of oxygen-18 and deuterium in river waters across the United States. *Hydrological Processes* 15:1363–1393.

Nissenbaum, A. 1974. Deuterium content of humic acids from marine and non-marine environments. *Marine Chemistry* 2:59–63.

Pataki, D.E., S.E. Bush, and J.R. Ehleringer. 2005. Stable isotopes as a tool in urban ecology. In L.B. Baker, J.R. Ehleringer and D.E. Pataki (eds.), *Stable Isotopes and Biosphere–Atmosphere Interactions: Processes and Biological Controls*. Elsevier, Amsterdam, pp. 199–216.

Sauer, P.E., T.I. Eglinton, J.M. Hayes, A. Schimmelmann, and A.L. Sessions. 2001. Compound-specific D/H ratios of lipid biomarkers from sediments as a proxy for environmental and climatic conditions. *Geochimica et Cosmochimica Acta* 65:213–233.

Sharp, Z.D., V. Atudorei, H. Panarello, J. Fernandez, and C. Douthitt. 2003. Hydrogen isotope systematics of hair: archeological and forensic applications. *Journal of Archaeological Science* 30:1709–1716.

Smith, B.N. and H. Ziegler. 1990. Isotopic fractionation of hydrogen in plants. *Botanica Acta* 103:335–342.

Sternberg, L., M.J. DeNiro, and H. Ajie. 1984. Stable hydrogen isotope ratios of saponifiable lipids and celluolose nitrate from CAM, C_3 and C_4 plants. *Phytochemistry* 23:2475–2477.

Taylor, H.P. 1974. The application of oxygen and hydrogen isotope studies to problems of hydrothermal alterations and ore deposition. *Economic Geology* 69:843–882.

Tzedakis, P.C., K.H. Roucoux, L. de Abreu, and N.J. Shackleton. 2004. The duration of forest stages in Southern Europe and interglacial climate variability. *Science* 306:2231–2235.

Whiticar, M.J. 1999. Carbon and hydrogen isotope systematics of bacterial formation and oxidation of methane. *Chemical Geology* 161:291–314.

Yakir, D. 1992. Variations in the natural abundance of oxygen-18 and deuterium in plant carbohydrates. *Plant Cell and Environment* 15:1005–1020.

Oxygen

Bowen, G.J., L.I. Wassenaar, and K.A. Hobson. 2005. Global application of stable hydrogen and oxygen isotopes to wildlife forensics. *Oecologia*, doi:10.1007/s00442-004-1813-y.

Hoffmann, G., M. Cuntz, C. Weber, P. Ciais, P. Friedlingstein, M. Heimann, J. Jouzel, J. Kaduk, E. Maier-Reimer, U. Seibt, and K. Six. 2004. A model of the Earth's Dole effect. *Global Biogeochemical Cycles* 18, GB1008, doi:10.1029/2003GB002059.

Koch, P.L. 1999. Isotopic reconstruction of past continental environments. *Annual Review of Earth and Planetary Sciences* 26:573–613.

Yakir, D. 1992. Listed above; see Section 3.1, Hydrogen readings.

Yakir, D. and M.J. DeNiro. 1990. Oxygen and hydrogen isotope fractionation during cellulose metabolism in *Lemna-gibba* 1. *Plant Physiology* 93:325–332.

Section 3.2

Balesdent et al. 1987. Listed above; see Section 3.1, Carbon readings.

Bellanger, B., S. Huon, F. Velasquez, V. Valles, C. Girardin, and A. Mariotti. 2004. Monitoring soil organic carbon erosion with $\delta^{13}C$ and $\delta^{15}N$ on experimental field plots in the Venezuelan Andes. *Catena* 58:125–150.

Cabana, G. and J.B. Rasmussen. 1996. Comparison of aquatic food chains using nitrogen isotopes. *Proceedings of the National Academy of Sciences of the United States of America* 93:10844–10847.

Costanzo, S.D., M.J. O'Donohue, W.C. Dennison, N.R. Loneragan, and M. Thomas. 2001. A new approach for detecting and mapping sewage impacts. *Marine Pollution Bulletin* 42:149–156.

Costanzo, S.D., J. Udy, B. Longstaff, and A. Jones. 2005. Using nitrogen stable isotope ratios ($\delta^{15}N$) of macroalgae to determine the effectiveness of sewage upgrades: Changes in the extent of sewage plumes over four years in Moreton Bay, Australia. *Marine Pollution Bulletin* 51:212–217.

Delegue, M.-A., M. Fuhr, D. Schwartz, A. Mariotti, and R. Nasi. 2001. Recent origin of a large part of the forest cover in the Gabon coastal area based on stable carbon isotope data. *Oecologia* 129:106–113.

Farrell, J.W., T.F. Pedersen, S.E. Calvert, and B. Nielsen. 1995. Glacial-interglacial changes in nutrient utilization in the equatorial Pacific Ocean. *Nature* 377:514–517.

Finlay, J., M.E. Power, and G. Cabana. 1999. Effects of water velocity on algal carbon isotope ratios: Implications for river food web studies. *Limnology and Oceanography* 44:1198–1203.

Hunt, J.M. 1970. The significance of carbon isotope variations in marine sediments. In G.D. Hobson and G.C. Speers (eds.), *Advances in Organic Geochemistry*, 1966. Pergamon, Tarrytown, NY, pp. 27–35.

Jennings, S. and K.J. Warr. 2003. Environmental correlates of large-scale spatial variation in the $\delta^{15}N$ of marine animals. *Marine Biology* 142:1131–1140.

Rogers, K.M. 2003. Stable carbon and nitrogen isotope signatures indicate recovery of marine biota from sewage pollution at Moa Point, New Zealand. *Marine Pollution Bulletin* 46:821–827.

Sackett, W.M. and R.R. Thompson. 1963. Isotopic organic carbon composition of recent continental derived clastic sediments of eastern Gulf coast, Gulf of Mexico. *Bulletin of the American Association of Petroleum Geologists* 47:525–528.

Savage, C. 2005. Tracing the influence of sewage nitrogen in a coastal ecosystem using stable nitrogen isotopes. *Ambio* 34:145–150.

Vitousek, P.M., H.A. Mooney, J. Lubchenco, and J.M. Melillo. 1997. Human domination of Earth's ecosystems. *Science* 277:494–499.

Wiesenberg, G.L.B., J. Schwarzbauer, M.W.I. Schmidt, and L. Schwark. 2004. Source and turnover of organic matter in agricultural soils derived from n-alkane/n-carboxylic acid compositions and C-isotope signatures. *Organic Geochemistry* 35:1371–1393.

Section 3.3

Ambrose, S.H. 2000. Controlled diet and climate experiments on nitrogen isotope ratios of rats. In S.H. Ambrose and M.A. Katzenberg (eds.), *Biogeochemical Approaches to Paleodietary Analysis*. Kluwer Academic, Hingham, MA, pp. 243–259.

Arrouays, D., J. Balesdent, A. Mariotti, and C. Girardin. 1995. Modelling organic carbon turnover in cleared temperate forest soils converted to maize cropping by using ^{13}C natural abundance measurements. *Plant and Soil* 173:191–196.

Beaudoin, C.P., W.M. Tonn, E.E. Prepas, and L.I. Wassenaar. 1999. Individual specialization and trophic adaptability of northern pike (*Esox lucius*): An isotope and dietary analysis. *Oecologia* 120:386–396.

Bowling, D.R., S.D. Sargent, B.D. Tanner, and J.R. Ehleringer. 2003. Tunable diode laser absorption spectroscopy for stable isotope studies of ecosystem-atmosphere CO_2 exchange. *Agricultural and Forest Meteorology* 188:1–19.

Branstrator, D.K., G. Cabana, A. Mazumder, and J.B. Rasmussen. 2000. Measuring life-history omnivory in the opossum shrimp *Mysis relicata*, with stable nitrogen isotopes *Limnology and Oceanography* 45:463–467.

Brenna, J.T. 2001. Natural intramolecular isotope measurements in physiology: Elements of the case for an effort toward high-precision position-specific isotope analysis. *Rapid Communications in Mass Spectrometry* 15:1252–1262.

Carleton, S.A., B.O. Wolf, and C.M. del Rio. 2004. Keeling plots for hummingbirds: A method to estimate carbon isotope ratios of respired CO_2 in small vertebrates. *Oecologia* 141:1–6.

Cerling, T.E., B.H. Passey, L.K. Ayliffe, C.S. Cook, J.R. Ehleringer, J.M. Harris, M.B. Dhidha, and S.M. Kasiki. 2004. Orphans' tales: Seasonal dietary changes in elephants from Tsavo National Park, Kenya. *Palaeogeography Palaeoclimatology and Palaeoecology* 206:367–376.

Culik, B.M. and R.P. Wilson. 1992. Field metabolic rates of instrumented Adelie penguins using double-labeled water. *Journal of Comparative Physiology B—Biochemical Systemic and Environmental Physiology* 162:567–573.

DeNiro, M.J. 1987. Stable isotopy and archaeology. *American Scientist* 75:182–191.

DeNiro and Epstein. 1976. Listed above; see Section 3.1, Carbon readings.

Estes, J.A., M.L. Riedman, M.M. Staedler, M.T. Tinker, and B.E. Lyon. 2003. Individual variation in prey selection by sea otters: Patterns, causes and implications. *Journal of Animal Ecology* 72:144–155.

Felicetti, L.A., C.C. Schwartz, R.O. Rye, M.A. Haroldson, K.A. Gunther, D.L. Phillips, and C.T. Robbins. 2003. Use of sulfur and nitrogen stable isotopes to determine the importance of whitebark pine nuts to Yellowstone grizzly bears. *Canadian Journal of Zoology* 81:763–770.

Fischer, H. and K. Wetzel. 2002. The future of ^{13}C-breath tests. *Food and Nutrition Bulletin* 23:53–56.

Fogel, M.L. and N. Tuross. 2003. Extending the limits of paleodietary studies of humans with compound specific carbon isotope analysis of amino acids. *Journal of Archaeological Science* 30:535–545.

Fogel, M.L., N. Tuross, B.J. Johnson, and G.H. Miller. 1997. Biogeochemical record of ancient humans. *Organic Geochemistry* 27:275–287.

Galloway, J.N., F.J. Dentener, D.G. Capone, E.W. Boyer, R.W. Howarth, S.P. Seitzinger, G.P. Asner, C.C. Cleveland, P.A. Green, E.A. Holland, D.M. Karl, A.F. Michaels, J.H. Porter, A.R. Townsend, and C.J. Vorosmarty. 2004. Nitrogen cycles: Past, present and future. *Biogeochemistry* 70:153–226.

Galloway, J.N., H. Levy, and P.S. Kashibhatla. 1994. Year 2020—Consequences of population-growth and development on deposition of oxidized nitrogen. *Ambio* 23:120–123.

Gotaas, G., E. Milne, P. Haggarty, and N.J.C. Tyler. 1997. Use of feces to estimate isotopic abundance in doubly labeled water studies in reindeer in summer and winter. *American Journal of Physiology-Regulatory Integrative and Comparative Physiology* 273:R1451–R1456.

Guiguer, K.R.R.A., J.D. Reist, M. Power, and J.A. Babaluk. 2002. Using stable isotopes to confirm the trophic ecology of Arctic charr morphotypes from Lake Hazen, Nunavut, Canada. *Journal of Fish Biology* 60:348–362.

Hadwen, W.L. and S.E. Bunn. 2004. Tourists increase the contribution of autochthonous carbon to littoral zone food webs in oligotrophic dune lakes. *Marine and Freshwater Research* 55: 701–708.

Hayes, J.M. 2001. Fractionation of the isotopes of carbon and hydrogen in biosynthetic processes. In J.W. Valley and D.R. Cole (eds.), *Stable Isotope Geochemistry, Reviews in Mineralogy and Geochemistry*, vol. 43. Mineralogical Society of America, Washington, D.C., pp. 225–278.

Hobbie, E.A., S.A. Macko, and H.H. Shugart. 1998. Patterns in N dynamics and N isotopes during primary succession in Glacier Bay, Alaska. *Chemical Geology* 152:3–11.

Hobbie, E.A., S.A. Macko, and H.H. Shugart. 1999. Interpretation of nitrogen isotope signatures using the NIFTE model. *Oecologia* 120:405–415.

Jackson, J.B.C, M.X. Kirby, W.H. Berger, K.A. Bjorndal, L.W. Botsford, B.J. Bourque, R.H., Bradbury, R. Cooke, J. Erlandson, J.A. Estes, T.P. Hughes, S. Kidwell, C.B. Lange, H.S. Lenihan, J.M. Pandolfi, C.H. Peterson, R.S. Steneck, M.J. Tegner, and R.R. Warner. 2001. Historical overfishing and the recent collapse of coastal ecosystems. *Science* 293:629–638.

Katz, J.J. 1960. Chemical and biological studies with deuterium. *American Scientist* 48:544–580.

Kelly, J.F. 2000. Stable isotopes of carbon and nitrogen in the study of avian and mammalian trophic ecology. *Canadian Journal of Zoology* 78:1–27.

Kidd, K.A. 1998. Use of stable isotope ratios in freshwater and marine biomagnification studies. In J. Rose (ed.), *Environmental Toxicology: Current Developments*. Gordon and Breach Science, Amsterdam, pp. 359–378.

Lee-Thorp, J.A., M. Sponheimer, and N.H. Van der Merwe. 2003. What do stable isotopes tell us about hominid dietary and ecological niches in the Pliocene? *International Journal of Osteoarchaeology* 13:104–113.

Los Gatos Research. www.LGRinc.com.

Mayntz, D., D. Raubenheimer, M. Salomon, S. Toft, and S.J. Simpson. 2005. Nutrient-specific foraging in invertebrate predators. *Science* 307:111–113.

McCutchan, J.H. Jr., W.M. Lewis Jr., C. Kendall, and C.C. McGrath. 2003. Variation in trophic shift for stable isotope ratios of carbon, nitrogen, and sulfur. *Oikos* 102:378–390.

Moseman, S.M., L.A. Levin, C. Currin, and C. Forder. 2004. Colonization, succession, and nutrition of macrobenthic assemblages in a restored wetland at Tijuana Estuary, California. *Estuarine Coastal and Shelf Science* 60:755–770.

Murnick, D.E. and B.J. Peer. 1994. Laser-based analysis of carbon isotope ratios. *Science* 263:945–947.

Pauly, D., V. Christensen, R. Froese, and M.L. Palomares. 2000. Fishing down aquatic food webs. *American Scientist* 88:46–51.

Ponsard, S. and R. Arditi. 2000. What can stable isotopes (δ^{15}N and δ^{13}C) tell about the food web of soil macro-invertebrates? *Ecology* 81:852–864.

Post, D.M. 2003. Individual variation in the timing of ontogenetic niche shifts in largemouth bass. *Ecology* 84:1298–1310.

Rice, S.K., B. Westerman, and R. Federici. 2004. Impacts of the exotic, nitrogen-fixing black locust (*Robinia pseudoacacia*) on nitrogen-cycling in a pine-oak ecosystem. *Plant Ecology* 174: 97–107.

Riddell, M.C., O. Bar-Or, H.P. Schwarcz, and G.J.F. Heigenhauser. 2000. Substrate utilization in boys during exercise with [C-13]-glucose ingestion. *European Journal of Applied Physiology* 83:441–448.

Schoeninger, M.J., J. Moore, and J.M. Sept. 1999. Subsistence strategies of two "savanna" chimpanzee populations: The stable isotope evidence. *American Journal of Primatology* 49:297–314.

Shearer, G. and D.H. Kohl. 1988. Estimates of N_2 fixation in ecosystems: The need for and basis of the ^{15}N natural abundance method. In P.W. Rundel, J.R. Ehleringer, and K.A. Nagy (eds.), *Stable Isotopes in Ecological Research*. Springer-Verlag, New York, pp. 342–373.

Sponheimer, M. and J.A. Lee-Thorp. 1999. Isotopic evidence for the diet of an early hominid, *Australopithecus africanus*. Science 283:368–370.

Sponheimer, M., T. Robinson, L. Ayliffe, B. Roeder, J. Hammer, B. Passey, A. West, T. Cerling, D. Dearing, and J. Ehleringer. 2003. Nitrogen isotopes in mammalian herbivores: Hair δ^{15}N values from a controlled feeding study. *International Journal of Osteoarchaeology* 13:80–87.

Stewart, A.R., S.N. Luoma, C.E. Schlekat, M.A. Doblin, and K.A. Hieb. 2004. Food web pathway determines how selenium affects aquatic ecosystems: A San Francisco Bay case study. *Environmental Science and Technology* 38:4519–4526.

Tieszen, L.L. and T. Fagre. 1993. Effect of diet quality and composition on the isotopic composition of respiratory CO_2, bone collagen, bioapatite, and soft tissues. In J.B. Lambert and G. Grupe (eds.), *Prehistoric Human Bone: Archaeology at the Molecular Level*. Springer Verlag, New York, pp. 121–155.

van der Merwe, N.J. 1982. Carbon isotopes, photosynthesis, and archaeology. *American Scientist* 70:596–605.

Vander Zanden, J.M. and J.B. Rasmussen. 2001. Variation in δ^{15}N and δ^{13}C trophic fractionation: Implications for aquatic food web studies. *Limnology and Oceanography* 46:2061–
2066.

Vander Zanden, M.J., J.M. Casselman, and J.B. Rasmussen. 1999. Stable isotope evidence for
the food web consequences of species invasions in lakes. *Nature* 401:464–467.

Vander Zanden, M.J., S. Chandra, B.C. Allen, J.E. Reuter, and C.R. Goldman. 2003. Historical
food web structure and restoration of native aquatic communities in the Lake Tahoe
(California-Nevada) Basin. *Ecosystems* 6:274–288.

Vander Zanden, M.J., J.D. Olden, J.H. Thorne, and N.E. Mandrak. 2004. Predicting occurrences
and impacts of smallmouth bass introductions in north temperate lakes. *Ecological Applications* 14:132–148.

Yoshinaga, J., T. Suzuki, T. Hongo, M. Minagawa, R. Ohtsuka, T. Kawabe, T. Inaoka, and T.
Akimichi. 1992. Mercury concentration correlates with the nitrogen stable isotope ratio in
animal food of Papuans. *Ecotoxicology and Environmental Safety* 24:37–45.

Section 3.4

Fry, B. 1981. Natural stable carbon isotope tag traces Texas shrimp migrations. *Fishery Bulletin*
79:337–345.

Fry, B. 1983. Fish and shrimp migrations in the northern Gulf of Mexico analyzed using stable
C, N, and S isotope ratios. *Fishery Bulletin* 81:789–801.

Hobson, K.A. 1999. Tracing origins and migration of wildlife using stable isotopes: A review.
Oecologia 120:314–326.

Hobson, K.A. 2002. Incredible journeys. *Science* 295:981–982.

Kelly, J.F., V. Atudorei, Z.D. Sharp, and D.M. Finch. 2002. Insights into Wilson's Warbler migration from analyses of hydrogen stable-isotope ratios. *Oecologia* 130:216–221.

Killingley, J.S. 1979. Migrations of California gray whales tracked by oxygen-18 variations in
their epizoic barnacles. *Science* 207:759–760.

Killingley, J.S. and M. Lutcavage. 1983. Loggerhead turtle movements reconstructed from ^{18}O
and ^{13}C profiles from commensal barnacle shells. *Estuarine Coastal and Shelf Science*
16:345–349.

Kline, T.C. Jr., J.J. Goering, O.A. Mathisen, P.H. Poe, and P.L. Parker. 1990. Recycling of
elements transported upstream by runs of Pacific salmon: I. δ^{15}N and δ^{13}C evidence in
Sashin Creek, southeastern Alaska. *Canadian Journal of Fisheries and Aquatic Science*
47:136–144.

Naiman, R.J., R.E. Bilby, D.E. Schindler, and J.M. Helfield. 2002. Pacific salmon, nutrients, and
the dynamics of freshwater and riparian ecosystems. *Ecosystems* 5:399–417.

Podlesak, D.W., S.R. McWilliams, and K.A. Hatch. 2004. Stable isotopes in breath, blood, feces
and feathers can indicate intr-individual changes in the diet of migratory songbirds. *Oecologia* DOI:10.1007/s00442-004-1737-6.

Rubenstein, D.R. and K.A. Hobson. 2004. From birds to butterflies: Animal movement patterns and stable isotopes. *Trends in Ecology and Evolution* 19:256–263.

Rubenstein, D.R., C.P. Chamberlain, R.T. Holmes, M.P. Ayres, J.R. Waldbauer, G.R. Graves,
and N.C. Tuross. 2002. Linking breeding and wintering ranges of a migratory songbird using
stable isotopes. *Science* 295:1062–1065.

Sapolsky, R.M. 2002. *A Primate's Memoir: A Neuroscientist's Unconventional Life Among the
Baboons*. Simon & Schuster Adult, New York.

Schaller, G.B. 1966. *The Year of the Gorilla*. University of Chicago Press, Chicago.

Schell, D.M., S.M. Saupe, and H. Haubenstock. 1989. Bowhead whale (Balaena-Mysticetus)
growth and feeding as estimated by δ^{13}C techniques. *Marine Biology* 103:433–443.

Taylor, H.P. 1974. Listed above; see Section 3.1, Hydrogen readings.

Wassenaar, L.I. and K.A. Hobson. 1998. Natal origins of migratory monarch butterflies at wintering colonies in Mexico: New isotopic evidence. *Proceedings of the National Academy of
Sciences of the United States of America* 95:15436–15439.

Section 3.5

Bender, M. 1968. Mass spectrometric studies of carbon 13 variations in corn and other grasses. *Radiocarbon* 10:468–472.

Boetius, A., K. Ravenschlag, C.J. Schubert, D. Rickert, F. Widdel, A. Gieseke, R. Amann, B.B. Jorgensen, U. Witte, and O. Pfaffkuche. 2000. A marine microbial consortium apparently mediating anaerobic oxidation of methane. *Nature* 407:623–626.

Boschker, H.T.S. and J.J. Middelburg. 2002. Stable isotopes and biomarkers in microbial ecology. *FEMS Microbiology Ecology* 40:85–95.

Boschker, H.T.S., S.C. Nold, P. Wellsbury, D. Bos, W. de Graaf, R. Pel, R.J. Parkes, and T.E. Cappenberg. 1998. Direct linking of microbial populations to specific biogeochemical processes by ^{13}C-labelling of biomarkers. *Nature* 392:801–804.

Bousquet, P., P. Peylin, P. Ciais, C. Le Quere, P. Friedlingstein, and P.P. Tans. 2000. Regional changes in carbon dioxide fluxes of land and oceans since 1980. *Science* 290:1342–1346.

Boutton, T.W., S.R. Archer, and A.J. Midwood. 1999. Stable isotopes in ecosystem science: structure, function and dynamics of a subtropical savanna. *Rapid Communications in Mass Spectrometry* 13:1263–1277.

Brooks, J.R., N. Buchmann, S.L. Phillips, B. Ehleringer, R.D. Evans, L.A. Martinmelli, W.T. Pockman, D. Sandquist, J.P. Sparks, L. Sperry, D. Williams, and J.R. Ehleringer. 2002. Heavy and light beer: A carbon isotope approach to detecting C_4 carbon in beers from different origins, styles, and prices. *Journal of Agricultural and Food Chemistry* 50:6413–6418.

Carlson, R.W., R.E. Johnson, and M.S. Anderson. 2000. Sulfuric acid on Europa and the radiolytic sulfur cycle. *Science* 286:97–99.

Chambers, R.M., J.W. Fourqurean, S.A. Macko, and R. Hoppenot. 2001. Biogeochemical effects of iron availability on primary producers in a shallow marine carbonate environment. *Limnology and Oceanography* 46:1278–1286.

Chimner, R.A. and D.J. Cooper. 2004. Using stable oxygen isotopes to quantify the water source used for transpiration by native shrubs in the San Luis Valley, Colorado USA. *Plant and Soil* 260:225–236.

Dawson, T.E., S. Mambelli, A.H. Plamboeck, P.H. Templer, and K.P. Tu. 2002. Stable isotopes in plant ecology. *Annual Review of Ecology and Systematics* 33:507–559.

Ehleringer, J.R. and T.E. Cerling. 2001. C_3 and C_4 photosynthesis. In H.A. Mooney and J. Canadell (eds.), *Encyclopedia of Global Environmental Change*, Vol. II. John Wiley and Sons, New York, pp. 186–190.

Fung, I., C.B. Field, J.A. Berry, M.V. Thompson, J.T. Randerson, C.M. Malmstroem, P.M. Vitousek, G.J. Collatz, P.J. Sellers, D.A. Randall, A.S. Denning, F. Badeck, and J. John. 1997. Carbon-13 exchanges between the atmosphere and biosphere. *Global Biogeochemical Cycles* 11:507–533.

Ghosh, P. and W.A. Brand. 2003. Stable isotope ratio mass spectrometry in global climate change research. *International Journal of Mass Spectrometry* 228:1–33.

Hoffmann, G., M. Cuntz, C. Weber, P. Ciais, P. Friedlingstein, M. Heimann, J. Jouzel, J. Kaduk, E. Maier-Reimer, U. Seibt, and K. Six. 2004. A model of the Earth's Dole effect. *Global Biogeochemical Cycles* 18:GB1008, doi:10.1029/2003GB002059.

Keeling, C.D. 1958. The concentration and isotopic abundances of atmospheric carbon dioxide in rural areas. *Geochimica et Cosmochimica Acta* 13:322–334.

Lammer, H., W. Stumptner, G.J. Molina-Cuberos, S.J. Bauer, and T. Owen. 2000. Nitrogen isotope fractionation and its consequence for Titan's atmospheric evolution. *Planetary and Space Science* 48:529–543.

Mortzazvi, B., J. Chanton, J.L. Prater, A.C. Oishi, R. Oren, and G. Kaul. 2005. Temporal variability in the ^{13}C of respired CO_2 in a pine and hardwood forest subject to identical climatic and edaphic conditions. *Oecologia* 142:57–69.

Nakagawa, F., U. Tsunogai, T. Gamo, and N. Yoshida. 2004. Stable isotopic compositions and fractionations of carbon monoxide at coastal and open ocean stations in the Pacific. *Journal of Geophysical Research-Oceans* 109:C06016.

Oakes, J.M. and R.M. Connolly. 2004. Causes of sulfur isotope variability in the seagrass *Zostera capricorni*. *Journal of Experiment Marine Biology and Ecology* 302:153–164.

O'Leary. 1988. Listed above; see Section 3.1, Carbon readings.

Orphan, V.J., C.H. House, K.-U. Hinrichs, K.D. McKeegan, and E.F. DeLong. 2001. Methane-consuming Archaea revealed by directly coupled isotopic and phylogenetic analysis. *Science* 293:484–487.

Ostrom, N.E., M.E. Russ, B.N. Popp, T.M. Rust, and D.M. Karl. 2000. Mechanisms of N_2O production in the subtropical North Pacific based on determinations of the isotopic abundances of N_2O and O_2. *Chemosphere—Global Change Science* 2:281–290.

Pataki, D.E., J.R. Ehleringer, L.B. Flanagan, D. Yakir, D.R. Bowling, C.J. Still, N. Buchmann, J.O. Kaplan, and J.A. Berry. 2003. The application and interpretation of Keeling plots in terrestrial carbon cycle research. *Global Biogeochemical Cycles* 17:doi:10.1029/2001GB001850.

Pearson, A., A.L. Sessions, K.J. Edwards, and J.M. Hayes. 2004. Phylogenetically specific separation of rRNA from prokaryotes for isotopic analysis. *Marine Chemistry* 92:295–306.

Pel, R., H. Hoogveld, and V. Floris. 2003. Using the hidden isotopic heterogeneity in phyto- and zooplankton to unmask disparity in trophic carbon transfer. *Limnology and Oceanography* 48:2200–2207.

Snover, A.K., P.D. Quay, and W.M. Hao. 2000. The D/H content of methane emitted from biomass burning. *Global Biogeochemical Cycles* 14:11–24.

Trolier, M., J.W.C. White, P.P. Tans, K.A. Masarie, and P.A. Gemery. 1996. Monitoring the isotopic composition of atmospheric CO_2: measurements from the NOAA global air sampling network. *Journal of Geophysical Research* 101:25897–25916.

Tyler, S.C. 1986. Stable carbon isotope ratios in atmospheric methane and some of its sources. *Journal of Geophysical Research* 91:13232–13238.

4
Isotope Chi

Overview

This chapter introduces a modeling approach to understand the interplay between isotope mixing and fractionation in the biosphere. The approach is named Isotope Chi (like Tai Chi) to represent the power in this new approach. An initial example deals with imaginary chocolate isotopes, to stimulate your appetite and to help you get in tune with how fractionation and mixing work together in isotope cycling. The next four sections use a more realistic example of photosynthesis and respiration in the sea to fully introduce the I Chi modeling. A final section switches to a terrestrial example with cows. The different examples develop the appreciation that isotope distributions in natural systems reflect the net balance between fractionation and mixing.

4.1. *Chocolate Isotopes.* This I Chi example is just for fun. Humans create inequality in the chocolate supplies via their fractionating pickiness for light chocolates versus heavy dark chocolates. Fortunately this fractionating selectivity is offset by a benevolent chocolate god who adds a good mix of chocolates to maintain a steady balance of light and dark.

4.2. *Oxygen in the Sea.* We keep track of oxygen isotopes in one of the eternal cycles of the planet, featuring photosynthesis that adds oxygen and mixes in isotopes to the dissolved oxygen pool in surface waters, versus respiration that subtracts oxygen and fractionates isotopes from this same pool.

4.3. *Equations for Isotope Chi.* We learn four equations for calculating isotope behavior during gains and losses that are at the heart of circulating isotopes in nature and in I Chi spreadsheet models.

4.4. *Building an I Chi Model, Step by Step.* Here is a step-by-step guide to constructing a gain–loss model with the equations of Section 4.3. Before starting this section, get a preview by opening I Chi Workbook 4.4 in the Chapter 4 folder on the accompanying CD. With the workbook open, spend five minutes clicking through the individual worksheets that give the individual I Chi steps.

4.5. *Errors in I Chi Models.* The I Chi spreadsheet models are not quite perfect, and do have minor errors that are evaluated here. This section and the next one are both quite technical, but included for how-to-do-it reference.

4.6. *Exact Equations for I Chi Models.* A usually minor source of error in the modeling concerns the isotope equations themselves. This section gives the exact equations and advice on when to use these more complex exact equations. This is a detailed technical section that beginning readers may skip, but is important for advanced readers who are writing their own spreadsheet models.

4.7. *Cows in a Pasture.* This everyday example shows that fractionation can give very different results depending on the balance between gains and losses, in this case, gains from grazing and losses via excretion. The isotope values of cows are not fixed by gains or by losses, but reflect the net balance between gains from the diet and losses via excretion. This balance explains trophic fractionations and enrichments commonly seen in food webs.

Main points to learn. If you grasp the main equations of I Chi presented in Section 4.3, you will be well on your way to following isotope cycling in the two big processes of mixing and fractionation. Fortunately, the equations are few and the algebra is easy. Learning how to use the equations is really the more interesting part, and that definitely needs practice. You will find spreadsheets for practicing I Chi with the examples from this and other chapters on the accompanying CD. As you view the joint action of mixing and fractionation in these spreadsheets, you may realize that if you want to understand source mixing, you will also need to develop a good understanding of fractionation. Interpreting isotopes in terms of source mixing generally means that you have to factor out or normalize for the effects of fractionation. This necessity arises because δ values reflect the net balance between fractionation and source mixing, not just source mixing alone.

4.1 Chocolate Isotopes

For most readers, the whole concept of isotopes is hard to grasp, seemingly simple, yet often mysterious. To start getting in tune with isotope cycling, here we consider a just-for-fun example, chocolate isotopes.

Now, there are no such things as chocolate isotopes, but then again, there is nothing to stop us from inventing some for our virtual entertainment. The isotope models that are the subject of this chapter are virtual spreadsheet models, so why not enlist chocolate as a virtual element? Having made that first step, let us also imagine that we humans actually could do the isotope selection, something that is normally impossible. Normally, isotope discrimination or fractionation occurs at the atomic level, where bonds are broken and formed. Down there at the atomic level, the probabilities of quantum theory dictate that heavy-isotope bonds are slower to react than the light-isotope bonds. But in this chocolate example, we create and enter an alternate universe where it is people who do the selection, not quantum probabilities. So, here you go. ⸍

In a yummy alternate universe, you have 100,000 pieces of chocolate and you eat 1000 pieces every day (with no consequence, because this is a virtual universe, one that is definitely worth visiting in your dreams). But already you may see the flaw: soon you would feel bad, because every day there would be 1000 fewer pieces of chocolate, and you know the day would come when—you guessed it—no more chocolate! To forestall this catastrophe, we have a Benevolent Chocolate God (BCG) who adds back 1000 pieces of chocolate every day, so the mountain of chocolate never grows smaller, never disappears, but always is there and waiting for our innermost desires. After all, we know that all is right with the universe when there is infinite chocolate.

"But oh," you sigh, "there are so many kinds of chocolate, and what kind will we get from our friendly local BCG?" It turns out that through earnest supplication, we can occasionally influence BCG on the mix of light and dark chocolate BCG supplies, and then of course we have our own eating preferences on this subject. So here is what happens in Two Years of Chocolate. We start a new each year from the same beginning point at $t = 0$ days, and with the same chocolate mixture that is dominated by light chocolates.

Year 1 (Figure 4.1). Getting started is great! We start with almost no dark chocolates, and tons of fluffy light chocolate cremes, but the BCG gives 500 pieces of dark and 500 pieces of light chocolate every day to replace the 1000 pieces we eat. As the year goes by, the total amount of chocolate does not change (very nice!), but we see that slowly but surely we get to the 50/50 mix of light and darks that our new supplier has granted us (Figure 4.1). In other words, we ate up all the old chocolate and now depend just on the new stuff. This example illustrates the maxim, "In with the new, out with the old," and the grand power of BCGs.

Year 2 (Figure 4.2). We start over this year with a fresh supply of chocolates that again are almost all light chocolates. This year we decide it is time to get picky. The BCG is still doing the 50/50 thing (BCGs take a while to change their minds), but we poor mortals finally get a little tired of those heavy dark chocolates, and make an important decision—to diet every day, deciding this diet means that we will choose more of the light chocolate pieces. Of course, this decision does not mean that we are actually eating less chocolate. No, we still eat the same total amount every day (so this would not qualify as a diet most places). But, instead of just taking whatever random mix of lights and darks is available, we get a

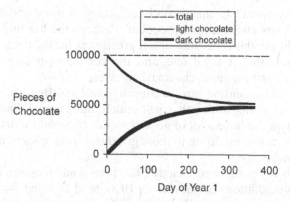

FIGURE 4.1. The changes in a chocolate mixture that is available for eating over the course of a year. Each day, a consumer eats 1000 chocolates total, with an equal preference for light and dark chocolates. This preference matches the supply dynamic that also replaces 1000 chocolates each day, 500 lights and 500 darks. In this case, daily additions resupply the eaten chocolate so that the amount of chocolate remains constant. However, the mix gradually changes composition from only light chocolates present initially to an equal mix of light and dark chocolates by the end of the year.

little selective in our new-found "diet", and discriminate against those big bad old heavy dark chocolates. But not too much, of course, for after all, it is still chocolate! So we decide to pick the lights a little faster than we pick the darks, twice as fast in fact, deliciously, delectably discriminating for the lights and against the heavies. *Et voilà*, the second year in chocolate (Figure 4.2).

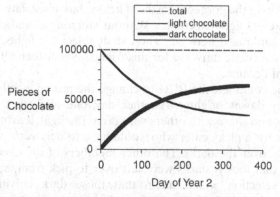

FIGURE 4.2. As Figure 4.1, but with a change in the preferences for chocolate eaten. Here the consumer prefers light chocolates twice as much as dark chocolates, leaving more dark chocolates to accumulate by the end of the year.

What did we learn? In this second year, there is more and more dark chocolate and less and less light chocolate. Perhaps life has turned against us. Some might say this is because of our greediness for the light, but others might point out that it is all one, and the composition doesn't matter, because it is all just the same chocolate calories.

You be the judge, and be sure to consult your local Benevolent Chocolate God about mixing ratios that will match and support your discriminating taste. You can have a lot of good designer chocolate years, whatever you choose, by paying attention to those gourmand principles of mixing and discrimination.

And what is the point here, exactly? The take-home message is twofold. First, chocolate additions or gains from BCG tend to bring the chocolate stocks to a characteristic baseline, 50% light chocolates and 50% heavy dark chocolates in this case as seen in the year 1 results. Mixing that occurs during gains is thus a stabilizing force that usually drives isotope dynamics towards characteristic baseline settings. Second, selection forces the chocolate stocks away from this baseline, as seen in the year 2 results with more heavies left behind in the overall stock versus the 50/50 starting stock. Selectivity or fractionation operates contrary to mixing, forcing isotope distributions away from steady baseline values. The interplay of mixing and fractionation thus leads to changing chocolate stocks (and isotopes) over time, even though the overall amounts remain absolutely constant in this chocolate universe. If you want to linger awhile in this delicious realm of chocolates for a gourmet's understanding of fractionation and mixing, work on the first three problems in the Chapter 4 folder on the CD. You can also open an I Chi workbook from that folder and practice mixing and fractionating chocolate isotopes until you are comfortable in this alternate yummy universe. Then go on to the next section for a more realistic example of how isotopes circulate in ecological systems, this time in the sea.

P.S. One friend read this essay on chocolate isotopes, and that friend is "isopicky," favoring the gorgeous tastes of heavy dark chocolate rather than those featherweight light chocolates. Without violating a fundamental rule of isotopes paraphrased here as "Thou shalt select for lights," could we rearrange the chocolate universe for this friend? Never fear, all is possible in this world of dreams.

One day, one could ask the BCG to change the mix to mostly dark, and presto, the day dawns brighter. Another day, one might get together a chocolate club, and share with others who favor the light, leaving the more abundant dark for a picky eater who would have to only very, very slightly favor the lights over the darks. The other members of the chocolate club would ensure that mostly darks were left over to pick from, and a nearly even-handed selection would mean that those dark cravings of our "isopicky" friend could be satisfied. So, a little manipulation and presto, while the world seemingly favors the light, the dark and heavy is also possible for those who wish upon a star, swinging to the beautiful tastes of chocolate isotopes.

P.P.S. As you leave this essay heading back towards the real world, don't forget to reset your thinking about how fractionation occurs. Leave behind the concept that people are busily selecting chocolate isotopes. Instead, think about our known reality, that the more difficult making and breaking of bonds at the atomic level leads to a probabilistic slower reaction of heavy isotopes. This atomic-level discrimination occurs completely without human intervention or control, so that we can expect it in all our biochemical and ecological systems, an implanted process from the beginnings of our universe.

4.2 Oxygen in the Sea

Imagine floating atop an ocean world, blue, deep pure blue, a seemingly empty ocean that is surprisingly alive, full of life that is producing and consuming again and again in ways that we macrohumans cannot easily perceive. Yet our instruments let us probe this cycle of the sea. Close examination shows that the microscopic algae of the ocean produce sugar and oxygen when the sun shines, and at night, this process runs in reverse, so that the stored algal sugar gets consumed, burned with aqueous oxygen. This may be stranger than the Greek fire of ancient ships, but true. The oxygen content of the water goes up with the sun and down after dark, over and over and over again following an ancient rhythmic sun–sea cycle.

There is isotope action inside these reactions, especially isotope mixing that is important for photosynthesis and isotope fractionation that is important for respiration. The isotopes cycle as behind-the-scene players in the oxygen dynamics, with spreadsheet models opening up the action to show what happens in the sea both day and night.

The goal of this section is to help you harmonize with the rhythms of this cycle, to view what is happening to two streams of data at once, the oxygen concentrations and the oxygen isotopes. There are four day–night examples to read and think about, but first let's consider some further introductory thoughts about the dynamics of dissolved oxygen, O_2.

We can write a generalized equation for photosynthesis, the chlorophyll-catalyzed reaction of water and CO_2 to produce O_2 and generic carbohydrates (CH_2O) for plant growth:

$$H_2O + CO_2 = O_2 + CH_2O$$

82   Chapter 4

The equation for respiration is just this same equation read in reverse, so that carbohydrates combine with oxygen to form CO_2 and water. The forward photosynthetic reaction occurs in the day, whereas respiration occurs both day and night.

Oxygen isotopes participate in these forward and reverse reactions, with the isotopes acting as components of the total oxygen fluxes. Because >99% of all oxygen is ^{16}O, most oxygen dynamics occur as ^{16}O fluxes. The $\delta^{18}O$ measurements help track the minor ^{18}O dynamics, showing which sources are involved in mixing and which respiratory sinks are involved in oxygen consumption. There is also a third, much more minor oxygen isotope involved, but we consider that ^{17}O isotope only briefly at the very end of this section. We keep the focus relatively simple, just on the total amount of oxygen and on the $\delta^{18}O$ isotope measurements.

We should also know that water is the source of oxygen produced during photosynthesis, but there is little or no fractionation in this reaction. Hence, O_2 produced during photosynthesis has the same $\delta^{18}O$ value as water, 0‰ in the case of seawater. Oxygen in CO_2 plays a negligible role in the dissolved oxygen dynamics because O_2 is formed from water, and because the amount of oxygen in water itself is very, very large compared to the amount of oxygen in dissolved CO_2. Dissolved oxygen in seawater has a baseline value of 24.2‰ $\delta^{18}O$, a value set by equilibrium with the atmosphere.

With these introductory thoughts, let's consider the ocean's day–night oxygen cycle for a few days, along with a little isotope spice so we can better tune in to this ancient harmony. Oxygen cycling (Figures 4.3 to 4.6) of this section all start at dawn and have 12 hours of light, 12 hours of dark (dark bars at the bottom of the graph indicate nighttime), and show four days of ocean oxygen cycling. The initial conditions represent near-surface water that has recently equilibrated with the large pool of atmospheric oxygen, but now is isolated and can develop its own oxygen dynamic (Figure 4.3).

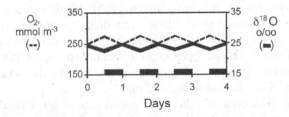

FIGURE 4.3. Oxygen dynamics in seawater during a four-day period. Black bars indicate nighttime conditions when oxygen concentrations decline due to respiration and the absence of photosynthesis. In the daytime, algal photosynthesis increases oxygen concentrations faster than respiration removes oxygen, leading to increasing oxygen concentrations. Isotope compositions of oxygen increase at night due to strong fractionation during respiration, removing light oxygen and leaving heavier oxygen behind with higher $\delta^{18}O$. In the daytime, photosynthesis adds back oxygen with low $\delta^{18}O = 0$‰, so that $\delta^{18}O$ declines.

The up and down of the oxygen concentration (top line, Figure 4.3) repeats each day and night, so there is no up or down trend in spite of the cycle. In each 24-hour period, there is a balance of production in the day and respiration at night, as is common in the sea. Now consider the related isotope action, shown in the bottom line. The isotopes are doing a similar dance, but inverted. The isotopes go down in the daytime and up at night. The daytime decrease reflects input of low isotope photosynthetic oxygen ($\delta^{18}O$ of 0‰, the same as the $\delta^{18}O$ value of the seawater source for this oxygen) so that isotopes go down in the day. At night, removal of that low-isotope oxygen by respiration drives the isotopes back up. So, concentration and isotope cycles are related in an inverse way. In the daytime, photosynthesis dominates, mixing in new oxygen that increases concentrations and decreases isotopes. But at night, respiration reverses these daytime excursions, reducing oxygen concentrations and also increasing and restoring oxygen isotope values with the secret weapon of isotope fractionation. The battle of oxygen additions and withdrawals, isotope mixing, and fractionation is at a draw or tie in this example, which probably occurs only very rarely in an unquiet sea.

This may be a little hard to get used to, thinking about concentrations and isotopes at the same time. But these same issues appear again and again when thinking about how materials cycle in the biosphere. And in fact this dual focus on concentrations and isotopes already appeared once in this book when we considered CO_2 dynamics in Section 3.5. Although it may seem confusing, the dual focus is actually an advantage that leads to better resolution of natural HCNOS cycles. For practice with this dual focus, we consider three more oxygen cycling examples, then stop.

First, let's shed this artificial shell of constant cycling. It never lasts long like that in the sea, saving the sea from endless repetition and boredom. So, let us suppose that there is a little more nutrient in the sea, and algal production and oxygen concentrations will climb over time during a small bloom of algae. What will the isotopes do? You may have guessed this, with more oxygen accumulating, the isotope values trend downwards towards the value of the new, low-isotope oxygen being mixed in (Figure 4.4).

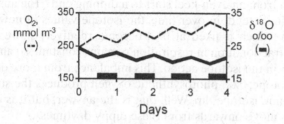

FIGURE 4.4. Oxygen dynamics, continued. As Figure 4.3, but there is stronger photosynthesis during a daytime algal bloom. Daytime oxygen isotope values decline when photosynthetic oxygen with $\delta^{18}O = 0‰$ is added.

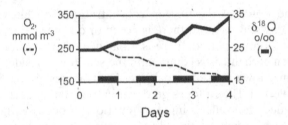

FIGURE 4.5. As Figure 4.3, but oxygen dynamics in deeper waters have balanced photosynthesis and respiration during the day. Strong respiration demands at night lead to progressively lower oxygen concentrations and higher and higher $\delta^{18}O$ values.

Moving right along to our third example, now think in a different direction, down in the deep where water stagnates near the bottom and little light penetrates. There is a lot of respiration and a small amount of photosynthesis, just enough to balance respiration during the daytime; what will happen to the oxygen dynamics?

And the answer for oxygen concentrations is this: overall, the oxygen concentration trends downwards when photosynthesis is weak and respiration is strong (bottom line, Figure 4.5). During the daytime, however, respiration and photosynthesis are just balanced and cancel out, so oxygen concentrations don't change during the day. That's the picture, neat and simple.

And what about the isotopes (top line, Figure 4.5)? Overall, respiration is winning so as concentrations fall, of course isotope values rise because . . . Why? Because respiration removes the light isotopes faster, leaving a heavy residue behind.

Enough said, except for one observant savant who wondered about the day-only results. You can see the problem too, if you look closely at Figure 4.5. At first, isotopes don't change during the day because of near-balance in the system. But by the last day, this isotope balance starts to break down, and isotopes dip downwards during the day. Why the switch in the daytime isotope action from an even-keel start to a dipping end? The answer is this: as isotope values trend up over time, the isotope values of new photosynthetic oxygen, which is fixed at 0‰, become relatively more deviant or extreme. But fractionation in respiration remains constant, so an imbalance starts to occur in the isotope action. This imbalance from more deviant (relatively low isotope, 0‰) photosynthetic oxygen produces the stronger and stronger daytime isotope dip. Well, that is the answer, but it is definitely a detail, so let's move onwards from these dippy daytimes.

For our final thought experiment, let's consider that the stagnating layer of algae in the bottom water has actually settled onto the sediment surface. There is still enough light for some photosynthesis during the day. But most

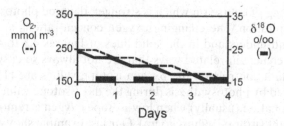

FIGURE 4.6. As Figure 4.5, but for water near the sediment surface. $\delta^{18}O$ values decline during the day when photosynthetic oxygen with $\delta^{18}O = 0‰$ is added, but at night, respiration occurs without fractionation in sediments, so that $\delta^{18}O$ values do not change.

of the respiration signal we observe now is from the sediment, rather than the algae, and this respiration has the unusual isotope characteristic that no fractionation is involved. This is because sediments contain high numbers of bacteria that easily consume all oxygen that diffuses in from the bottom waters. In this case, there is no opportunity for any fractionation because all oxygen is used. Another way to think about this is that the isotope flavor is irrelevant to the oxygen-starved bacteria. These bacteria use any and all oxygen that arrives to the sediments, regardless of isotope content. With this special case of limiting oxygen delivery and consequent lack of fractionation during respiration, what will happen to the day–night oxygen dynamics we are considering?

The concentrations are doing the same thing as in the previous example, just going down and down (Figure 4.6). But now the isotopes are also drifting down. Why? When we stop fractionating isotopes during respiration, the overall isotope trend is down because we continued adding low-isotope oxygen ($\delta^{18}O = 0‰$) every day to the system with photosynthesis, and mixing in this new low-isotope oxygen is slowly replacing the initial high-isotope oxygen ($\delta^{18}O = 24.2‰$). So, in the absence of fractionation, we see the clear mixing dynamic of the daytime additions (Figure 4.6). This illustrates one of the interesting uses of models: turn off the fractionation, and you see clearly the effects of mixing alone.

Conclusion

The day–night oxygen cycles in the sea are governed by photosynthesis and respiration that add and subtract oxygen from our lovely life-giving pool of OO (that is O_2 or dissolved oxygen to you who are not true aficionados of OO). When photosynthesis and respiration are balanced, we start one place and can cycle up in the day and back down at night in a stately steady state, without any overall long-term directional change in oxygen concentrations. But when there is an imbalance in the strengths of photosynthesis and respiration, the oxygen concentrations can trend up or down as the world turns, depending of course on which is stronger, daytime photosynthesis or nighttime respiration. The changing oxygen concentrations are relatively easy to see and understand in the solid lines of Figures 4.3 to 4.6, but the isotopes also change in related ways that are not always so easy to follow.

The important controls on the oxygen isotope values are (1) mixing as oxygen is added by photosynthesis during the day (isotope values go down) and (2) fractionation (usually) when low-isotope oxygen is removed by respiration at night (isotope values go up). Our last example showed a special case with respiratory oxygen removal in sediments without any fractionation (we eliminated fractionation, just like that!), and what did we see? Well, without fractionation, what do you think might appear? Because the isotope choices are mixing and fractionation, if you get rid of one, the other should show up. And sure enough, with fractionation kicked out of the picture, it was the mixing dynamic that emerged.

Scientists working with oxygen dynamics are beginning to use these dual focus approaches to help calculate the gross rates of oxygen production and consumption. Without the extra isotope insight, scientists using the oxygen concentrations alone can really only measure net fluxes, differences between photosynthesis and respiration. Adding the second dimension of oxygen isotope measurements resolves the individual strengths of photosynthesis and respiration, the gross rates. The combined concentration plus isotope measurements should help answer long-standing questions about the relative roles of bacteria and algae in oxygen cycling, and whether respiration rates are higher during the day than at night.

Scientists can study these questions in bottle incubations, but there are many possible experimental artifacts associated with those incubations. Extracting oxygen from natural water samples and measuring oxygen concentrations and isotopes is a satisfying alternative way to study what happens in the field, and provides at least a reality check for the labor-intensive bottle incubations.

Besides help with these rate and process estimates, $\delta^{18}O$ measurements can help distinguish oxygen sources and sinks in the sea. The measurements readily distinguish source inputs of atmospheric oxygen versus oxygen produced by in situ marine photosynthesis. Small $\delta^{17}O$ anomalies may also help in these source estimates (Luz and Barkan 2000), although at this time, there is some controversy about the origins of these anomalies (Young et al. 2002). Oxygen isotopes can also help budget sediment versus water column sinks for oxygen, especially where respiratory demand in sediments occurs with little fractionation (Brandes and Devol 1997).

In the end, it is either elegant and informative to view the oxygen cycles from the two different, but related perspectives, the oxygen concentrations and the oxygen isotopes, or else you feel you are going cross-eyed trying to keep up with two streams of information that converge and diverge in mysterious ways. A goal of this book is to help you unfocus your eyes to make these cross-eyed, double-vision experiences routine and understandable at an intuitive level. Practice is the key to acquiring this dual-focus ability. On the accompanying CD, you will find the I Chi model used for these oxygen examples in the Chapter 4 folder, workbook 4.2. Open that workbook and change master variables such as the amount of oxygen added or the fractionation factor, to see if you can re-create some of the results shown here. Also, Problems 4 and 5 for this chapter on the CD can help you work on a better understanding of these oxygen cycles and the spreadsheet models of this section.

4.3 Equations for Isotope Chi ("I Chi")

Isotope models are built around the mathematics of gains and losses. Simple addition and subtraction account for the changes in the total amount during gains and losses, but what equations are needed for isotopes? As it turns out, two simple equations govern isotope cycling, one equation for isotope mixing during gains and one equation for isotope fractionation during losses. Overall, gains involve adding mass and mixing isotopes. Losses involve subtracting mass and fractionating isotopes.

Mastering the equations for gains and losses will enable you to cycle isotopes through ecological systems in graceful and surprising ways. This is the art of Isotope Chi (like Tai Chi) that will give you unexpected power, isotope power, power that appears in the virtual reality of spreadsheets. (Technical note: the knowledgeable reader will see that the equations of this section are approximations, but in fact they are surprisingly good approximations. Section 4.5 considers errors associated with these approximations and Section 4.6 presents the exact equations.)

For the biologists who dislike equations, you should know that you are not alone as you read the following section which emphasizes mathematics. So, here is an introductory word from Bond, James Bond. In the book, *Facts of Death* by Raymond Benson, Bond is investigating a secret society devoted to mathematics. The secret society is inspired by a famous Greek from the past, Pythagoras. As you read this passage from page 162 of the Penguin Putnam edition, keep in mind this exposition of Pythagoras comes from a really evil person:

'The Master – that is, Pythagoras – demanded that those desiring instruction should first study mathematics. The Pythagoreans reduced everything in life to numbers because you can't argue with numbers. We usually don't get upset about multiplying two (and two) and getting four. If emotions were involved, one might try to make it five and quarrel with another who might try and make it three, all for personal reasons. In maths, truth is clearly apparent and emotions are eliminated. A mind capable of understanding mathematics is above the average, and is capable of rising to the higher realms of the world of abstract thought. There, the pupil is functioning closest to God.'

'I should have studied harder in school,' Bond said.

(From *The Facts of Death* by Raymond Benson, copyright © 1998 by Glidrose Publications Ltd. Used by permission of G.P. Putnam's Sons, a division of Penguin Group (USA) Inc.)

Here are the important equations.

1. Gains

$$m_{t+1} = m_t + m_{GAIN}$$

$$\delta_{t+1} * m_{t+1} = \delta_t * m_t + \delta_{GAIN} * m_{GAIN},$$

so that

$$\delta_{t+1} = (\delta_t * m_t + \delta_{GAIN} * m_{GAIN})/m_{t+1}$$

For gains or additions, the total mass (m) increases each time step by a mass gain, m_{GAIN}, from m_t to m_{t+1}. The isotope compositions changes according to mass balance, from the initial isotope value δ_t towards the new isotope value δ_{GAIN}, with the material with the largest mass or fractional contribution dominating the weighted average. The isotope balance derives from the realization that a δ value is a very good proxy for "% heavy isotope" (see Figure 2.1), so that multiplying mass $\times$ δ gives the amount of heavy isotope, and the above isotope equations just become accounting equations for the amount of heavy isotope.

Let's stop and restate this idea. If δ values are stand-ins or proxies for % heavy isotope, then terms such as $\delta_{t+1} * m_{t+1}$ could be rewritten as

$$\delta_{t+1} * m_{t+1} = (\% \text{ heavy isotope}) * (\text{mass}) = (\text{amount of heavy isotope})_{t+1}$$

so that the overall equation for $\delta_{t+1} * m_{t+1}$ becomes

$$\delta_{t+1} * m_{t+1} = (\text{amount of heavy isotope})_{t+1}$$
$$= (\text{amount of heavy isotope})_t + (\text{amount of heavy isotope})_{GAINED}.$$

With this restatement, it is clearer that these complex-looking isotope equations are actually simple sums that budget the amounts of heavy isotope. The equations just add amounts of atoms, heavy-isotope atoms, and this is what is happening when we use the δ values as proxies or stand-ins for % heavy isotope.

For completeness, a detail here is that because the correlation between δ and % heavy isotope is not completely exact, there are some very minor inaccuracies in using δ as a proxy for % heavy isotope in most cases. This leads to very slight errors in the mixing and fractionation calculations for most natural abundance δ values, as evaluated in Sections 4.5 and 4.6. An important cautionary note is that these equations are inexact for samples that have been highly enriched with added heavy isotope; exact equations presented in Section 4.6 are necessary for such samples.

2. Losses

$$m_{t+1} = m_t - m_{LOSS}$$

$$f = \text{fraction lost} = m_{LOSS}/m_t = 1 - m_{t+1}/m_t$$

$$\delta_{t+1} = \delta_t + \Delta * f = \delta_t + \Delta * (1 - m_{t+1}/m_t),$$

where f is the fraction lost and Δ is the fractionation factor expressed in positive ‰ units. In the first of these loss equations that concerns only the amount lost, the total mass m_t decreases at each time step by an amount, m_{LOSS}. The second equation for f gives the fraction lost, and the third equation concerns isotope changes, splitting the overall mass into a part m_{LOSS} that is lost and a part m_{t+1} that is carried forward.

Technical Note: There are important approximations involved in using this last equation for isotope loss, and in some cases when working with hydrogen isotopes or with enriched samples that have δ values very different (>100‰ different) from 0‰, a more exact equation should be substituted. The substitute equation is

$$\delta_{t+1} = (\delta_t + 1000) * (f * \alpha + 1 - f) - 1000,$$

where $\alpha = (1000 + \Delta)/1000$; Section 4.6 explains this more complex substitute equation.

But generally speaking, how do we understand these equations for isotope changes during losses? First, we recognize that the part lost typically has a lower δ value according to the general formula for isotope fractionation:

$$\delta_{LOST} = \delta_{SOURCE} - \Delta.$$

But this simple isotope difference between source and lost materials only partly accounts for the observed fractionation during I Chi loss reactions. A second step is considering the more complex result when the source substrate pool splits into two parts, a part that forms product and a residual substrate pool. This type of reaction is known as a split or open system

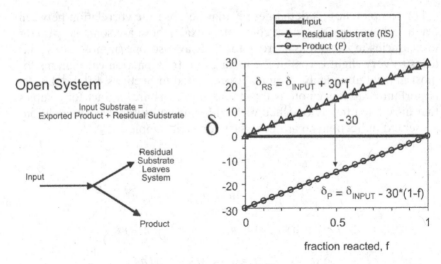

FIGURE 4.7. Isotope dynamics in open systems where reactions are split or branched. In this example, a fractionation factor of $\Delta = 30‰$ always gives the difference between substrate and product δ values, as discussed in the text.

reaction, where substrate constantly enters the reaction center. Unused substrate and newly formed product continually leave the reaction (Figure 4.7).

In these open systems, the entering substrate pool splits into two exit streams, product and residual (unused) substrate. The key feature is that the isotope fractionation also splits. For example, suppose that an overall isotope fractionation factor involved in the formation of a product from substrate is $\Delta = 30‰$ and the initial substrate has a δ value of 0‰. Then this 30‰ fractionation will appear as a −30‰ δ value in the product when only a very small amount of product is split out from the reactant pool, and the isotope value of the remaining substrate pool will stay near, but slightly above 0‰. This situation occurs when f is near 0 in Figure 4.7. Now consider a split loss of half of the overall substrate pool ($f = 0.5$, Figure 4.7), when the reactor is more efficient and converting a much larger fraction of substrate to product. In this case, the isotope value of the product becomes −15‰, and the half of the fractionation is split to the substrate pool which obtains a value of 15‰. The fractionation factor of 30‰ has been split between a −15‰ product and a +15‰ residual substrate, retaining a constant fractionation difference of 30‰ between the two pools. In summary, the open system splits the reaction streams into product and residual substrate, and also splits the fractionation between product and substrate.

Overall, the important consequence for I Chi modeling is that when losses are accompanied by fractionation, part of the fractionation is expressed in the residual substrate pool. Small losses of substrate are accompanied by small increases in δ values of the surviving substrate (Figure 4.7, top line). Fractionation is also important in products (Figure 4.7, bottom line), and accounting for this fractionation becomes important in multiple-pool models that include eventual recycling of products into substrate pools. Most of the models in this book are simple, single-pool models that only need consider gains and losses from substrate pools, without explicitly tracking products.

These are the technical details, so now let's stop and step back for perspective. The reader may have noticed that in the above equations, isotopes are presented together with amounts and masses. Isotopes are not really independent of amounts, but are just an associated property, like the color of a rock is an associated property of that rock. To understand and budget isotopes, one has to involve amounts. This involvement of amounts with isotopes pertains generally for isotope modeling of laboratory and field results (see readings below), and also accounts for the dual focus on isotopes and amounts noted in the oxygen cycling example presented in the preceding section.

This also concludes the math. The rest of the modeling is really only learning to diagram gains and losses in box-and-arrow diagrams, and learning to use spreadsheets to write these equations over and over to keep track of losses and gains. Finally, there is also a simpler class of models that involve no net changes in mass and isotopes over time, the steady-state models. In steady-state models, the stocks remain constant even as mass and isotopes are cycling. Chapter 7 considers these steady-state models alongside the time-dynamic I Chi models.

4.4 Building an I Chi Gain–Loss Model, Step by Step

Let's look at how to build an I Chi isotope model, step by step. In reality, the I Chi models are elaborations of two kinds of steps, gains and losses. Gains involve adding mass and mixing isotopes. Losses involve subtracting mass and fractionating isotopes. Here we construct an I Chi model of oxygen cycling in the sea, as portrayed in Section 4.2.

The first step is to make a diagram of the processes that interest you. In this case, gains are due to photosynthesis adding oxygen, and losses occur when respiration consumes oxygen. The gain–loss dynamic is depicted in a box-and-arrow diagram centered on a pool of dissolved oxygen (Figure 4.8). There are fluxes in and out of the central storage compartment, with inputs representing photosynthetic oxygen gains and outputs representing respiratory losses. We need gain equations for addition and isotope mixing, and loss equations for subtractions and isotope fractionation. Sometimes the system will be balanced, but more often it will be unbalanced as photosynthesis seldom exactly equals respiration.

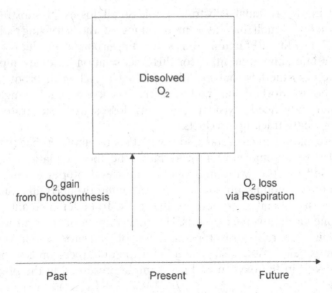

FIGURE 4.8. A simple model of oxygen dynamics in the sea. Photosynthesis adds oxygen to a central pool and respiration removes oxygen from the pool.

Also, we want our model to capture the timing of oxygen build-up and oxygen loss, so that oxygen changes over time, with continuity from the previous time step to the current time, and continuity forwards to the next time step. We add an arrow of time at the bottom of our model to denote continuity from the previous time and continuity towards the next time (Figure 4.8).

A second important beginning step is to think out a good test for the model, something simple that the model should depict accurately, a model result that will tell us that the equations we are writing in fact give a correct answer. One test would be this: suppose the oxygen gains equal the oxygen losses, then the amount of oxygen in the central pool should not change in a balanced system. Another test would be this: if oxygen gains are much stronger than losses, then oxygen amounts in the central pool should increase. Many such simple tests are possible, and it is good to think of a few tests before you really get started with the modeling. If you make a mistake here or there in the next steps of actually writing the model, having some simple final tests will alert you to those mistakes. Then you can find and fix the mistakes before using a flawed model that would produce confusing results.

Armed with these initial ideas, a diagram of the process, and some simple tests for the model, we proceed to the individual steps involved in gains and losses over time. We write equations for the gains and losses in a convenient spreadsheet program, Excel. As it turns out, it does not matter for the overall dynamics which comes first within a time step, the gains or losses, just that both are included in each time step.

The individual steps are given as numbered worksheets on the accompanying CD in the Chapter 4 folder, I Chi Spreadsheets, workbook 4.4. Take a minute to look at the workbook and its worksheets; much is self-explanatory in the sequentially numbered worksheets. Click through those worksheets to see the progression of how to build an I Chi model. To do this, first open the workbook in the Excel program, and set your computer cursor on the first tab on the bottom left. This will bring up the first worksheet. The next tab to the right opens the next worksheet with the next step, and so on through 18 sequential worksheets. Table 4.1 gives an overview of results obtained from the basic model, and the following text gives a guide to each numbered worksheet.

1. *Establish time intervals.* The first task is to set up the time sequence. Use the leftmost column for this, start with time 0 at the top, and add time steps downwards, 1, 2, 3, 4, and so on. This gives each row a consecutive number that will represent hours in this model, although in some applications, the intervals are spatial, for example, meters or kilometers. In any event, gains and losses will play out across each row in the following worksheets. For this oxygen model, each row will have a complete cycle of oxygen gains and losses before the next time step.

TABLE 4.1. I Chi Spreadsheet Model for the Oxygen Dynamics.

			Gain		Loss	
	Initial amount	250	Amount gained	10	Amount lost	10
	Initial δ	24.2	δ gained	0	Fractionation factor, Δ	18
Time (Hours)	Amount	δ	Amount	δ	Amount	δ
0	250	24.200	260	23.269	250	23.962
1	250	23.962	260	23.040	250	23.732
2	250	23.732	260	22.819	250	23.512
3	250	23.512	260	22.607	250	23.300
4	250	23.300	260	22.404	250	23.096
5	250	23.096	260	22.208	250	22.900
6	250	22.900	260	22.019	250	22.711
7	250	22.711	260	21.838	250	22.530
8	250	22.530	260	21.664	250	22.356
9	250	22.356	260	21.496	250	22.188

[a] The model has initial conditions of 250 mmol oxygen m^{-3} with an isotope value of 24.2‰. Each hourly time step is represented by one row for the 9 hours of this model. Each hour, photosynthesis adds 10 units of oxygen with an isotope value of 0‰, and respiration removes 10 units of oxygen with a fractionation factor of 18‰. Equations developed in Section 4.3 account for gains and losses during each time step, using the master variables at the top of the table.

Source: From Chapter 4 folder on the accompanying CD, I Chi workbook 4.4, worksheet 8.

2. *Write initial conditions.* The fundamental quantities of the isotope models are amounts and isotopes, so that there are always dual columns for each process: the first column for the amounts, and the second, adjacent column for the δ value of that material. Here we use representative (average) values for oxygen concentrations and $\delta^{18}O$ in air-equilibrated seawater, $250\,mmol\,m^{-3}$ and 24.2‰, respectively, to fill in the beginning columns. These special values are the initial conditions, and we also write these values at the top of the columns. These and later variables written at the top of the worksheet will become master variables that are linked to values in the worksheet. Typing in a change to a master variable will lead to updates throughout the whole worksheet, so that you can type in a value as a thought experiment, then view the results.

3. *Gain equations.* Oxygen gain occurs as new photosynthetic oxygen enters the central oxygen pool, changing both concentrations and isotope compositions via mixing. The increment amount (10) and isotope value of added oxygen (0‰) are given at the top of the spreadsheet. The oxygen addition occurs via simple addition: old amount + increment = new amount, so $250 + 10 = 260$. The isotope addition occurs via mass balance or weighted average mixing, that is ((previous amount * previous δ) + (amount of new material added * δ of new material added))/new amount, $(250 * 24.2 + 10 * 0)/260 = 23.269‰$, the updated δ value of the oxygen pool. The 0‰ value for new oxygen represents conditions for seawater ($\delta^{18}O = 0‰$) and the assumption that there is no fractionation during the photosynthetic production of oxygen from this seawater.

4. *Loss equations.* Loss occurs via respiration, affecting both amounts and isotopes of oxygen left behind in the central pool. The loss amount (10) and isotope fractionation during respiration (18‰) are given at the top of the spreadsheet. The oxygen loss occurs via simple subtraction, old amount − increment = new amount, $260 - 10 = 250$. The isotope loss occurs with fractionation, and the new value after loss = previous $\delta + \Delta * f$ where Δ is the fractionation factor in positive ‰ units, and f = fraction reacted = (1 − (new amount/previous amount)). Completing the end of the first row, δ following loss $= 23.269 + 18 * (1 - (250/260)) = 23.962‰$. Believe it or not, this is the end of the hard part of the modeling!

5. *$ signs.* Add $ signs to the equations of the first row so that terms reference the master variables (gain and loss terms) at the top of the spreadsheet. For example, in workbook 4.4 cell C7 becomes = D2 instead of = D2, cell D7 becomes = D3 instead of = D3, and so on for each equation in row 1. This $ addition will ensure that formulas in all rows copied from row 1 will refer to the master variables at the top of the sheet; for example, the gain term of "10" in cell G2 will always be referenced appropriately if in the equations it appears as G2. Don't worry if you forget the $ signs; various nonsense indicators will appear after a while in the worksheet columns, to remind you that you left out the $ signs.

6. *Wraparound.* Copy the last two entries of the first row, the final amount and final isotope values, down and left so that they form the starting values of next row. This wraps around the values from the end of one time step into the beginning of the next time step, allowing a steady progression of oxygen dynamics from one hour to the next.

7. *Copy and Drag.* Drag down formulas for gain and loss (or copy and paste them) from the first row of the model into the second row.

8. *Copy row 2 downwards.* If you copy row 2 downwards, all the equations of gain, loss, and wraparound will propagate over time, 9 hours in the example worksheet. Don't pull down the row 1 equations as a whole, because there are initial conditions in the first row that you don't want in the rest of the spreadsheet. Instead, copy row 2 equations downwards.

9. *Graphs.* Use the initial values for each hour, and graph up results for amounts and isotopes versus time. The initial variables used in this example give no change in concentration, which stays steady at $250 \, mmol \, m^{-3}$.

10. *Changing master variables.* You are finished building your model, and now you can change master variables in workbook 4.4 on the accompanying CD, to see what might happen if . . . for example, you increase photosynthesis from 10 to 20. Satisfyingly, the oxygen concentration goes up when hourly gains of 20/h from photosynthesis exceed hourly losses of 10/h from respiration. Isotopes also change, declining because of the addition of photosynthetic oxygen with a low 0‰ value. *Note*: you may have to collect and move your master variables to the left upper corner of the worksheet and place graphs below these variables, so that it is convenient and easy to see the graphical consequences of any changes you make to the master variables.

Checking the Model

We set out two tests before starting the model, and now before going any further, it is time to check and see if the model is working correctly. Test 1 was that if oxygen gains and losses were balanced, then oxygen concentrations should not change. Worksheets 8 and 9 in workbook 4.4 show this to be the case, so the model passes this first test. Test 2 was that if gains in photosynthetic oxygen exceed respiratory losses, then oxygen concentrations should increase. Worksheet 10 shows this also to be the case. But we should add a few more tests, for the isotope portion of the model. These are more complex tests, but here is one of the simpler ones. Isotopes should not change if the isotope values of added oxygen are 24.2‰ and there is no fractionation in respiration; that is, $\Delta = 0$‰. If you type in the values "24.2" into cell G3 and "0" into cell J3 of worksheet 10, you can verify that there is a flat horizontal line for the oxygen isotopes. So, all in all, the model passes tests for amounts and isotopes.

A few thought experiments follow in worksheets 11 to 13 of workbook 4.4 on the accompanying CD, and these can also function as additional tests of the model if you approach them with the thinking that results should make sense for isotopes as well as amounts.

11. *Scenario A*. Turn off photosynthesis. Type in 0 in cell G3, turning off oxygen gain, and note the following results. Amounts: Oxygen concentrations drop without new oxygen additions from photosynthesis and respiration remains turned on. Isotopes: Isotopes increase as isotopically lighter oxygen reacts faster during respiration, leaving heavier oxygen behind accumulating in the central oxygen pool.

12. *Scenario B*. As previous worksheet with scenario A, but increase the rate of respiratory oxygen loss from 10 to 20. Amounts: Oxygen concentrations drop faster. Isotopes: Respiration effects are stronger, with residual oxygen more enriched in heavy isotopes and reaching higher values than in scenario A.

13. *Scenario C*. As previous worksheet with scenario B, but double the fractionation factor, Δ, from 18 to 36‰. Amounts: No change. Isotopes: An even steeper increase in respiration effects with high, heavy isotope values exceeding 50‰.

These three tests indicate oxygen isotope dynamics that are consistent with general principles, especially that when respiration effects increase, isotope values increase. These tests thus suggest that the model is also adequate for oxygen isotopes as well as oxygen amounts.

Before leaving this step-by-step approach to modeling, let's consider some final useful modifications of models: (1) switches, (2) more complex functions for gains and losses, and (3) a more exact equation for fractionation during losses.

1. *Switches*. Sometimes processes are active only at certain times, and need to be switched on or off. In our oxygen model, photosynthesis switches on in the day, but switches off at night. Switching is readily possible using 1 and 0 as multipliers. For example, make a column with entries of 1 for daytime, and 0 for nighttime, then multiply appropriate equations by this column, in effect switching on only reactions that happen during the day. You can also use "if" statements to create the column of 1s and 0s using "if" statements (e.g., in worksheet 14, cell C7 translates "day" into a value of "1" by this logic: If B7 = "day"1,0). These if statements are generally useful in other ways, because you can also use if statements to generate columns of 1s and 0s around threshold values (e.g, If B7 < 250, 1,0). Worksheets 14 and 15 illustrate using if statements to switch photosynthetic oxygen gains on during the day and off at night, and the basic worksheet is reprinted as Table 4.2 for reference (see especially the first three columns of Table 4.2 for the day–night switching).

TABLE 4.2. Table 4.1 with a Day–Night Dynamic Added.[a]

	Initial amount	250		Gain		Loss	
	Initial δ	24.2		Amount gained	10	Amount lost	10
				δ gained	0	Fractionation factor, Δ	18
Time (Hours)	Switch	Amount	δ	Amount	δ	Amount	δ
0 day	1	250	24.200	260	23.269	250	23.962
1 day	1	250	23.962	260	23.040	250	23.732
2 day	1	250	23.732	260	22.819	250	23.512
3 day	1	250	23.512	260	22.607	250	23.300
4 day	1	250	23.300	260	22.404	250	23.096
5 night	0	250	23.096	250	23.096	240	23.816
6 night	0	240	23.816	240	23.816	230	24.566
7 night	0	230	24.566	230	24.566	220	25.349
8 night	0	220	25.349	220	25.349	210	26.167
9 night	0	210	26.167	210	26.167	200	27.024

[a] Dynamic added by using a "switch," with 1 representing day and 0 representing night. This switch controls photosynthetic oxygen gains that occur during the day, but not at night.

Source: From Chapter 4 folder on the accompanying CD, I Chi workbook 4.4, worksheet 14.

2. *Complex gain functions.* The models thus far are based on fixed amounts for mass gains and losses. These fixed amounts or increments can be modified in many ways, for example, making part of the photosynthetic oxygen gain dependent on the pool size of oxygen, with algal growth during the day leading to increasing amounts of oxygen produced each hour. Worksheet 16 gives a formula for complex gain during daytime photosynthesis, with oxygen increments increasing exponentially over time. This is accomplished by multiplying daytime oxygen concentrations by a fixed fraction (see cell H3 in spreadsheet 16). Worksheets 16 and 17 implement this multiplicative strategy for increments for both losses and gains, rather than using the fixed increments of the previous worksheets. Using the multiplier strategy allows for first-order (exponential) reactions based on percentage changes. Using fixed amount removals, similar to a flat sales tax regardless of the amount involved, is termed a zero-order reaction. These two different types of reactions, zero-order reactions and first-order reactions, are discussed again and illustrated more fully in Section 7.2.

3. *Exact fractionation during loss.* The main error in I Chi models lies in the equation used for fractionation during loss. A more complex equation can be substituted in spreadsheets after they are completed (or starting from scratch), as detailed in Section 4.6. Spreadsheet 18 makes this substitution in column N; click on cell N7 to see the equation algebra. Overall, results usually don't change much with this substitution, but for exactness, it is good to make this substitution at the very end, when you are satisfied with the overall worksheet model.

Conclusion

Now you have the behind-the-scenes view of how to make an isotope cycling model based on gain and loss steps. Generally each row represents one time unit, with several processes happening in that time unit. Processes usually have two columns associated with them, one column for the amounts, and a second column for isotopes. The simple model for oxygen gives details of only two processes, photosynthetic gains and respiratory losses. Models can be expanded to include more processes. For example, this simple aquatic oxygen model could be modified to include oxygen gains from the atmosphere or oxygen losses to the atmosphere. But a word to the wise is to keep only two to five processes in each time step in your models. That will yield plenty of complexity in the end.

Problem 5 in the Chapter 4 folder on the accompanying CD asks you to develop your own I Chi photosynthesis model that is similar to the oxygen model of this section, but focused on carbon dioxide rather than on oxygen.

4.5 Errors in I Chi Models

The connoisseur is often more interested in the problems than the successes, just as much interesting literary material is often found in footnotes. Also, considering errors usually illuminates material from a different angle, leading to a different perspective about complex processes. With these thoughts in mind, let's consider several sources of errors potentially at work in the I Chi models: (1) time intervals are too large, (2) choice of equations for loss reactions is flawed, and (3) the δ notation used in I Chi models is inexact in some details, and can lead to flawed isotope accounting.

Time Intervals Are Too Large

Ideally, reactions such as gain and loss occurring in models are instantaneous, so that time intervals between reactions are infinitesimal. Realizing this ideal is possible in many situations, and involves writing and solving differential equations. But solutions to differential equations are not always easy to find, nor are they typically within reach for students with average mathematical capabilities. The alternative method, adopted for I Chi, is to separate steps and solve sequentially, keeping intervals as narrow (infinitesimal) as is practical. In effect, we split the ongoing reaction into several discrete, time-sequential steps (Figure 4.9).

This splitting into discrete steps leads to time intervals with finite breadths, so this is known as the finite difference method. There are several helpful methods that can be embedded in finite difference calculations to make equations more simultaneous. Those methods such as the fourth-order Runge-Kutta algorithm are not employed here, but can be accessed through commercial modeling packages.

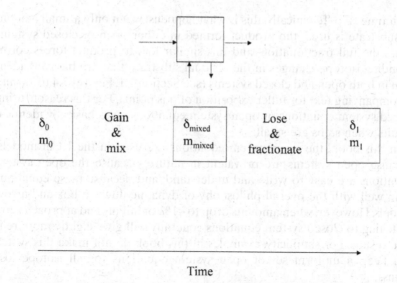

FIGURE 4.9. A generic box model with explicit representation of processes involved in input gains and output losses. The gain and loss steps are coupled in that they occur during each time interval in the I Chi spreadsheet models.

For the I Chi models of this book, practical tests show that it is wise to keep the stepwise gains and losses small relative to pool sizes, <10%. If you suspect the increments are too large at each time step, divide the master gain and loss terms each by 10 (reduce them by an order of magnitude), and see if this makes a strong difference in the results. This test makes the mass gain or loss per time step 10 times smaller, moving in the direction of infinitesimal mass changes at each time step. When results stop changing at smaller and smaller gain and loss steps, the intervals are small enough. Typically, this process of testing smaller changes in gains and losses rapidly leads to finding acceptable time steps for the I Chi models, with <500 time steps (rows) sufficing for simpler models.

Choice of Equations for Loss Terms

Although calculating gains and accompanying isotope mixing is very straightforward, it is not always obvious which equations should be used for loss reactions. In Chapter 7, we learn that there are two systems of equations used to understand isotope loss dynamics, equations applying to open and closed systems. But with the finite difference approach adopted here, the open and closed systems are actually similar and both really only consider a loss step, stopping any flow-through dynamic to complete the loss as a single reaction step. And as it turns out, both the open and closed system equations give essentially identical answers when losses are small at each time step. Technically, this is what happens: when only a small amount of substrate is used, the product formed in either open or closed systems bears the full fractionation, and this similar loss to product forces corresponding isotope changes in the residual substrate that are basically identical in both open and closed systems (see Section 7.1, Figures 7.4 to 7.6 and accompanying text for fuller exposition of this point). For this reason, using closed system equations or open system equations gives basically identical results when steps are small.

In this book, the open system equations are used in the I Chi models, because open systems are prevalent in nature, because the open system equations are easy to write and understand, and because these equations fit in well with the overall philosophy of dynamic flows in box-and-arrow models. However, when amounts drop to <1% of initials and approach zero, switching to closed system equations generally will give slightly more reliable results. For simplicity, examples in this book do not make this switch and keep a uniform set of open system equations for all isotope loss terms.

The δ Notation Can Lead to Flawed Isotope Accounting

Ideally, one would like to account for all the heavy isotope atoms and all the light isotope atoms, not misplacing any in a good accounting scheme. Unfortunately, the δ notation is not this kind of truly exact accounting notation (see Section 2.2). But δ values can be recalculated exactly into heavy and light isotope fractions, and flows of these isotope components can be tracked separately, occasionally recombining the results for a δ value. Although this sounds a little confusing, it is exact and is illustrated in the next section. Overall, results from the simpler I Chi approach are usually very close to accurate, within acceptable limits of errors. However, significant errors do appear in some cases, especially when amounts drop to <1% of initials and approach zero, and when working with hydrogen isotopes and enriched samples in ranges of δ values that are more than 100‰ different than 0‰ (Hayes 2004; Sessions and Hayes 2004). Section 4.6 considers these problems and recommends a relatively simple solution for these cases. The simple solution is to write I Chi models as outlined in the step-by-step approach of Section 4.4, then go back when the model is otherwise complete, and substitute a more exact equation for isotope changes during loss. That is, the normal equation for isotope change during loss is:

$$\delta_{t+1} = \delta_t + \Delta * f$$

and the recommended substitution is:

$$\delta_{t+1} = (\delta_t + 1000) * (f * \alpha + 1 - f) - 1000 \quad \text{where}$$
$$\alpha = (1000 + \Delta)/1000.$$

Section 4.6 also gives exact equations for all the gain and loss steps of isotope accounting, and the sophisticated reader may wish to use those equations from the start when constructing I Chi models.

Conclusion

There are several kinds of computational errors in the I Chi models. A first concern is for interval errors that are common to all finite difference models. The largest errors are associated with large intervals, gains or losses of 10% or more relative to affected pools. Gains and losses in the 1 to 10% range seem quite robust compared to results generated with much smaller increments, and are recommended for these simple gain–loss models. Other computational errors also occur, but in most cases these errors are minor or, as the next section shows, can be avoided by using a slightly more complex equation for fractionation during loss steps.

4.6 Exact Equations for I Chi Models

The δ-based notation used in this book for I Chi modeling is a slightly inexact notation in some contexts, and small errors can accumulate in multistep models. However, exact equations are readily available for the gain and loss reactions of I Chi, so that you can gain exactness by amending models built with the simpler, δ-based I Chi equations. The main source of error lies in the equation for fractionation during losses, so one amendment is a simple substitution of a more exact fractionation equation. The other source of error occurs during gain and mixing, so the second amendment is to change equations governing mixing. Changing the mixing equations is more elaborate, although still straightforward once you have done it once or twice. Overall, the choices about gaining exactness are (a) no change, (b) changing equations for fractionation only, or (c) changing equations for both fractionation and mixing. In fact, experience shows that no change and use of the simple I Chi equations of Section 4.3 are adequate for most applications considered in this book. But when working with hydrogen isotopes and enriched samples, there is a significant gain in exactness when equations for fractionation are modified (option b). There is little further advantage to the truly exact solutions (option c), but for reference these are presented first.

Exact Equations for Mixing Gains and Fractionation Losses in I Chi Models

Gains

When mass is being added, the δ values can be recalculated into exact fractions of heavy and light isotopes that together sum to the whole. This calculation of fractions derives from the δ definition:

$$\delta = ((H/L_{\text{SAMPLE}})/R_{\text{STANDARD}} - 1)*1000,$$

where R is the known isotope composition of a standard (see Table 2.1), and H and L are the fractional heavy and light isotope components of the sample. Rearranging this equation, one obtains:

$$H/L = (R_{\text{STANDARD}}/1000)*(1000+\delta).$$

A second equation states that the fractional heavy and light isotope components add to 1,

$$H + L = 1.$$

(Technical Note: This equation becomes slightly more complex for oxygen and sulfur which have three (oxygen) or four (sulfur) stable isotopes instead of just two stable isotopes present in hydrogen, carbon, and nitrogen. But the fractions of all O and S isotopes can be calculated from the δ values and robust assumptions made about proportional "mass dependent" fractionations of the isotopes (Hulston and Thode 1965). Here we continue assuming that there is only one heavy isotope of interest for all the HCNOS elements, in effect aggregating the other minor isotopes of O and S with the most abundant, lightest isotope. But strictly speaking, the exact mathematics given here apply only to the HCN elements that have only two stable isotopes). Substituting and solving, one obtains:

$$L = 1 - H \quad \text{and} \quad H = (\delta + 1000)/((\delta + 1000) + 1000/R_{\text{STANDARD}}).$$

With these last two equations, the isotope composition of initial and gained materials can be recalculated in terms of fractions of heavy and light isotopes, H and L. These fractions are multiplied by the initial amounts or gained amounts to obtain the amounts of heavy and light isotope, and sums are performed with these exact isotope amounts. After addition, δ values can be recalculated from the new H and L amounts via the δ definition above.

Losses

Fractionation typically occurs during the loss reactions, and exact equations for fractionation can be written using the ratio (R) notation and α fractionation values (Figure 4.10). The equation of concern for the I Chi loss steps is the upper equation of Figure 4.10 for residual unused substrate:

$$R_{t+1} = R_t *(f*\alpha + 1 - f).$$

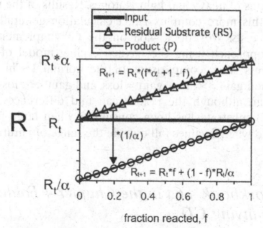

FIGURE 4.10. Fractionation in an open system, with exact equations for fractionation in residual substrate (top line) and product (bottom line). R is the isotope ratio and α is the fractionation factor for the reaction, as defined in the text.

In this equation, R_t is the H/L ratio at the start of the loss process, and α is the fractionation factor (see Box 2.1 in Chapter 2 for definition), and f measures the extent of reaction or fraction reacted:

$$f = (1-(\text{amount after loss/amount before loss})).$$

Now recalling that $H + L = 1$, $L = 1 - H$ and $R = H/L$,

$$H = R/(1+R).$$

The equation for unused substrate can be rearranged to yield the fraction of heavy isotope, H:

$$H_{t+1} = R_t *(f*\alpha+1-f)/(1+R_t *(f*\alpha+1-f)).$$

This solution gives the fraction of heavy isotope H after fractionation has occurred (and also, by difference, the fraction light isotope $L = 1 - H$). Multiplying these fractions by the amount left after loss gives the amounts of H and L isotopes after the loss reaction, and the ratio of these amounts can be used to calculate δ values from the first equation of this section.

(Note that similar approaches can be used with closed system equations to derive exact budgets for heavy and light isotopes. The closed system equations are given in the appendix and derived in Technical Supplement 7B in the Chapter 7 folder on the accompanying CD).

Overall, this complex but exact algebra makes detailed calculations at the level of fractional abundances and occasionally recalculates δ values from the exact fractions of heavy and light isotopes. Results of the simpler I Chi approach and this more complex exact calculation generally agree well in side-by-side tests. Workbook 4.6a on the CD implements equations for this exact approach using the oxygen cycling model of Section 4.2. Three worksheets show comparisons of the normal I Chi versus exact calculations when gain >>> loss, gain = loss, and gain <<< loss. Results are generally similar, although the reader can find differences at extremes when almost all substrate has been consumed, when large fractionations >50‰ apply, and when δ values fall outside the range of natural abundance values.

Guide to Workbook 4.6a in the Chapter 4 Folder on the Accompanying CD

The worksheet "short-form simple" shows the calculations for exact budgeting of heavy and light isotopes in columns A–U, then columns AA–AK show the normal I Chi calculations based on δ. Differences between the two methods in calculated δ values are given in column AM. Graphs show the δ values versus time and the differences between the two methods versus time. The exact budgeting is based on fractional abundances, or "F" values for heavy (H) and light (L) isotopes, and the title "F-based" calculations refers to the exact budgeting approach.

In the remaining worksheets, the equations are extended from 10 time steps to about 2000 time steps in a "long-form" comparison. Different scenarios are compared where gain >>> loss, gain = loss, and gain <<< loss. You can change master variables highlighted in bold (see cells K2, K3, and P2 and P3) to explore similarities and dissimilarities between the two methods of calculating δ values in these dynamic gain–loss models.

Exact Equations for Loss in I Chi Models

The other option to gain exactness in the I Chi models is to change only the fractionation equation for loss, leaving the mixing equations in simple form. Most error arises in the fractionation equation, and can be avoided rather simply by substituting an exact fractionation equation for the basic I Chi equation. Derivation of this exact equation follows.

The first steps concern calculating the isotope ratio of the substrate at time t (R_t) by rearranging the definition of δ:

$$\delta = (R_t / R_{STANDARD} - 1) * 1000,$$

so that

$$R_t = R_{STANDARD} * (\delta + 1000)/1000.$$

This is the starting ratio value, and fractionation changes this R_t ratio according to the top equation in Figure 4.10. For residual substrate, the post-fractionation result is:

$$R_{t+1} = (R_{STANDARD} * (\delta + 1000)/1000) * (f * \alpha + 1 - f).$$

The final step is to calculate δ_{t+1} values from R_{t+1} again using the δ definition:

$$\delta_{t+1} = (\delta_t + 1000) * (f * \alpha + 1 - f) - 1000.$$

This equation can be substituted for the simpler I Chi equation

$$\delta_{t+1} = \delta_t + f * \Delta$$

used throughout the I Chi spreadsheets of this book, where Δ is the fractionation factor expressed in positive permil units. Those spreadsheets use Δ values rather than α values to express fractionation, but α values are derived readily from Δ values:

$$\alpha = (1000 + \Delta)/1000.$$

Thus, when amending spreadsheets of this book to ensure exact isotope changes during loss reactions, the changes needed are twofold, calculating α from Δ, $\alpha = (1000 + \Delta)/1000$, and substituting a new equation for fractionation during loss processes, $\delta_{t+1} = (\delta_t + 1000) * (f * \alpha + 1 - f) - 1000$. When constructing new spreadsheet models, the reader may prefer to work with these last two equations throughout, or alternatively, follow the lead of this book and use the simpler I Chi equations, then substitute these last two equations at the end to check if more exactness is really needed.

Guide to Workbook 4.6b in the Chapter 4 Folder on the Accompanying CD

Workbook 4.6b is parallel to workbook 4.6a, but gives comparisons between the I Chi model with exact equations for both fractionation and mixing versus a model with the exact fractionation equations but the simpler I Chi mixing equations of Section 4.2. Inspection of the various scenarios presented in workbook 4.6b shows very little difference at all in any scenario. This simple substitution for fractionation is therefore recommended to readers interested in gaining exactness generally, those interested in hydrogen isotopes that have large fractionations, and those working with highly enriched samples.

Conclusion

This book provides a beginning approach to modeling isotope circulation using write-your-own desktop models. The simple I Chi equations given in earlier in this Chapter in Section 4.3 suffice primarily because the box models are simple, usually just one box with a few processes of gain and loss associated with this box. There are errors, but they are small because the models are still relatively small and simple. More complex models with many box-and-arrow diagrams could have larger errors that accumulate, and in these cases, the exact equations for both mixing and fractionation should be implemented. Future modeling software will undoubtedly automatically write these exact equations for the user, so that the emphasis will be on the structure and logic of the models, rather than on the isotope equations themselves.

4.7 Cows in a Pasture

Having seen the progression of math steps through the last four sections, it is perhaps wise to end this chapter by returning to an example that shows how I Chi models can be used to understand ecological problems. Also, after the chocolate isotopes and oxygen isotopes introduced at the start of the chapter, it is perhaps good to focus on a more everyday example, in this case a simple gain–loss model for cows grazing in a pasture. For the next few minutes, let's think about nitrogen (N) dynamics for cows that lead a peaceful life in a pasture. We use I Chi modeling to explore how isotopes can indicate dietary status of cows, and show how isotope trophic enrichments arise in food webs. Most of this essay considers a simple gain–loss model, but the end of the essay considers an amendment to the simple model.

We begin with a cow that grazes and gains weight and nitrogen mass over time. The amount of nitrogen starts low and increases, with the visible net increase in N supported by larger N fluxes in and out. The N influx is, of course, N gained from the diet, and the N efflux is a combination of N losses in urine and feces. Including loss of N in our thinking makes sure that N is turning over even while it is being added via the diet, so that N pools are dynamic and not static. In our thinking, we aggregate the fluxes into just one input flux and just one output flux, recognizing that future models could split the fluxes into multiple streams, for example, the output flux could become the urinary N flux plus the fecal N flux.

We find that the isotope value of pasture foods averages 0‰, the same value as most fertilizers and also the same value as atmospheric nitrogen that is the standard for $\delta^{15}N$ measurements. Indeed, some nitrogen-fixing plants such as clover are present in the pasture and have $\delta^{15}N$ values of −2‰. Other plants have higher $\delta^{15}N$ values of 1 to 3‰, because source soil

N has these higher values. All these foods average to 0‰, and so we would expect the cow to have that isotope value too, if it were not for fractionation.

Where would we expect fractionation as the cow is both gaining N from the diet and losing N via urine and feces? Typically, fractionation occurs during the making or breaking of chemical bonds, so we might not expect fractionation during food assimilation that involves uptake of large molecules without breaking N bonds. Once the large molecules are inside the cow, we assume all the N in these molecules is used for growth. In this case, all N isotopes are used for growth as well, and there is no opportunity for an overall fractionation between diet and cow during the N gains.

Without fractionation in the feeding process, you might think that a cow would have the same isotope value as the diet. But in addition to N gains during feeding, we also have to consider the loss side of the reactions, and whether isotope fractionation might occur during N loss from the cow. The answer here is definitely yes. While the cow is gaining N from the diet, N is also leaving, and this loss occurs when N in amino acids is "deaminated" with cleavage of N bonds. This loss does fractionate isotopes, and laboratory studies suggest that the maximum fractionation possible for deamination reactions when amino acid substrates are abundant is about 9‰ (Macko et al. 1986). Thus, we might expect N losses to be 9‰ lower than the amino acid substrates from which they are formed in the cow. As the cow loses this N which is low in $\delta^{15}N$ value, predictably the cow itself will become enriched in ^{15}N. This preserves the isotope balance, low $\delta^{15}N$ in excreta, and, by difference, high $\delta^{15}N$ in the cow.

In summary, the cow is growing and adding food at a $\delta^{15}N$ value of 0‰, but it is also losing N with fractionation accompanying the loss. We might guess that overall, the isotope consequences of adding dietary N might be stronger than those of losing N, because the growing cow is gaining more N than it is losing. So, let's see if this really checks out, using an I Chi model (see CD, workbook 4.7a in the Chapter 4 folder for the cow model if you want to type in some of the values given in the following).

Here is a prediction for 400 days of growth. The cow starts life at birth as a young calf, with N inherited from its mother. It spends the first few months drinking mother's milk. We start to make measurements of cow hair once the young cow is weaned and begins grazing exclusively. We follow the life of this cow for 400 days after weaning (Figure 4.11).

The cow is growing and gaining N at a rate of 80 g N per day but losing 30 g N per day, for a net N growth of 50 g N/day. The growth rate is steady and unremarkable but the isotope dynamics are more interesting (Figure 4.11). Initially high $\delta^{15}N$ values of 10‰ are associated with feeding on mother's milk, but after 100 days of grazing the isotope values decrease and stabilize at 3.4‰ (Figure 4.11). This value of 3.4‰ is intermediate between the 0‰ value of the diet and the 9‰ value associated with fractionation during loss. And because the 3.4‰ value is closer to the 0‰ value of the

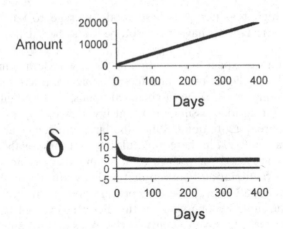

FIGURE 4.11. Model results for a cow growing 400 days in a pasture, gaining 80 g N each day while also losing 30 g N to various forms of excretion. The net growth is 50 g N each day, leading to a constant increase in cow N each day. During this growth, cow isotope values change from initial values to reflect the new pasture diet at 0‰. Final isotope values in the cow reflect a balance between dietary isotope values that pull values towards the 0‰ diet values and fractionation during excretion losses that pushes isotope values of the cow up and away from the dietary values. Cow values approach the 0‰ value of the mixed pasture diet by the end of 400 days, but are still offset 3.4‰ higher than the diet due to fractionation operating during loss. The fractionation factor for total losses is set at $\Delta = 9‰$ in these and the remaining examples of this chapter, except for Figure 4.12.

diet than it is to the 9‰ value associated with fractionation, we might conclude (correctly) that the diet is more important than fractionation in determining the isotope value. However, even if diet is more important, clearly fractionation is still a strong modifier of the observed values, because if only diet were important then. . . . Well, let's see, shall we? Let's turn off fractionation in our model by setting it to 0‰ instead of 9‰ (Figure 4.12). What happens is that now the diet clearly dominates, and the cow $\delta^{15}N$ values are 0‰, the same as the diet. This simple experiment also tells us that fractionation is the agent elevating the isotope values above those of the diet.

But what if the cow grows a little slower or faster, so that the balance between gains and losses differs from the 80/30 ratio depicted above? To explore this, let's turn the fractionation back to 9‰, and look at what happens when we have a cow that is a super-grower, with 100 g N/day gained but only 10 g N/day lost. This might be a feed-lot cow that eats a lot and does not exercise much (Figure 4.13). The amount of growth is higher after the 100 days, as you might expect, but the isotope values for the cow are still low, near those of the diet. It seems as though we turned off fraction-

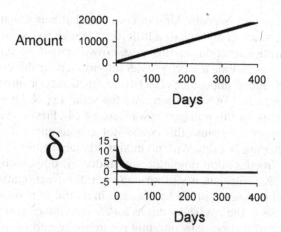

FIGURE 4.12. As Figure 4.11, but fractionation during loss has been changed from $\Delta = 9‰$ to $\Delta = 0‰$ (no fractionation). In this case, only diet influences isotopic compositions, and cow isotopes conform to the simple maxim, "You are what you eat;" that is, the cow has the same isotopes as the diet.

ation, and in a way, this perception is accurate. But what really happens is this: the 9‰ fractionation remains active, but isotope effects associated with loss are small because losses are minor compared to the very large gains. Fractionation is thus active during loss, but all loss-associated effects, including fractionation, are swamped by the massive gains.

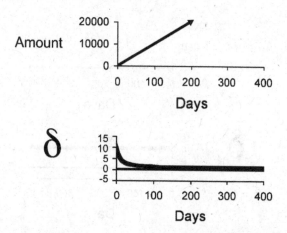

FIGURE 4.13. As Figure 4.11, but the cow is growing more rapidly and losing less N every day, gaining 100 g N and losing only 10 g N each day. With this strong growth, cow isotopes still reflect mostly diet, even though fractionation during loss has been switched back on versus Figure 4.12, from 0‰ to 9‰. Because there is little net loss, only 1 N atom lost per 10 N atoms gained, fractionation that occurs during loss still is not very important. The result is that cow isotopes are still close to diet isotopes.

But what about an opposite kind of cow, one that was unhappy and not really growing much at all? It eats a little, just enough to balance the losses it still has during excretion. In this case, growth might be balanced, and because the losses assume a much greater importance for this cow, we might expect a much more important effect from fractionation involved in the losses. So we type in 15 g N/day gains and the same 15 g N/day for losses, to see what happens for this unhappy cow (Figure 4.14). First we see a big contrast in the amounts, because this cow is not gaining mass, a consequence of N losses equaling N gains. With no mixing (dietary) gains of N, we might expect to see fractionation dominate, and indeed it does, with cow values stabilizing at 9‰. Nitrogen is still turning over, the fractionation works to produce excreta depleted in ^{15}N by 9‰, and in a kind of reverse or inverse way, the isotopes in the cow show this as a +9‰ fractionation versus its diet. We come back to this perhaps puzzling result at the end of this essay, but let us proceed to two final cases before ending.

Another unhappy cow example would be a starving cow, one that doesn't eat and only loses N. This cow is still holding on after 400 days, perhaps improbably, but we have the following model results (Figure 4.15). The amount of N declines as expected during starvation, and without any feed to add nitrogen with low δ^{15}N, values just increase as fractionation ensures that low δ^{15}N nitrogen continues to be excreted.

Finally, before we leave this starving cow, let's put it in a field of clover where it can eat all by itself and happily regain some weight, the resurrection cow. The δ^{15}N value of the clover food is a little lower, −2‰, and the

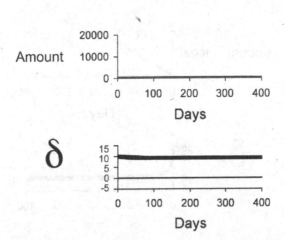

FIGURE 4.14. As Figure 4.11, but now the cow is gaining and losing N at the same rate, so that there is no net growth. In this case, the 9‰ fractionation during loss reactions plays an important role in pushing cow isotopes away from the diet isotopes. The maxim, "You are what you eat," no longer applies simply, and clearly needs amendment to something like this, "You are what you eat less excrete." Cow isotopes are 9‰ higher than those of the diet due to the excretion losses.

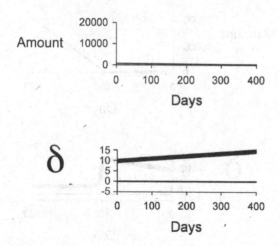

FIGURE 4.15. As Figure 4.11, but now the cow is starving and not gaining any N, only losing N. Fractionation during loss is unchecked by new dietary inputs, and fractionation leads to higher and higher values as the cow loses more and more N.

cow gains 80 g N per day while losing 30 g N per day. Observing the next 400 days, and putting it together with the first 400 days of starvation, we have 800 days of observations (Figure 4.16). Yes, we see that starving cow syndrome in the first 400 days, with declining amounts and rising isotopes. But after transferring the cow to the new field of clover, growth sets in once again, and isotopes drop to reflect a dominant role of dietary inputs. We note that the final values of this cow have shifted from 3.4‰ to 1.4‰, reflecting the shift in baseline dietary isotope values from 0‰ in the first pasture to −2‰ in the field of clover. Baseline corrections or shifts are important in many real-world examples with isotopes.

Summarizing, the cow isotope values rise and fall in two ways, first if the baseline food values change, but also if the net balance changes between gains and loss. This latter point derives from the fact that effects of fractionation are split between the cow and its excreta, or, as we learned in Section 4.3, effects of fractionation are split between substrates and products. In simpler words, when the cow is gaining weight rapidly and most of the diet is retained for growth, the cow has isotopes similar to that of the diet. In this case, the effects of fractionation are mostly expressed in the excreta, not in assimilated cow tissue. On the other hand, when the cow is eating just a maintenance ration, neither gaining nor losing weight, most N is lost to excreta that has isotopes similar to that of the diet, and the effects of the fractionation are mostly expressed in the cow. Overall, the fractionation effects in the cow are proportional to the fraction (f) lost to excretion where f = loss/dietary input = loss/gain. This is the dynamic at work in the pasture food web, consistent with the basic equation of fractionation

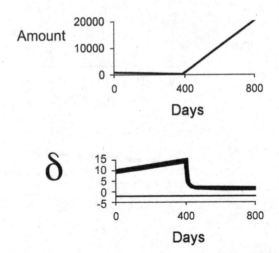

FIGURE 4.16. As Figure 4.15 for the first 400 days for a starving cow, but then the cow is moved to a new clover pasture where it starts to grow again at the rate assumed in Figure 4.11, that is, 80 g N gained from the diet and 30 g N lost each day. The cow values approach the new −2‰ value of the clover diet by the end of 800 days, but are still offset 3.4‰ higher than this new diet due to isotope fractionation during excretion.

during loss presented in Section 4.3 and in Figure 4.7. (Note that although those equations do not directly suggest the link between observed fractionation and the relative strengths of loss versus gain, it is because of turnover that the fraction lost to excretion does eventually determine the expressed fractionation). In the end, however, an important conclusion is that the ways that gains and losses are partitioned typically create divergent results in even seemingly simple I Chi gain–loss isotope models. Even simple models create complex results.

In future chapters, we find that fractionation actually creates balanced effects that are split between substrates and products. The above essay about cows is incomplete in this regard, for the cows are the substrate, and what is missing is a presentation of the amounts and δ values of the excreted products. Putting in the extra graphs of amounts and isotopes for products is deferred to Chapter 7, where we try to build an expert-level understanding of the balanced effects of fractionation. So, help is on the horizon for these fractionation effects.

Also, although fractionation can seemingly move isotopes in mysterious ways, it is also true that ecology and ecosystems usually process elements in routine modes, so isotope compositions become predictable in larger contexts. Many ecological studies share the observation that animals are enriched in ^{15}N by an average of 3.4‰ versus their diets, as presented in the first example above (Figure 4.11). This average enrichment that is especially observed among vertebrates likely derives from the balance between gains and losses. Perhaps the fraction N retained is strongly selected through the course of evolution, because it would seem a characteristic growth dynamic that makes isotopes predictable in field studies. These models give a variety of fractionation expectations, such that smaller ^{15}N fractionations noted especially for invertebrates (McCutchan et al. 2003) could denote higher assimilation efficiencies, and ^{13}C enrichment observed in food webs may also be linked to loss reactions of CO_2 or fecal carbon excretion. The models can explore odd and unusual behavior such as starvation, but the presence of such extremes should not obscure strong averaging that often prevails in the natural world and results in characteristic patterns of isotope enrichment in food webs.

Models characteristically need testing and refinement before yielding accurate results, and it is good to start with simple models such as the one outlined thus far. We can compare the model results to field data that show urinary N losses in cows with substantial ^{15}N depletions of up to 7.8‰ versus cow tissues (Steele and Daniel 1978). This 7.8‰ value is close to the 9‰ fractionation value used in the modeling above and documented by Macko et al. (1986) for deamination losses. However, much of the N loss from herbivores is via feces and not via urine, with fecal N typically having δ^{15}N values equal to or higher than those of the diet, instead of showing the substantial ^{15}N depletions typically found in urine (Steele and Daniel 1978; Sponheimer et al. 2003a,b). Problem 4 in the Chapter 7 folder on the accompanying CD asks you to develop a revised fractionation estimate for total losses that includes fecal N losses. For your current reference, the answer is about 5.5‰ for the total combined losses. This 5.5‰ fractionation can be substituted for the 9‰ value used above for the cow model, leading to slightly revised quantitative results, but very similar qualitative results.

Finally, you might want to check out a basic gain–loss model that is similar to this cow example, but a little more abstract. The basic gain–loss model can be thought of in terms of not only cows and animals, but also in terms of processes such as formation and loss of soil organic matter, of plant biomass, or of atmospheric gases. By changing the initial values to suit the example, you can use the I Chi model 4.7b labeled "Abstract Gain–Loss" (see accompanying CD, Chapter 4 folder, I Chi Spreadsheets) to investigate isotope dynamics in a system that interests you. That model has a tutorial aspect, because it suggests different values you can type in to assess effects of the different master variables. One set of master parameters controls the amounts (initial amount, amount of gain, amount of loss), and the second set of master variables controls isotope compositions (initial δ, δ of the material gained, and a Δ fractionation factor for loss steps). Type in a list of variables then use back and forward arrows to review the sequence of effects that accompany each change. Generally if you change amounts, isotopes also change, but not vice versa; that is, isotopes can change enormously without affecting overall amounts. This basic observation makes sense because isotope parameters are just controlling relative amounts of heavy and light isotopes, leaving the total amounts the same. Working through this idea and typing in the various parameters of the Abstract Gain–Loss workbook 4.7b (see accompanying CD, Chapter 4 folder, I Chi Spreadsheets) may help give a better appreciation of how fractionation and mixing combine in these I Chi models. But working with just an abstract model can be tedious. If so, try returning to the cow model after the abstract model and see if the cow model now makes more sense – hopefully the answer is yes!

Problem 7 in the Chapter 4 folder on the accompanying CD asks you to reconsider these gain–loss models in terms of carbon cycling, to estimate trophic enrichment factors for carbon in food webs.

4.8 Chapter Summary

This chapter introduces spreadsheet models that circulate stable isotopes at your fingertips, in any virtual reality biosphere or ecosphere you choose. There is always an element of unreality or fantasy in any model or concept, and the first example in this chapter makes this explicit by considering chocolate isotopes (Section 4.1). Here you are the "fractionator," and light and dark chocolates are the "fractionatees" that receive the benefit of your picky selectivity. This example captures the attention of chocolate lovers, and gives you the opportunity to pit your skills against a formidable opponent, a benevolent chocolate god who tries to keep things in balance. Working with the chocolate isotopes can be fun and instructive if you take the time to open up a spreadsheet on the accompanying CD and try manipulating mixing and fractionation. A surprising array of outcomes awaits your efforts and will challenge your isotope understanding in this delicious model domain of chocolate isotopes.

The next five sections of Chapter 4 (Sections 4.2 to 4.6) deal with a more realistic example, oxygen cycling in the sea. Oxygen is produced during photosynthesis and consumed during respiration in one of the great biogeochemical cycles of our planet. There is isotope action inside these reactions, especially isotope mixing that is important for photosynthesis and isotope fractionation that is important for respiration. The isotopes cycle as behind-the-scene players in the oxygen dynamics, with spreadsheet models opening up the action to show what happens in the sea both day and night. Section

4.2 shows four characteristic patterns of oxygen dynamics in the sea: a balance between photosynthesis during the day and respiration at night, a strong daytime algal bloom, mild oxygen consumption during decomposition of an algal bloom, and strong oxygen consumption in the sediments. You can explore these and other scenarios by typing in values in spreadsheet model 4.2 found on the accompanying CD (Chapter 4 folder, I Chi Spreadsheets), simulating the living dynamics of oxygen in the sea.

With chocolates and oxygen behind you, Section 4.3 introduces the equations you need for all this isotope modeling. These equations boil down to this: add and mix, subtract and fractionate, just four equations in all. These are the I Chi equations, named for Isotope Power or Isotope Chi. Study and learn these four equations that are simple algebra, so you can write your own models. Section 4.4 does just that, guiding you through how to write a spreadsheet model in ten easy steps. Just like karaoke, singing along with the words, you click along on the spreadsheets, and pretty soon you have an isotope model up and running. Amazing but true!

Section 4.5 reviews all this modeling from another point of view, locating the mathematical errors in these models. Errors are unfortunately facts of life in dynamic models that employ a stepwise or finite difference approach to simplify mathematical calculations. Section 4.5 recommends keeping time steps small and using exact equations for fractionation as the best ways to minimize math errors. Exact equations are presented and discussed in Section 4.6.

Chapter 4 concludes on a peaceful note, with cows grazing in a pasture. These virtual cows exist courtesy of a generic gain–loss model that predicts their growth and isotope dynamics. Cows gain nitrogen (N) and mix in isotopes from their diets, while losing N and fractionating isotopes during excretion. The balance between gains and losses can work out in several ways, from the starving cow with higher and higher isotope values to the resurrection cow that finds a field of clover in the end, and returns to a normal diet and low isotope values.

A central concept of this last section and for the chapter as a whole is that isotope values are net values, reflecting the balance between gains and losses. This realization takes a while to sink in, but becomes very useful when trying to understand things such as trophic enrichments (higher δ values) commonly seen in food webs. The cow example of Section 4.7 shows that trophic enrichments arise from the interplay between isotope mixing during dietary gains and isotope fractionation during the loss processes of excretion and respiration. The net balance is the trophic enrichment. Similar balances explain the isotope compositions of many atmospheric gases and apply generally for isotope distributions throughout the biosphere.

An ending challenge is for the reader to open a prefabricated, generic gain–loss workbook 4.7b on the CD (Chapter 4 folder, I Chi Spreadsheets) and click along, following how the four basic equations of I Chi let you model isotope dynamics correctly and creatively for any system that interests you. The basic gain–loss model can be thought of in terms of not only cows and animals, but also in terms of processes such as formation and loss of soil organic matter, of plant biomass, or of atmospheric gases. With the examples of this chapter to lead you, you should be able to think out your own isotope models and actualize them in your own spreadsheet gain–loss models.

Further Reading

Section 4.2

Bender, M.L. and K.D. Grande. 1987. Production, respiration, and the isotope geochemistry of O₂ in the upper water column. *Global Biogeochemical Cycles* 1:49–59.

Brandes, J.A. and A.H. Devol. 1997. Isotopic fractionation of oxygen and nitrogen in coastal marine sediments. *Geochimica et Cosmochimica Acta* 61:1793–1801.

Epstein, S. and L. Zeir. 1998. Oxygen and carbon isotopic compositions of gases respired by humans. *Proceedings of the National Academy of Sciences USA* 85:1727–1731.

Lane, G.A. and M. Dole. 1956. Fractionation of oxygen isotopes during respiration. *Science* 123:574–573.

Luz, B. and E. Barkan. 2000. Assessment of oceanic productivity with the triple-isotope composition of dissolved oxygen. *Science* 288:2028–2031.

Marra, J. 2004. Phytoplankton and heterotrophic respiration in the ocean. *U.S. JGOFS Newsletter* 12:6,19.

Marra, J. and R.T. Barber. 2004. Phytoplankton and heterotrophic respiration in the surface layer of the ocean. *Geophysical Research Letters* 31:L09314, doi:10.1029/2004GL019664, 2004.

Quay, P.D., D.O. Wilbur, J.E. Richey, A.H. Devol, R. Benner, and R. Forsberg. 1995. The ^{18}O: ^{16}O of dissolved oxygen in rivers and lakes in the Amazon Basin: determining the ratio of respiration to photosynthesis rates in freshwaters. *Limnology and Oceanography* 40:718–729.

Roberts, B.J., M.E. Russ, and N.E. Ostrom. 2000. Rapid and precise determination of the $\delta^{18}O$ of dissolved and gaseous dioxygen via gas chromatography-isotope ratio mass spectrometry. *Environmental Science and Technology* 34:2337–2341.

Sarma, V.V.S.S., O. Abe, S. Hashimoto, A. Hinuma, and T. Saino. 2005. Seasonal variations in triple oxygen isotopes and gross oxygen production in the Sagami Bay, central Japan. *Limnology and Oceanography* 50:544–552.

Young, E.D., A. Galy, and H. Nagahara. 2002. Kinetic and equilibrium mass-dependent isotope fractionation laws in nature and their geochemical and cosmochemical significance. *Geochimica et Cosmochimica Acta* 66:1095–1104.

Section 4.3

Benson, R. 1998. *The Facts of Death*. Glidrose Publications Ltd, London.

Hart, E.A. and J.R. Lovvorn. 2002. Interpreting stable isotopes from macroinvertebrate food-webs in saline wetlands. *Limnology and Oceanography* 47:580–584.

Herman, D.J. and P.W. Rundel. 1989. Nitrogen isotope fractionation in burned and unburned chaparral soils. *Soil Science Society of America, Journal* 53:1229–1236.

Mariotti, A., J.C. Germon, P. Hubert, P. Kaiser, R. Letolle, A. Tardieux, and P. Tardieux. 1981. Experimental determination of nitrogen kinetic isotope fractions: some principles; illustration for the denitrification and nitrification processes. *Plant and Soil* 62:413–430.

Martinez del Rio, C. and B. Wolf. 2005. Mass-balance models for animal isotopic ecology. In J.M. Starck and T. Wang (eds.), *Physiological and Ecological Adaptations to Feeding in Vertebrates*. Science Publishers, Enfield NH, pp. 141–174.

O'Reilly, C.M, R.E. Hecky, A.S. Cohen, and P.-D. Plisnier. 2002. Interpreting stable isotopes in food webs: Recognizing the role of time averaging at different trophic levels. *Limnology and Oceanography* 47:306–309.

Pennock, J.R., D.J. Velinsky, J.M. Ludlam, J.H. Sharp, and M.L. Fogel. 1996. Isotopic fractionation of ammonium and nitrate during uptake by *Skeletonema costatum*: Implications for $\delta^{15}N$ dynamics under bloom conditions. *Limnology and Oceanography* 41:451–459.

Rastetter, E.B., B.L. Kwiatkowski, and R.B. McKane. 2005. A stable isotope simulator that can be coupled to existing mass-balance models. *Ecological Applications* 15:1772–1782.

Shearer, G., J. Duffy, K.H. Kohl, and B. Commoner. 1974. A steady-state model of isotopic fractionation accompanying nitrogen transformations in soil. *Soil Science Society of America, Journal* 38:315–322.

Section 4.5

Hayes, J.M. 2004. *An introduction to isotopic calculations*. http://www.nosams.whoi.edu/docs/IsoCalcs.pdf

Sessions, A.L. and J.M. Hayes. 2004. Calculation of hydrogen isotopic fractionations in biogeochemical systems. *Geochimica et Cosmochimica Acta* 69:593–597.

Section 4.6

Hulston, J.R. and H.G. Thode. 1965. Variations in the S^{33}, S^{34} and S^{36} contents of meteorites and their relation to chemical and nuclear effects. *Journal of Geophysical Research* 70:3475–3484.

Section 4.7

Macko, S.A., M.L. Fogel Estep, M.H. Engel, and P.E. Hare. 1986. Kinetic fractionation of stable nitrogen isotopes during amino acid transamination. *Geochimica et Cosmochimica Acta* 50:2143–2146.

McCutchan, J.H. Jr., W.M. Lewis Jr., C. Kendall, and C.C. McGrath. 2003. Variation in trophic shift for stable isotope ratios of carbon, nitrogen, and sulfur. *Oikos* 102:378–390.

Minagawa, M. and E. Wada. 1984. Stepwise enrichment of ^{15}N along food chains: Further evidence and the relation between $\delta^{15}N$ and animal age. *Geochimica et Cosmochimica Acta* 48:1135–1140.

Minson, D.J., M.M. Ludlow, and J.H. Troughton. 1975. Differences in natural carbon isotope ratios of milk and hair from cattle grazing tropical and temperate pastures. *Nature* 256: 602.

Olive, P.J.W., J.K. Pinnegar, N.V.C. Polunin, G. Richards, and R. Welch. 2003. Isotope trophic-step fractionation: A dynamic equilibrium model. *Journal of Animal Ecology* 72:608–617.

Post, D.M. 2002. Using stable isotope methods to estimate trophic position: Models, methods, and assumptions. *Ecology* 83:703–718.

Sponheimer, M., T. Robinson, L. Ayliffe, B. Roeder, J. Hammer, B. Passey, A. West, T. Cerling, D. Dearing, and J. Ehleringer. 2003a. Nitrogen isotopes in mammalian herbivores: hair $\delta^{15}N$ values from a controlled feeding study. *International Journal of Osteoarchaeology* 13:80–87.

Sponheimer, M., T.F. Robinson, B.L. Roeder, B.H. Passey, L.K. Ayliffe, T.E. Cerling, M.D. Dearing, and J.R. Ehleringer. 2003b. An experimental study of nitrogen flux in llamas: is ^{14}N preferentially excreted? *Journal of Archaeological Science* 30:1649–1655.

Steele, K.W. and R.M. Daniel. 1978. Fractionation of nitrogen isotopes by animals: a further complication to the use of variations in the natural abundance of ^{15}N for tracer studies. *Journal of Agricultural Science*, Cambridge 90:7–9.

Sutoh, M., Y. Obara, and T. Yoneyama. 1993. The effects of feeding regimen and dietary sucrose supplementation on natural abundance of ^{15}N in some components of ruminal fluid and plasma of sheep. *Journal of Animal Science* 71:226–231.

Vander Zanden, J.M. and J.B. Rasmussen. 2001. Variation in $\delta^{15}N$ and $\delta^{13}C$ trophic fractionation: implications for aquatic food web studies. *Limnology and Oceanography* 46:2061–2066.

West, A.G., L.K. Ayliffe, T.E. Cerlig, T.F. Robinson, B. Karren, M.D. Dearing, and J.R. Ehleringer. 2004. Short-term diet changes revealed using stable carbon isotopes in horse tail-hair. *Functional Ecology* 18:616–624.

5
Mixing

Overview

As elements circulate in the biosphere, mixtures arise when two or more sources contribute materials. Isotopes are excellent tracers for mixing processes and indicate which sources dominate the mixtures. This chapter considers isotope mixing in ecological systems.

5.1. *Isotope Mixing in Food Webs.* This review of 30 years of estuarine research shows how isotope mixing models are used in practical ways to solve a food web problem. The review sets the stage for the next four sections 5.2 to 5.5 that consider mixing from more theoretical viewpoints. After these sections, future food web studies are considered in extended problem 10 for Chapter 5, on the accompanying CD.

5.2. *Isotope Sourcery.* Isotope mixing proves rather simple, like mixing blue and yellow colors to make an intermediate green color, or black and white to make grey. Mixing models let you go backwards, calculating the contributions of sources from the isotope colors. This is the magic part of isotope sourcery.

5.3. *Mixing Mechanics.* Here we derive the equations for isotope mixing in an easy-to-follow way, and give an example of how isotope mixing helps solve a laboratory problem.

5.4. *Advanced Mixing Mechanics.* Isotope mixing can be more complex than you might think, especially when concentrations play a role, or when there are too many sources.

5.5. *Mixing Assumptions and Errors.* This is a lecture about the problems and pitfalls involved with isotope mixing models. The take-home message is to look for well-poised mixing systems that have good signals and little statistical noise. This section also marks the end of general comments about mixing.

5.6. *River Sulfate and Mass-Weighted Mixing.* This is the first of four hypothetical mixing examples that help introduce you to common problems encountered during isotope mixing. Each example has a spreadsheet in the Chapter 5 folder on the accompanying CD to practice these mixing skills with your new-found I Chi powers. The sulfate story concerns work in the Mississippi River basin, and has a mistake in the mixing equations that the researcher has to find.

5.7. *A Special Muddy Case and Mixing Through Time.* This second hypothetical example concerns nitrogen pollution detected in a core from muddy sediments.

5.8. *The Qualquan Chronicles and Mixing Across Landscapes.* Mixing occurs at many levels, and this example concerns mixing marsh materials across an estuarine landscape, with both quality and quantity of materials important in "qualquan" isotope mixing budgets.

5.9. *Dietary Mixing, Turnover, and A Stable Isotope Clock.* During a diet change experiment, isotopes shift more quickly or slowly, depending on the metabolic activity of tissues involved. Keeping track of this turnover and mixing leads to an isotope clock for the experiment, a clock based on stable isotopes rather than radioisotopes.

Main points to learn. Isotope mixing is generally easy to understand and model, once you understand isotope budgeting, the weighted averaging of isotopes by amounts. Mixing is also the most common use of isotopes, so it is good to learn all you can about the art and equations of isotope mixing. You need to pay special attention to Section 5.5 on mixing problems if you want to become a successful isotope mixmaster. The examples in Sections 5.6 to 5.9 and the Problems for Chapter 5 on the accompanying CD will help test your newfound mixing skills. Can you really follow the mixing action in different venues that range from forests to rivers to the ocean, and mud to tuna to eagles?

5.1 Isotope Mixing in Food Webs

To begin this chapter on mixing, we start with an example that shows how isotope mixing dynamics have been important in food web research. This essay traces evolution of several mixing approaches across thirty years of seagrass research, and ends with some summary advice about how to conduct isotope investigations in food web studies.

As a beginning graduate student, I worked on the Texas coast where my research focused on the role of seagrasses for estuarine consumers. These marine grasses form underwater meadows in nearshore areas, and small fish and crustaceans are exceptionally abundant in these meadows. So what was the secret of the seagrasses? Were these plants fish food, or were the meadows just convenient places to hide from predators? To partially address this question, we began investigating the possible dietary importance of seagrasses for the abundant consumers.

We knew some things about seagrasses. These plants grew quickly, but were heavily colonized by small algae (epiphytes). After a month or so the seagrasses would senesce, breaking off at the base to sink or float away. Not many animals ate the live seagrasses, although in the past sea turtles had been very abundant (before humans ate most of them). Turtles had probably eaten a lot of the seagrass production, a cows-(turtles)-in-the-sea kind of scenario. Also, other vertebrates such as dugongs, manatees, and some ducks were active seagrass grazers, feeding on the leaves and belowground roots (rhizomes) of the plants. These animals, like the turtles, are no longer nearly as abundant as they once were. So, with little grazing, most seagrass today contributes to a hard-to-trace pool of marine detritus, breaking down after plant death into fine particles that are common in the estuarine shallows. Bacteria and fungi (the microbes) colonize this material. Although the plant material itself is not very nutritious, the microbes are nutritious. Various "detritivore" animals can ingest the particles, digesting off the microbes and defecating the plant substrate that will be colonized by the next round of microbes. Sometimes I think of this when eating breakfast cereal rich in fiber: the bananas on the top are easy to assimilate, but much of the fiber goes on through undigested.

The idea of a detrital food web based on seagrass makes sense in some ways. We know most seagrass is not grazed while alive, and dead seagrass does not accumulate over the longer term, so something must be breaking down the dead seagrass. But it is hard to follow the fate of seagrass after it dies, because first of all, everything occurs underwater where it is hard to see what is happening, and secondly, because most of the action is in very fine particles that are also hard to observe. Said another way, estuarine ecologists believe that the detrital food web exists, but they weren't (and aren't) sure how important it is. Are the fine particles in the stomachs of estuarine animals really nutritious, or are they just filler fiber, of no real consequence for growth of the common inhabitants of the seagrass meadows? You can draw diagrams that show that detritus contributes strongly to estuarine food webs (Figure 5.1), but are the diagrams really correct?

One group of ecologists maintains the right answer is that seagrasses are important because seagrasses generally have high productivities relative to other foods such as epiphytes or phytoplankton. Estuarine productivities for seagrasses often average $1200\,g\,C\,m^{-2}\,yr^{-1}$ whereas productivities for epiphytes and phytoplankton are much lower, 150 and $300\,g\,C\,m^{-2}\,yr^{-1}$, respectively. A counterview among ecologists is that the quality of the detrital seagrass food is relatively low, so that you should discount the 1200 seagrass number by a factor of 10 or so, in which case the 1200 value becomes 120 and seagrass importance declines dramatically. So, with the answer depending on which numbers you accept, using tracers to resolve the dispute seemed a great idea. The idea was that isotopes would provide a color code showing the importance of detrital seagrasses (Figure 5.2).

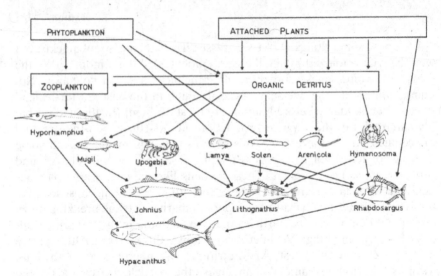

FIGURE 5.1. A generalized estuarine food web for the Knysna ecosystem of South Africa. Note the two major sources of organic matter, phytoplankton and attached plants (shown at the top of the figure), with attached plants contributing to food webs via a pool of organic detritus. Arrows show trophic links from foods to consumers. (Reprinted with permission from Day, J.H. 1967. The biology of Knysna estuary, South Africa, pp. 397–407. In G.H. Lauff (ed.), *Estuaries*. Copyright 1967, AAAS.)

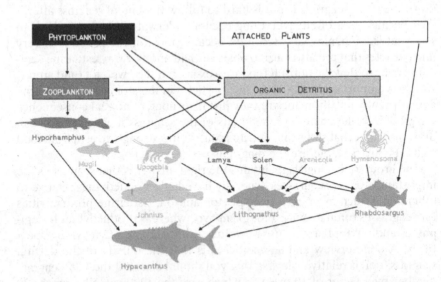

FIGURE 5.2. A two-source estuarine food web as a coloring (dye) experiment. Isotope labeling could trace the flow of color (tracer) through the major branches of the food web. As diagrammed, fisheries production is more linked to (grey) detritus rather than (black) phytoplankton. A better color version of this figure is on the CD. (Diagram reprinted with permission from Day, J.H. 1967. The biology of Knysna estuary, South Africa, pp. 397–407. In G.H. Lauff (ed.), *Estuaries*. Copyright 1967, AAAS.)

The first isotope ecology study actually weighed in on this point (Parker 1964). A study of plants and animals from a seagrass meadow at Redfish Bay, Texas showed that animal carbon isotopes were arrayed between those of algae and seagrass (Figure 5.3). These early data were consistent with a strong trophic or feeding importance for the detrital seagrass. In the late 1970s, we began studying these Texas sites again, combining laboratory experiments with field surveys. One set of laboratory experiments tested for isotope fidelity in the seagrass food web, finding little change (<1‰) in carbon isotope values of seagrass during decomposition and also close isotope similarity (1‰ or better) between two kinds of seagrass herbivores (amphipods and sea urchins) and their seagrass foods (Fry et al. 1987). Armed with these laboratory results, we developed a conceptual model of seagrass food webs that was straightforward. Given seagrass carbon isotope values that normally averaged near −10‰, we expected seagrass detritus and consumers of seagrass to also average about −10‰, with lower values nearer −20‰ indicating reliance on the major alternative food, planktonic algae (Figure 5.4). Thus, the carbon isotopes could trace mixing of seagrass and planktonic carbon in seagrass ecosystems (Figure 5.4). Fractionation during plant photosynthetic carbon fixation resulted in differential isotope labeling of seagrass versus plankton, and the question was how the isotope color from detrital seagrass would mix through the food web.

Armed with these laboratory results and using this simple conceptual model, we investigated seagrass influences in the field by focusing on land-scape areas with lots of seagrass versus those that were too shady for the seagrasses. This comparison involved the shallow seagrass meadows of the Upper Laguna Madre versus the deeper waters of the region (Baffin Bay, Corpus Christi Bay, and the nearby offshore Gulf of Mexico). Surveys of carbon isotopes in sediments and fauna of the Upper Laguna Madre

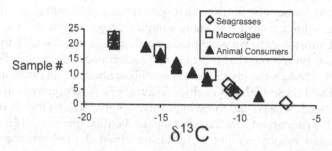

FIGURE 5.3. Carbon isotopes in the food web of a Texas seagrass meadow (Parker 1964). Animals are a mixture of algal and seagrass carbon. (Adapted from *Geochimica et Cosmochimica Acta*, v. 28. P.L. Parker. The biogeochemistry of the stable isotopes of carbon in a marine bay, pp. 1155–1164. Copyright 1964, with permission from Elsevier.)

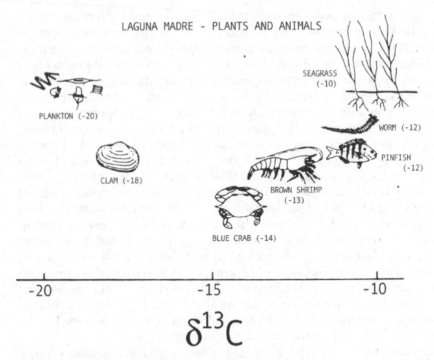

FIGURE 5.4. Conceptual model of carbon flow in the Texas seagrass meadows, with only two carbon sources present, seagrass and phytoplankton (P.L. Parker, personal communication, ca. 1976).

showed a shift towards higher $\delta^{13}C$ values characteristic of seagrasses versus samples collected in the deeper planktonic systems (Figures 5.5 and 5.6). The Upper Laguna Madre system seemed to be supported by seagrass.

But these studies also showed that sometimes you should quit while you are ahead—don't collect too many samples or you might get the wrong answer! Unfortunately, we were plodding along and found such a fly in the ointment. It turned out that not only were seagrasses enriched in ^{13}C (had higher $\delta^{13}C$), but also many kinds of macroalgae abundant in estuarine shallows also had these high $\delta^{13}C$ values (Figure 5.7). And epiphytes from the seagrass meadows also had these high values (Table 5.1), so that in the end, we could write about the importance of benthic plants (seagrasses + macroalgae + epiphytes), but were unsure about the importance of seagrasses per se. We had learned that carbon flow in the seagrass meadows was strongly based on benthic plants and not on phytoplankton (some progress here), but were not sure exactly which benthic plants were most important (and a lack of progress here). We could not cleanly partition the effects of seagrass and planktonic carbon when there was a third source (benthic macroalgae and microalgae) in the middle (Figure 5.8). This was the dreaded trap of the "mixing muddle." However, we pressed on to find some interesting answers about seagrass food webs by combining several different approaches detailed below and eventually escaped the mixing muddles.

Three lines of evidence made us think harder about epiphytic algae rather than seagrass detritus as the likely most important benthic food source. First, a simple experiment was to put common "arrow" shrimp

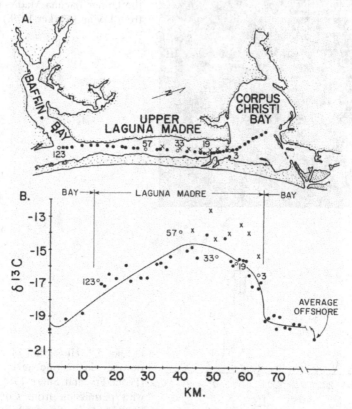

FIGURE 5.5. Carbon isotope values for organic matter in sediments from bays and lagoons of the south Texas coast, near the border with Mexico. Highest $\delta^{13}C$ values occur in the seagrass meadows of the Upper Laguna Madre, consistent with high inputs of ^{13}C-enriched seagrasses ($\delta^{13}C = -10‰$). In other deeper bay sites that lack seagrass, $\delta^{13}C$ values are consistent with strong inputs from phytoplankton, not seagrass. X = sample collected in shallow water inside seagrass meadows; dot = sample collected in deeper bays or in the deeper Intracoastal Waterway that traverses the seagrass meadows in the Upper Laguna Madre. (Reprinted from *Geochimica et Cosmochimica Acta*, vol. 41, B. Fry, R.S. Scalan, and P.L. Parker. Stable carbon isotope evidence for two sources of organic matter in coastal sediments: Seagrasses and plankton, pp. 1875–1877. Copyright 1977, with permission from Elsevier.)

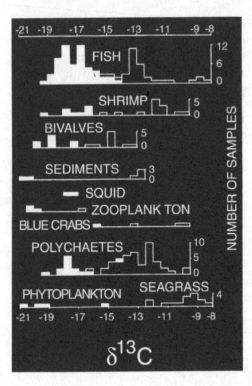

FIGURE 5.6. Histogram of carbon isotopes in plants and consumers from seagrass meadows of the Upper Laguna Madre (dark) and from the offshore Gulf of Mexico (white). Phytoplankton inputs dominate in the offshore ecosystem, whereas values are shifted away from the phytoplankton values towards seagrass values in the Upper Laguna Madre. (Data from Fry and Parker 1979.)

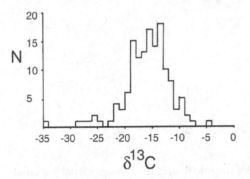

FIGURE 5.7. Histogram of carbon isotopes in marine macroalgae. (From Fry and Sherr 1984; used with permission from Contributions in Marine Science.)

TABLE 5.1. Carbon Isotope δ^{13}C Values of Epiphytes.[a]

Epiphyte Species	On *Halodule wrigthii* Seagrass	On Plastic Strips
Dermatolithon sp.	−12.7	−14.8
Heteroderma lejolisii	−10.4	−15.4

[a] Taken from seagrasses and from artificial seagrass (plastic strips) from the same site (Redfish Bay, Texas, ca. 1980). Epiphytes were common coralline algae.
Source: Fry, unpublished.

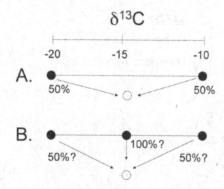

FIGURE 5.8. Conceptual mixing models for carbon isotopes. Our seagrass research started with a two-source model (Model A) with −20‰ phytoplankton and −10‰ seagrasses contributing 50/50 to −15‰ consumers (open circles; closed circles are sources). But further work changed the picture, especially discovery that marine macroalgae had intermediate −15‰ isotope values. This complicated interpretation of the isotope results (model B), creating a "mixing muddle" with no unique solution; that is, source contributions of 50/0/50 and 0/100/0 were both logically possible. To resolve this muddle, we turned to observational studies, comparative isotope surveys, and more tracers, as explained in the text. (Adapted from Fry and Sherr 1984; used with permission from Contributions in Marine Science.)

(*Tozeuma carolinense*) from the seagrass meadows in a dish with seagrass + epiphytes, and look at what got eaten. The answer was unequivocal: epiphytes! Arrow shrimp fed on the epiphyte flora in differentiated ways, grooming diatoms from larger encrusting *Acrochaetium* epiphytes and leaving the underlying seagrass substrate intact (Fry, unpublished observations). These experiments did not really rule out the possible importance of fine detrital seagrass particles that might be embedded in the epiphyte complex, but did show that shrimp were selecting for the epiphyte matrix and avoiding feeding on live and dead seagrass when seagrass blades were still whole. In the field, we also listened to and photographed common shrimp consumers feeding on epiphytes during undisturbed nighttime conditions (see Figure 2 in Kitting et al. 1984), thus supporting the laboratory results that epiphytes were important foods in seagrass meadows.

A second approach relied on carbon isotopes, surveying multiple seagrass meadows to see if consumer isotopes tracked values of seagrasses or epiphytes. In this isotope tracking or shift experiment, it turned out that animal consumer isotopes did not shift along with those of seagrass but stayed close to those of epiphytes (Figure 5.9, sites I–III), or were intermediate between seagrasses and epiphytes (Figure 5.9, sites IV–VI). Our conclusion was that where we had good separation between epiphyte and seagrass isotope values, consumers tracked the epiphyte values more closely. We repeated this sampling in Florida where the natural carbon isotope difference between

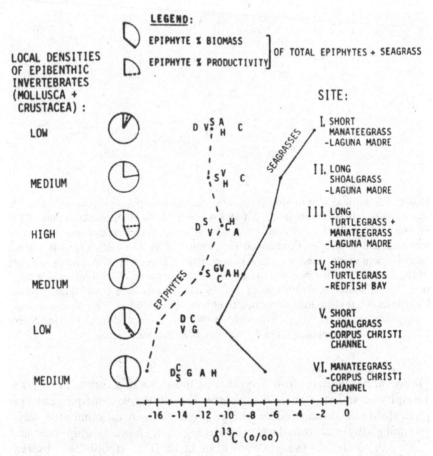

FIGURE 5.9. Carbon isotopes in plants and animals from six Texas seagrass meadows. Dashed line connects values of epiphytes and solid line connects seagrass values. Letters denote common consumers. Note that consumer values usually fall close to those of epiphytes or are intermediate between those of seagrasses and epiphytes, consistent with strong epiphyte inputs to the local food webs. (From Kitting, C.L., B. Fry, and M.L. Morgan. 1984. Detection of inconspicuous epiphytic algae supporting food webs in seagrass meadows. *Oecologia (Berlin)* 62:145–149. This is a reprint of Figure 4, p. 148 from the article and is used with permission from Springer.)

seagrasses and epiphytes was very large, about 14‰, and found the same result that consumers tracked epiphyte carbon, not seagrass carbon (Fry 1984). The emerging picture was that epiphytes were often equally or more important nutritional sources than seagrasses in the seagrass meadows. We did not fully understand the factors controlling the carbon isotope differences between seagrasses and their epiphytes, but these differences were probably due in part to differences in nutrient loading across sites.

Our third and final tests used more tracers. The isotope shift experiments of Figure 5.9 had and have a flaw. They assume that only the isotope values shift, but the rest of the system stays the same; that is, the physical setting, the species composition, and the productivities don't change from system to system. In fact, our observations showed that although these seagrass systems were located relatively near to one another in shallow protected bays and had the same species composition, there were some marked differences between systems in invertebrate densities and in epiphyte productivities (Figure 5.9). Although some authors recommend the isotope shift experiment (see Finlay et al. 2002; McCutchan and Lewis 2002; Melville and Connolly 2003), and the potential for many interesting shift experiments exists in many other seagrass systems (Mumford 1999; Anderson and Fourqurean 2003; Yamamuro et al. 2003; Fourqurean et al. 2005), we realized that these shift results should be treated with caution. Mr. Polychaete wondered, "Was it a simple shift in only plant isotopes, or was it more complex, a little 'shifty' or a little untrustworthy?" With this in mind, we also looked at using more tracers, in addition to these cross-system comparisons of isotope shifts.

In the late 1970s, only the carbon isotope measurements were routine, but we also investigated using hydrogen, nitrogen, and sulfur isotopes as additional tracers in seagrass food webs. Comparisons between seagrass and open water systems indicated little hydrogen or nitrogen isotope contrast for the few shrimp samples we analyzed (Fry 1981; Fry et al. 1987), so we did not pursue these measurements that in any case were quite difficult at that time. (Marilyn Estep, now Marilyn Fogel, made those initial hydrogen isotope measurements, and went on to publish the first observations on hydrogen isotopes in natural food webs; Estep and Dabrowski, 1980). But sulfur isotopes appeared promising for resolving sources of nutrition at Redfish Bay, the original Texas seagrass site studied by Parker (1964). Here the combined C + S isotope values were consistent with seagrass rather than epiphytes being the most important food resource at this particular site, with isotope values for most consumers most closely resembling those of seagrasses (Figure 5.10).

However, as we were progressing in our thinking about algal and microbial food sources in the seagrass meadows, we also realized that we were likely undersampling important foods. In particular, we could not isolate benthic algae (diatoms) that grew on sediment surfaces. These benthic microalgae might have low $\delta^{34}S$ values and be important in food webs. Without more sampling and estimates of plant productivities that could help us sort and estimate likely food web inputs, we felt we could not reliably interpret the dual-isotope S + C isotope food web diagrams, and so based our published interpretations on the carbon isotopes. Later studies of seagrass food webs began to incorporate N and S isotope results along with the C isotope results (Harrigan et al. 1989; Loneragan et al. 1997; Moncreiff and Sullivan 2001; Melville and Connolly 2003), although there

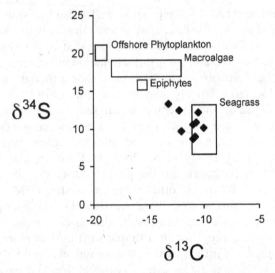

FIGURE 5.10. Dual-isotope, carbon–sulfur isotope diagram for the food web in sea-grass meadows at Redfish Bay, Texas, sampled in 1980 (Fry, 1981). Rectangles indicate ranges of measured plant values in the case of seagrasses, macroalgae, and epiphytes; offshore plankton values are estimates (Fry et al. 1987). The diamond symbols indicate isotope values for common consumers, including four shrimp species, blue crabs, snails, toadfish, pipefish, and anchovies.

are still relatively few S isotope data. Overall, combining the isotope studies with detailed field observations of feeding activities and gut contents gave the most powerful combined approach for tracing seagrass food webs (Harrigan et al. 1989).

Conclusion

Although seagrass food web isotope studies have advanced in several ways, they remain anchored in the isotope mixing models. Our early isotope mixing model (Figure 5.4) turned into a mixing muddle (Figure 5.8) when we discovered that there were too many sources in our system, but experimentation, cross-system comparisons, and multiple tracer studies sometimes helped us out of the mixing dilemmas. We did not perform tracer addition experiments, although such experiments are underway now (Mutchler et al. 2004). There the major challenge is introduction of isotope label to seagrasses while not simultaneously labeling other food resources.

So, what did we learn from all of this? There were several general findings. First, the most important finding was that inconspicuous epiphytic algae could be the most important foods in seagrass meadows. Second, even if seagrasses themselves were of poor nutritional quality and not the dominant food resources, seagrass plants were still important substrates for growing the nutritionally important epiphytes. Lastly, we found that sea-grass meadows were not all the same in terms of trophic dynamics, but instead seemed to represent a wide continuum of systems where seagrass detritus and algal foods vary in their relative importance; seagrass detritus dominates here, algae there, and so on.

This last finding is a good puzzle for future seagrass ecologists; what controls the trophic importance of algae versus seagrasses in different seagrass meadows? One idea is that nutrients are the key to understanding the algal–seagrass switch, as follows. In undisturbed seagrass meadows, nutrients available to algae in the water are often low, but rooted seagrasses can still flourish because they access nutrients in the sediments. In this case, sea-grass detritus would be the dominant available food and be widely used. The hypothesis continues that as nutrient levels increase in the water due to human inputs and eutrophication, algal foods would become more abundant and important, with an endpoint that seagrasses decline in abundance as algal blooms overgrow and shade out the benthic seagrasses. With this explanation in mind, the next research steps could test if low nutrient supply is a good predictor of seagrass nutritional dominance across different types of seagrass systems, and to investigate low-level contributions (<40%) that seagrass detritus makes at different times and places, for example, in winter or on adjacent sand flats when other foods may be scarce. Looking back, we realize that the isotope studies were only a chapter in the ongoing inves-tigation of detrital contributions to estuarine food webs (Currin et al. 2003; Hyndes and Lavery 2003; Lepoint et al. 2004; Moore et al. 2004). Also, the seagrass meadows we studied in Texas and on the east coast of Florida may have been relatively rich in epiphytes (Harrigan et al. 1989; P. Ostrom, per-sonal communication), so that low-nutrient, epiphyte-poor sites still need testing for possible seagrass dominance of local food webs.

In a more general way, the seagrass studies also showed that isotope tracing sometimes works and sometimes fails. Multiple tracers and tracer addition experiments were of some help, but not decisive. We obtained clearest results with just the carbon stable isotope measurements when working in a comparative fashion across systems, finding that seagrass epi-phytes were often the important food resource, not seagrasses themselves. This idea became much more believable when supported by follow-on lab-oratory and field experiments that did not involve isotopes.

As we turn now to a more abstract consideration of isotope mixing models, it is good to keep in mind such empirical experience, especially that most isotope investigations benefit greatly if isotopes are only part of the evidence gathered to address scientific questions. For those interested in food webs specifically, Box 5.1 gives a list of ten practical lessons for con-ducting food web research based on isotope mixing models. An extended problem on the accompanying CD (Chapter 5, problem 10) leads you through a modern exercise in interpreting carbon and nitrogen isotope results for a food web, using a terrestrial food web as an example.

Box 5.1. Ten Practical Suggestions for Using Isotopes to Study Food Webs

1. Define your central question (e.g., "Are seagrasses important in the food web?"), then read widely about this question. You are unlikely to be the first scientist to have thought about your topic. Read widely to find out what are the informed opinions and also what are the wild guesses. Then think about what you need to do to test these ideas. If isotopes were no help at all, what kinds of other things would you do to help address your question? (For example, in the seagrass studies described in this section, we thought about establishing a meadow of artificial plastic seagrass that would be colonized by epiphytes, then comparing animal communities in this artificial meadow to those in a nearby natural seagrass community. If there were no difference in animal communities, then seagrasses would not be likely to be important as food resources. We decided that although this would be a good test, it would be difficult to perform convincingly, and so left it to the next generation of investigators.) Once you have thought broadly, think about balancing your isotope approach with one or two other approaches (such as simple observation), because it may be that isotopes will not provide a magic bullet answer. Although you may be lucky and get the magic bullet answer from isotopes, this actually happens in only about 25% of the isotope investigations. Building in isotopes as only one part of your overall approach is much safer, and generally leads to much better and more believable science.

2. In your actual isotope field work, proceed stepwise. If you can, plan a three-visit approach, and collect twice as many samples each time as you think you will analyze. Analyze some of the samples from the first visit and see what you have, then analyze some more of those saved, extra samples to make sure you are getting the right picture. After this first round of interpretation, use the second visit to follow up with more focused sampling, polishing off the study with mop-up sampling during the third visit. Often the first visit is rather brief, the second very extensive, and the third is brief again: remember that one day in the field can translate into many months in the lab! The stepwise approach lets you check out ideas and follow the isotope trail to best advantage, without spending too much effort in places that don't need it.

3. Think about the advantages of compositing samples versus using individuals. It is always better to analyze individuals, but this always costs more in terms of time, effort, and money. If what you really need is the average isotope value, pooling 30 individuals into three groups and measuring those three samples can be as effective as measuring all 30 individuals, and will be much more cost-effective. Some balance of composite and individual samples is usually wise. Long-lived organisms

such as larger fish tend to integrate their isotopes over years, so they are good candidates for analysis of individuals. Smaller short-lived organisms (which you may need more of anyway) may be good to pool into composite "pre-averaged" samples.

4. Animals average much of the food web action, with top carnivores expected to be essentially invariant in isotope compositions (Parker 1964). Most of the isotope action is at the lowest trophic levels. To see the true dimensions of the isotope variation in your food web, sample intensively at the base of the food web. Also, you should realize that your data for food resources at the base of the food web are often quite aggregated and averaged. After all, in a one-liter water sample, there are usually tens to hundreds of plankton species, yet you collect them all onto a filter for a single, final pooled isotope value. So sampling more kinds of plants and basal food resources often pays off as you seek to keep track of sources of isotope variation. Detailed isotope analyses of signature or biomarker compounds are also possible (see Further Reading for Section 3.5), allowing researchers to investigate isotopes of microbes that also may be important food sources.

5. There are other reasons for paying special attention to sampling low in the food web. A disconcerting possibility is that animals might be consuming most of the important foods, keeping those foods at such low abundances that they are very hard to sample. The animals are much more expert at this kind of (gourmet) sampling than you are; bet on it! If you think this applies, you may have to plan some very special sampling to get those important but hard-to-catch foods.

6. Sampling directly from stomach contents often works convincingly in food web studies (Fry 1988; Grey et al. 2002) because what you are sampling is definitely what an organism has chosen to eat, and not just some "potential" food item.

7. A good way to see if you missed an important food in your sampling is to work backwards from a consumer isotope composition, for example, subtracting the respective trophic enrichment factors of 0.5 and 3.4‰ for ^{13}C and ^{15}N from consumer isotope values, and using the resulting inferred isotope compositions to see if they match any of the measured food isotope values. If you work backwards from top consumers in this top-down approach, do you find an important food at the base of the food web? Or, can you find a mixture of foods that will form this inferred composite or average food? If you don't find the pot-of-gold food source at the end of the inference chain, then your sampling was likely inadequate, or your use of fractionation factors was inappropriate. Agreement within 1‰ between measured and inferred values is generally good agreement, but larger disagreements may need your more detailed attention.

8. So, treat your food isotope values with caution. As you look at your food web from the bottom-up or basal source point of view, think that

your foods are representative rather than giving the whole array of possible foods, and realize that your sampling may have missed something
important. If indeed something seems to be missing, don't be surprised.
If you can guess what food was missing, you may want to go back to see
and get a sample of it; that is what the second or third trip is for.

9. Once you think you have the basic outlines of the isotope food web,
there are nice statistical packages to help you calculate contributions
from important food sources and to deal with mixing muddles that arise
when there are too many sources and not enough tracers. Extended
Problem 10 for Chapter 5 on the accompanying CD gives a step-by-step
guide to using these statistical approaches pioneered by Donald Phillips
and colleagues.

10. But in the end, and even if you have gone to the extra trouble to
sample isotopes across systems, use multiple isotopes, or add isotope
tracers, nonetheless you will almost certainly have to use your biological insight and intuition to really understand the isotope results. And so,
closing here at point 10, it is just a reminder that you should have pro-
foundly considered point 1 before embarking on the isotope path.

5.2 Isotope Sourcery

The main use of isotopes involves magic. You can't see, feel, touch, hear,
smell, or taste isotopes with our normal senses, yet there they are, magical
scraps of information fluttering gently all around us. Pluck a piece of isotope
out of the air and read the invisible scrip; see what your fortune will be. If
you are lucky, you will find that Nature has kindly painted your experiment
in contrasting shades of isotopes, so that you only need to consult your
friendly (isotope) paint color chart to read what mixtures pertain all
through the hills and valleys of your investigations. However, Nature has
also painted many a landscape in isotope monochromes, so the brilliant
contrasting hues of a South Sea Gaugin are not to be found, no, just a
drab background full of disappointment. Except—we come to these
exceptions—later.

But first, the mixing magic. Consider two sources, good versus evil, right versus wrong, asphalt versus grass; you name it. These sources mix it up and isotopes let you follow the action. Say the two sources are colors, black and white. The whiter the mixed sample, the more the white source dominates the mix; the blacker the sample, the more the black source wins. The color in the middle gives a color chart guide to which source is more important, the white or the black. If we set up a scale with 0% as pure black and 100% as pure white, then we can use the colors to move along the scale, like moving an abacus bead along a string from black (0%) to white (100%). The color exactly in the middle is gray, the perfect blend of black and white, with 50% contributions from each source or end member (Figure 5.11, top). (Color versions of these figures in the Chapter 5 folder on the accompanying CD give this mixing in terms of blue and yellow, with green as the intermediate color.)

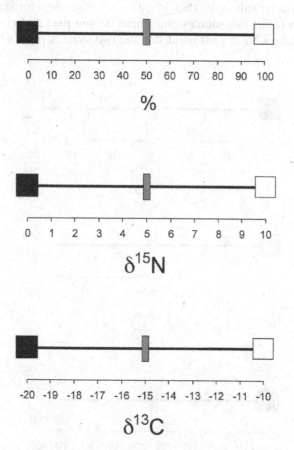

$$\delta^{15}N$$

$$\delta^{13}C$$

FIGURE 5.11. Mixing models for percentages, nitrogen and carbon isotopes. Black and white sources at the ends of the scales yield a grey sample in the middle; colors and isotopes index the % contributions of the sources, 50%–50% in these cases.

This concept of isotopes as colors is not hard to follow, with the color (isotope) telling us the proportions of the two sources, like following an isotope cream swirl that ends up as an even blend in your morning coffee. The harder part is getting used to all the odd number scales for the isotope colors and flavors. These δ scales were developed long ago when geochemists chose standards. The standards provide the central δ value reference point at 0‰ for each of the HCNOS elements. But samples are not uniformly higher or lower in δ values than the standards, so both negative and positive δ values occur for the different HCNOS elements. Thus, the often confusing reality is that the isotope numbers are positive and >0‰ for some elements such as nitrogen (Figure 5.11, middle), negative and <0‰ for carbon (Figure 5.11, bottom), negative and positive for sulfur and oxygen (Figure 5.12, top and middle), and usually very negative for hydrogen (Figure 5.12, bottom). Here the secret is not to pay much attention to the actual isotope values, be they negative, positive, large or small, but to pay attention to the two sources, and where the sample falls between those two source values. So, in your mind, find the two sources, paste a black label

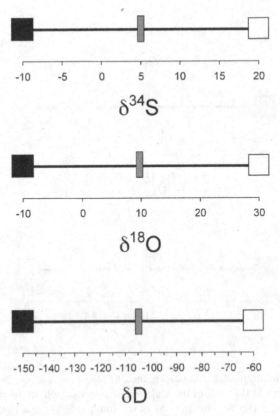

FIGURE 5.12. As Figure 5.10, but for sulfur, oxygen, and hydrogen stable isotopes.

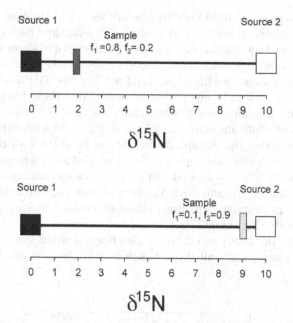

FIGURE 5.13. Mixing models: two sources at the ends and a sample in the middle; sources contribute unequally to the sample in the top and bottom case, so the split is not 1:1, but 2:8 (top, black source is larger contributor) and 1:9 (bottom, white source dominates). In these two-source mixing problems, source 1 contributes fraction f_1 and source 2 contributes fraction f_2 to the mixed intermediate sample so that $f_1 + f_2 = 1, f_2 = 1 - f_1$ and as derived in Section 5.3, $f_1 = (\delta_{SAMPLE} - \delta_{SOURCE2})/(\delta_{SOURCE1} - \delta_{SOURCE2})$.

onto one and a white label onto the other, divide up the intervening isotope scale into colors or percentage points, *et voilà*, you are ready to interpret the isotope mixes. It isn't hard at all; it is just the odd units that so distract the first-time users. The advice here from mixmaster Mr. Polychaete is to ignore those units; just pay attention to the overall separations between sources and sample.

Let's look a little closer on how this mixing works. There are different ways to look at mixing beyond the simple color (or flavor) analogies, and for your viewing pleasure, here are three more examples.

In the first example, we move our isotope marker off the grey middle of the isotope scale. The color changes towards black or white, depending on which way we move the marker and for now we simply note that there seems to be a close linear correspondence between the color and isotope value (Figure 5.13). In the next section, we write an equation that gives the color or isotope value in terms of source contributions. But for now, just keep in mind that as the isotope marker shifts, isotope colors are shifting because source contributions are shifting. This simple colorized sourcery works as long as samples truly represent mixtures of the sources.

Approaching this all from another point of view, let's think about chocolate candy available at the store as a mixture of white and dark chocolates. We like to know how much candy we have, but also like to know what kinds we have. Those isotope colors can come in handy, as you might suspect. We go to the store and get two bags of candy. Each bag has 100 pieces of chocolate, but you can tell right away from the colors that these bags are really different, one containing mostly white chocolates and the other containing an equal mix of white and dark chocolates (Figure 5.14, upper two panels). So you can see that the isotope coding lets us keep track of things happening even when total amounts are the same, just as you might be interested in someone's family, and ask not only how many children there were, but also how many girls and boys. Yes, proportions and percentages are of great interest in everyday situations including family histories, chocolates, and a thousand other things.

Not to leave the candy too soon, we also linger a while, visualizing a sad hungry day when nearly all those delicious white chocolates have been

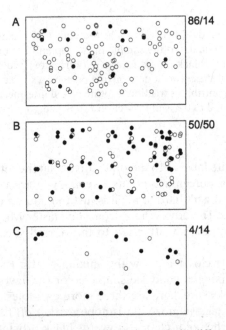

FIGURE 5.14. Mixing white and dark chocolate candy in different proportions: the color gives the mix proportions, without actually having to count all the candy. Paired numbers are no. of white chocolates/no. of dark chocolates. Chocolate proportions in bottom panel were calculated assuming a starting 50/50 mix (middle panel) then removing white chocolates twice as fast as dark chocolates. (Precise calculations for these removals followed the closed system fractionation rules explained in Section 7.2.)

eaten, leaving mostly the dark chocolates (Figure 5.14, bottom panel). This unhappy day arrived because although we had started with the 50/50 mix (Figure 5.14, middle panel), we ate white chocolates twice as fast as the dark ones. But fortunately there is a friendly supplier of white chocolates nearby, and smart shopping soon fixes the problem and recreates a preponderance of white chocolates (Figure 5.14, top panel). We keep track of these acquisitions by the color of the chocolate mix, until those undesired dark chocolates are rare. The isotopes strike again, tracking proportions and percentages, not the totals.

The last example is actually a flock of thoughts. We are using the isotopes as a scale between two sources, and we do this kind of thing constantly in everyday life. For instance, our grading system in the United States sets the scale between knowledge and ignorance with grades of A to F, our Olympic sports judges rate high to low for excellent to poor, and as you turn down the sound on your radio, you move between loud and soft. You can find an analogy to isotopes in any of the senses that are set up to detect contrasts: sight, taste, smell, hearing, touch. Beyond the immediate senses, the brain is also integrating, divining which is more important in the situation at hand, the yin or the yang. There is a weighting of individual items going on in all this, but sometimes it is hard to keep an account of what each particular component is doing. Isotopes provide a quick way to assay the integrated picture, the fractions, proportions, and percentages, without having to count all the items at hand.

5.3 Mixing Mechanics

We have been learning that isotopes record proportions and percentages. When two sources combine to form a mixture, isotopes will indicate the relative contributions of the sources. The first way to think about this is with percentages or fractions, with percentages of the two sources adding to 100% and fractions of the two sources adding to 1. Fractions are a little simpler in the equations, so we use fractions below.

Let's give our sample and two sources isotope values of δ_{SAMPLE}, $\delta_{SOURCE1}$, and $\delta_{SOURCE2}$, then conceptualize the mixing dynamic as the sample consisting of two fractions that sum to a whole, fraction (f_1) for source 1 and fraction (f_2) for source 2.

$$\delta_{SAMPLE} = (\delta_{SOURCE1}) * f_1 + (\delta_{SOURCE2}) * f_2.$$

Realizing that

$$f_1 + f_2 = 1 \quad \text{so that} \quad f_2 = 1 - f_1$$
$$f_1 = (\delta_{SAMPLE} - \delta_{SOURCE2})/(\delta_{SOURCE1} - \delta_{SOURCE2}).$$

A more fundamental way to arrive at this same result is by mass balance, an accounting of the isotopes. Here we write an equation for the total amounts or masses (the m terms), followed by an equation for the isotopes (the δ terms) in these amounts:

$$m_{\text{SAMPLE}} = m_{\text{SOURCE1}} + m_{\text{SOURCE2}}$$

$$\delta_{\text{SAMPLE}} * m_{\text{SAMPLE}} = \delta_{\text{SOURCE1}} * m_{\text{SOURCE1}} + \delta_{\text{SOURCE2}} * m_{\text{SOURCE2}}.$$

This second equation holds because δ values are excellent indicators of % heavy isotope, and so really just gives the isotope accounting for the total amount of heavy isotope (if needed, review Section 4.3 on "gains" for further details of this accounting). Dividing terms by m_{SAMPLE},

$$f_1 = m_{\text{SOURCE1}} / m_{\text{SAMPLE}}$$

$$f_2 = m_{\text{SOURCE2}} / m_{\text{SAMPLE}} = 1 - f_1,$$

substitution yields

$$\delta_{\text{SAMPLE}} = \delta_{\text{SOURCE1}} * f_1 + \delta_{\text{SOURCE2}} * (1 - f_1)$$

and

$$f_1 = (\delta_{\text{SAMPLE}} - \delta_{\text{SOURCE2}}) / (\delta_{\text{SOURCE1}} - \delta_{\text{SOURCE2}}),$$

the same result obtained above.

Now that we have derived this mixing formula in two ways, an intuitive way based on illustrations (Figures 5.11–5.13 of the previous section), and a more formal way based on accounting for all the masses and isotopes, let's practice using this formula in different ways in some numerical examples. Suppose you have two sources and a sample that have isotope values of 0‰ (source 1), 10‰ (source 2), and 2‰ (the sample; Figure 5.13, top). You can already see that source 1 contributes most because the sample is closest to source 1 in δ values. Does the formula for f_1 give us the right answer? Well, $f_1 = (2 - 10)/(0 - 10) = -8/-10 = 0.8$, or 80% for the source 1 contribution, so that the source 2 contribution is 0.2 or 20%. (Interestingly, the source 2 contribution satisfyingly corresponds to the numbers on the δ scale of Figure 5.13, another indication that our equation is correct.) Let's try another example. The two sources and the sample have respective values of 0, 10, and 9‰ (Figure 5.13, bottom), so we can see that this is going to be a 10%/90% split: $f_1 = (9 - 10)/(0 - 10) = -1/-10 = 0.1$ or 10% and f_2 is 0.9 or a 90% contribution. The equations worked again, very nicely.

Now, for our final test, let's use some of those often troublesome isotope numbers that are negative δD values (although we could have used $\delta^{13}C$

values that are also typically negative). The δ values of our two sources and sample are respectively –150, –60, and –105‰. Here it is not really immediately obvious what the answer is, so you really have to use (and trust) the formula: $f_1 = [(-105) - (-60)]/[(-150) - (-60)] = -45/-90 = 0.5$, or a 50% contribution from source 1. You can see that this is the right answer when you set it up on a graph (see Figure 5.12, bottom), but this particular example really reinforces the idea that you need to keep an eye on all those negative signs when using this formula. Altogether there are six minus signs in this hydrogen isotope example, and that is quite a lot to keep track of. Experience shows that it is all too easy to neglect a minus sign in some part of the calculations. To remind you that it is essential to carefully use the minus signs, Mr. Polychaete advises this seeming bit of nonsense, "Remember to minus the minus," something that you will find yourself doing quite a bit as you work with these mixing equations.

Another point of confusion lies in deciding which source is source 1 and which source is source 2. The formula solves for the contribution of source 1, but does it make a difference which source you assign as source 1? The answer is no, it doesn't make a difference, but experience shows that it does pay to be consistent. The examples given here consistently designate the source with the lowest δ value as source 1. Thus in graphs, source 1 will appear at the left and source 2 at the right, following the normal left-to-right progression of written numbers.

But why, you may ask, are we spending so much time on these very small details? It is because the majority of all isotope applications use these simple types of two-source mixing models, or variations on this theme. The simplicity of the mixing models is one reason why isotopes are popular, for novices and experts alike. So it is good to make sure you understand all these simple details.

With many great examples to choose from, Mr. Polychaete decided to make up a brand new example that shows how to use the mixing models to avoid getting mixed up. Here it is.

You are studying fish metabolism and get interested in lipids and lipid turnover. While isolating lipids in the laboratory, you begin to suspect that your reagents are dirty and are contributing some contaminant lipid to your samples. This is bad, but you want to know how bad. "Hmmm," you say; "Hmmmm," says your yoga instructor as you cogitate your next steps; "Hmmm," says your major professor, "pray tell what will you do, oh you budding independent thinker." Suddenly, you remember this book and isotopes pop into your head. That old mixing magic should get me the proportions of the sources; the isotope book was very definite about that, wasn't it? So, you go down the hall, pull out clean lipid standard and run it through your procedures. You can just measure an ever-so-slight increase in the weight at the end of your procedures, but you aren't sure everything went perfectly (it never does, you know). The small change in weight could be due to something else besides contamination. So, you measure the lipid isotopes before and after the purification, as well as the potential contaminating solvent. You find these isotope values for source 1 (pure lipid), source 2 (solvent), and sample: −45, −25, and −35‰ $\delta^{13}C$, respectively. You can see immediately that the −35‰ sample is halfway between the −45‰ source 1 starting lipid and −25‰ source 2 solvent, so what you are isolating is 50% solvent contamination. You shake your head in disgust, and take this result into your major prof's office. He looks at you, and you both start nodding your heads up and down in unison (the bobble head theory of scientific accord applying here), saying, "Now we know."

Suddenly, that is quite enough work for the whole next month. Every day you and your professor see each other, and begin bobbing heads. Yes, "Now we know," but also, "What next?" begins to penetrate and dissolve the unanimity of those bobbing heads. And later at night, you think back on those weights of the lipid that did not show much contamination problem, and think you may have to work on those weighing procedures as well. Maybe you have been losing a lot of material through the extractions, without really realizing it. In the end, your yoga teacher advises "Attentive Neglect" of this problem, and in due course (over the next six months) you find and eliminate the contaminant (using a cleaner triple-distilled solvent that is a lot more work), while also reading about, then implementing a new extraction procedure that is much more quantitative and doesn't lose sample at each step. You also try correcting for the contaminant effect using the mixing equation of this section, but that does not work out because the blank is oddly variable, for reasons unknown. With inconsistent blanks, you had to develop the cleaner, more troublesome procedures. But in the end, you have great procedures and great results, thanks to—isotopes!

You might not think so, but using this isotope mixing approach takes practice, even for these seemingly simple problems that deal with blanks (or are they cursed blankety-blank-blanks?). Problems 2 and 3 for this chapter on the accompanying CD give you a chance to practice mixing with laboratory blanks, and Problem 4 shows that mixing in other settings such as the classroom also can produce results that resemble the effects of blanks. Section 5.7 of this chapter shows how an ecologist uses these same "blank" mixing equations to work out a field ecology problem. These examples show that isotope mixing calculations can provide robust quantitative answers in carefully conducted research.

5.4 Advanced Mixing Mechanics

Here we revisit mixing dynamics of the previous section in two ways, emphasizing the role of amounts in two-source mixing, and emphasizing outcomes when there are more than two sources involved in mixing.

Two-Source Mixing with Weighted Averages

When two sources mix, both isotopes and masses are involved. In the last section, we used fractions to indicate the role of mass, but here we step back from fractions, and explicitly use mass instead. We show that when mass becomes involved, isotope mixing magic becomes more complex but also more interesting.

Let's start with the algebra. If the total amount of mass (m_T) in a sample comes partly from source 1 (m_1) and partly from source 2 (m_2) then

$$m_T = m_1 + m_2.$$

Now let the fractions f_1 and f_2 represent the amounts of source 1 and source 2 divided by the total amount so that:

$$f_1 + f_2 = 1$$
$$f_1 = m_1/m_T = m_1/(m_1 + m_2)$$
$$f_2 = m_2/m_T = m_2/(m_1 + m_2).$$

Recalling the mass balance equation from Section 5.3,

$$\delta_{SAMPLE} = f_1 * \delta_{SOURCE1} + f_2 * \delta_{SOURCE2}$$

and substituting for f_1 and f_2, one obtains:

$$\delta_{SAMPLE} = (m_1 * \delta_{SOURCE1})/(m_1 + m_2) + (m_2 * \delta_{SOURCE2})/(m_1 + m_2),$$

which rearranges to

$$\delta_{SAMPLE} = (m_1 * \delta_{SOURCE1} + m_2 * \delta_{SOURCE2})/(m_1 + m_2).$$

In this last equation, it is clearer that the isotope value of the sample reflects the mass-weighted average of the two sources. So, if source 1 makes a large mass contribution, the sample will have an isotope color (δ value) more like the isotope color of source 1, but where source 2 makes the large mass contribution, the isotope colors shift towards source 2. This is the math behind the visual mixing models (Figures 5.11–5.13). In fact, the mass contributions are the main driving force in mixing dynamics, with mixing of mass from the two sources dragging along the isotope colors.

Because of the underlying mass contributions, there are many varieties of two-source isotope mixing. Roy Krouse compiled some of these possibilities that show a surprisingly broad scope of isotope variation arising from "simple" two-source mixtures (Figure 5.15). These diverse outcomes occur because four parameters are being varied across the various scenarios of Figure 5.15, the two δ values of the sources, and the two mass contributions from the sources. Understanding these scenarios was important for tracking pollution from industrial gas plants in western Canada. Krouse and colleagues used these models to estimate effects of acid rain pollution and sulfur deposition in terrestrial ecosystems downwind of the gas facilities (Winner et al. 1978; Case and Krouse 1980; Krouse 1980; Krouse et al. 1984).

These examples indicate that mass contributions are important in the isotope mixing equations. In real life, there are often additional weighting factors that enter into these isotope mass balance equations. These factors include concentrations in the sources, so that sources with higher concentrations will still dominate a 50:50 mix. Oddly enough, isotopes do not directly record concentration effects. This can be confusing, so let's consider an example, the case of the black bear.

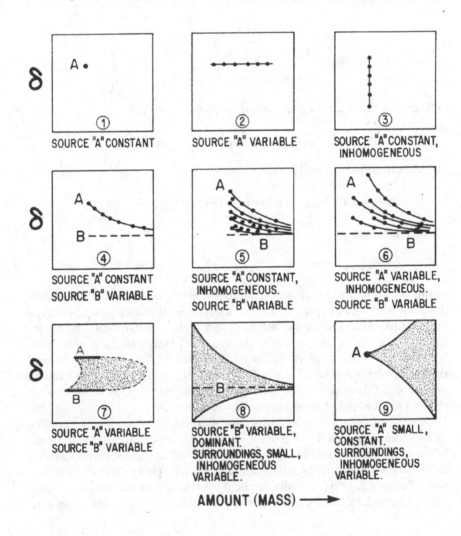

Here our goal is to use isotopes to understand whether a bear feeds on plants or on fish. This particular black bear eats nitrogen-poor plants that contain 1% N. It also eats nitrogen-rich salmon that contain 14% N in the fish fillet muscle tissue. What is it eating more frequently, plants or fish? We turn to nitrogen isotope analyses to help answer this question.

Nitrogen isotope analysis of the bear and its potential food sources shows $\delta^{15}N$ values of 8.4‰ for the bear (using hair tissue), 0‰ for plants (source 1), and 10‰ for salmon (source 2). We know from Sections 3.3 and 4.7 that animals usually have higher $\delta^{15}N$ values than their diets, so we subtract 3.4‰ from 8.4‰ to account for this trophic fractionation. The subtraction results in a value of 5‰ as an estimate for the $\delta^{15}N$ value of the bear's diet. Using the isotope mixing equations for fractional contributions of the two sources,

$$f_1 = (\delta_{SAMPLE} - \delta_{SOURCE2})/(\delta_{SOURCE1} - \delta_{SOURCE2})$$

$$f_2 = 1 - f_1,$$

FIGURE 5.15. Isotope mixing between two sources is governed by a combination of two things: isotope compositions of the sources, and also amounts (mass) of sources. So, when mixing only two sources, there are actually four things to keep track of: source A isotopes, source A mass, source B isotopes, and finally source B mass. This figure shows that when one or more of these four quantities is not fixed, but can vary, mixing of even two sources can get complex. (From Krouse (1980). Used with the author's permission.) Guide to the examples distinguished by the numbered circles: (1) Only source A is present, and source A has fixed isotope value and mass. (2) Source A has a constant isotope value, but can vary in mass. (3) Source A can vary in isotope value, but has a constant mass. (4) Sources A and B have fixed isotope values, source A has a fixed low mass, but source B increases in mass. Mixing results will approach the isotope value of source B as the mass of source B increases. (5) Source A can vary in isotope value, but has a constant low mass. Source B has a constant isotope value, but can increase in mass. The resulting family of mixing curves all approach the isotope value of source B as the mass of B increases. (6) Source A can vary in isotope value and has a variable, but low mass. Source B has a fixed isotope value, and can increase in mass. As mass increases, results approach the isotope value of source B. (7) Sources A and B have fixed isotope values, and can each vary in narrow, but fairly similar, ranges of mass. The shaded envelope gives the range of possible mixing outcomes. (8) Source A is not really depicted, but has a near-zero mass and widely variable isotope values. Source A is perhaps better thought of not as a single source, but as a family of sources. Source B has a fixed isotope value and can increase in mass. At higher mass, isotope values converge on the dominant source, source B. (9) Source A is fixed in isotope value and has a low mass, and mixes with various other B sources that all have higher mass, but different isotope values. The shaded envelope gives possible mixing solutions.

this 5‰ translates to equal nitrogen contributions from the plant (0‰) and salmon (10‰) sources, a 50–50 or 1:1 result. So, this result indicates the bear is eating an equal mix of plants and salmon.

But as we think about this result over the next few days, we find that we are not really comfortable with this answer. Eventually we pinpoint the problem. Although the 1:1 result may be strictly accurate for the source of nitrogen, it is nonetheless misleading for understanding the total amounts of material being ingested, the total amounts of plants plus fish. We think this through carefully, and finally realize that because the plants have much lower nitrogen contents, the plants must dominate the diet at a 14:1 ratio even when nitrogen comes equally from the two sources. We talk this over with a bear biologist. As it turns out, the biologist is most interested in the total diet, the 14:1 ratio of plants versus fish, because it shows the bear's overall feeding strategy. Nonetheless, the 1:1 result is also very interesting because it shows that a little feeding on fish provides the bear with a very important part (50%) of its nitrogen needs. Overall, we agree with the bear biologist that it is valuable to calculate mixing results from both perspectives, the relative amounts of fish and plants as well as the narrower focus on the relative amounts of N in the bear's diet.

Generally, the isotopes·give the narrower result centered on the HCNOS element tracked by the δ value being used. The ecologist must then use this result with concentration data to infer total amounts of material involved. Involving concentrations in mixing equations makes these equations more realistic, but also more complex.

Here is the algebra involved when concentrations or other weighting factors differ in the sources. First, the initial mass balance equation must be amended. So instead of

$$m_T = m_1 + m_2,$$

the mass balance becomes

$$m_T * W_T = m_1 * W_1 + m_2 * W_2,$$

when concentrations affect or weight (W) the mass contributions. In the bear example, W_1 and W_2 represent the nitrogen concentrations in the source plant and fish materials, but as we show in Sections 5.6 and 5.8, weights can also be assigned from other factors such as loadings or even relative importance values that you estimate for modeling purposes. Many different kinds of weightings are permissible, as long as they help accurately budget the amounts of isotope involved. Weighting factors are related to the mass balance of elements and their isotopes, and are "mass-related" weighting factors.

In any event, weighting factors for the present example insert into the isotope mixing equation as follows.

$$\delta_{SAMPLE} = (m_1 * W_1 * \delta_{SOURCE1} + m_2 * W_2 * \delta_{SOURCE2})/(m_1 * W_1 + m_2 * W_2).$$

This is a more general isotope balance equation with both mass (m) and weighting (W) factors. It allows realistic tracking of the amounts and masses in mixing problems that start with sources and calculate forwards to predicted δ values (Figure 5.15a; Fry 2002; Phillips and Koch 2002). We need these more complex weighted average equations for the forward calculations.

But ecologists are actually interested in the reverse process, using the δ values to calculate the source contributions. This is what we did in the bear example above. The following equations first calculate the source contributions by element from the δ values, then use a ratio-based approach to calculate the total amount of material contributed by each source. Weighting factors appear again in the second part of the calculations.

Where isotopes show the fractional contributions in two-source mixing as

$$f_1 = (\delta_{SAMPLE} - \delta_{SOURCE2})/(\delta_{SOURCE1} - \delta_{SOURCE2}) \quad \text{and} \quad f_2 = 1 - f_1$$

and the weighted values for the HCNOS element in consideration are W_1 and W_2 (e.g., W_1 and W_2 are different % N values for the sources), then the ratio of the two fractional contributions of total material (f_{TOTAL}) from the two sources is

$$f_{TOTAL1}/f_{TOTAL2} = (f_1/f_2)/(W_1/W_2) \quad \text{so that}$$

$$f_{TOTAL1} = f_1 * W_2/(f_1 * W_2 + f_2 * W_1) \quad \text{and} \quad f_{TOTAL2} = 1 - f_{TOTAL1}.$$

For the bear example, $f_1 = 0.5$, $f_2 = 0.5$, $W_1 = 1$, $W_2 = 14$, $f_{TOTAL1} = 0.93333$, and $f_{TOTAL2} = 0.06667$. To check this result, we divide f_{TOTAL1} by f_{TOTAL2}, obtaining the correct 14:1 ratio for plants:fish in the total diet.

Note that the isotope equations calculate the relative contributions that the sources make to the HCNOS element being budgeted in the mixture. But to calculate the relative contributions of the total mass, not just the element in the mixture, the second ratio step is needed. Phillips and Koch (2002) outline calculations of this type when more than two sources are involved in the mixing.

There are two important concluding thoughts for this section. The simple two-source mixing equations

$$f_1 = (\delta_{SAMPLE} - \delta_{SOURCE2})/(\delta_{SOURCE1} - \delta_{SOURCE2}) \quad \text{and} \quad f_2 = 1 - f_1$$

are quite reliable, giving the source contributions of the HCNOS elements via δ values. In addition, these simple mixing results can be extended to have multiple meanings when concentrations and weightings are involved, so it is good to keep the ecological context in mind as you interpret even simple mixing results.

Multisource Mixing

We end this mixing math by considering the case when there are more than two sources. Having three, four, or more sources involved in mixing problems is not uncommon, but your goal remains to calculate or back out the contributions of each source. Here is the solution: each source contributes some fraction to the total mass of the sample, and isotope fractions also have to add up to the observed total isotope amount. Summing fractions gives the total sample, and summing mass-weighted or fraction-weighted δ values gives the total isotope amounts. Mathematically, this is straightforward. For example, if you measure a sample for carbon and nitrogen isotopes and three potential sources that can contribute to the sample, you can write three equations:

$$f_1 + f_2 + f_3 = 1,$$

$$f_1 * \delta^{13}C_1 + f_2 * \delta^{13}C_2 + f_3 * \delta^{13}C_3 = \text{observed } \delta^{13}C \text{ of the sample,}$$

$$f_1 * \delta^{15}N_1 + f_2 * \delta^{15}N_2 + f_3 * \delta^{15}N_3 = \text{observed } \delta^{15}N \text{ of the sample,}$$

where the three sources are denoted by the subscripts 1–3, and f is the fractional contribution of a source. Because there are three equations and three unknowns (f_1, f_2, and f_3; all the δ values are measured knowns), you can do the algebra and solve for the three fractions.

This approach can be extended to multiple sources when you measure more isotopes. If the three-source problem can be solved by measuring two sets of isotope tracers, as above, a four-source problem will require measurement of three isotope tracers, for example, S in addition to C and N isotopes. Solving all the algebra is time-consuming, but still straightforward.

However, in the real world there are often too many sources (>5–10) and not enough tracers (Phillips and Gregg 2003). What to do? One new approach is to use statistical models to solve for the constellation of potential or feasible solutions (this is the IsoSource software available on the Web at www.epa.gov/wed/pages/models.htm). However, when using this elegant software to give these solutions, you should realize that these feasible solutions are not the same as the true actual solution, and no one knows where in the constellation of feasible solutions the actual solution really will reside. Be aware that the only really reliable results from this statistical modeling are the minimum and maximum (minmax) contributions calculated for each source.

But you should also realize that these minmax models can be amended in interesting ways. If you can gather more information about source contributions, even some seemingly simple information such as "one of the potential sources contributes more than another one of the potential sources," this will help constrain the equations you are seeking to solve. For food web studies, this further information often comes from natural history observations or gut content analyses (Harrigan et al. 1989; Peterson 1999). There is a danger here of course. The danger is that this additional information is genuinely hard to obtain reliably and often why you started isotope studies in the first place. But in many cases, you can find reliable additional information that helps you in simple but powerful ways with the IsoSource modeling. Added information that is credible will rapidly shrink the gap between minimum and maximum values, closing in on the unique solution. In the future, using this IsoSource model along with a good field-testing program will likely be a powerful way to accurately estimate source contributions in the difficult but common cases where there are too many sources and not enough tracers.

Problems 5 through 7 for this chapter on the accompanying CD consider these more advanced aspects of mixing, focusing on weighted averages related to concentrations and on multisource mixing.

5.5 Mixing Assumptions and Errors or the Art and Wisdom of Using Isotope Mixing Models

If you are skeptical by nature, you have been wondering about this isotope mixing. It all seems too easy somehow. We just add one source to another and bingo, there is the answer. Is it really that simple, you keep wondering? You might start voicing your concerns as this: "Let's see some hard thinking and statistics, where the rubber meets the road." And while you are at it, you might go on objecting, "Maybe yes, the math is that simple, but look, there are bound to be errors. That is what statistics are for, to help us track the errors. Where are the statistics? And how about the interpretation, and are there any questionable assumptions? So, Mr. Polychaete, what will you show us when we get you under the close scrutiny of the statistoscope and the assumptoprobe?" This all puts Mr. Polychaete on the spot, but he tries to gather his thoughts to respond to these valid criticisms. We listen in on his rambling lecture that is just starting. Fortunately, he begins with an overview.

Overview

In the end, the experienced scientist knows that one should not expect too much from the isotope mixing models. The sage advice is that if large errors in the final mixing models are probable, then it is probably better to start your scientific program with multiple lines of inquiry. Don't rely on isotopes

alone, but see how the isotope results fit in. You can see this kind of thinking in a quote:

Warning! Stable isotope data may cause severe and contagious stomach upset if
taken alone. To prevent upsetting reviewer's stomachs and your own, take stable
isotope data with a healthy dose of other hydrologic, geologic, and geochemical
information. Then, you will find stable isotope data very beneficial.

This quote is from Marvin O. Fretwell, USGS (1983), reprinted from
Isotope Tracers in Catchment Hydrology by Carol Kendall and Jeffrey J.
McDonnell (1998).

Lecture

With the overview idea in mind that isotopes will never be a perfect form
of evidence in scientific inquiries, let us investigate some of the errors and
assumptions associated with these mixing models, starting with the basic old
two-source mixing model. How much can we really trust the results from
this model? The model has three isotope terms (source 1, source 2, and
sample), each of which has a measurement error associated with it. For the
final result that we are interested in, the fractional contribution of source
1, our confidence in the answer should somehow reflect all three sources of
error. Statisticians have worked this out (Phillips and Gregg 2001), and generally call this propagation of errors. Box 5.2 gives an example calculation
for propagation of errors for a two-source isotope mixing problem. Working
through this and other similar example problems, one rapidly reaches the
following sensible conclusion. The most favorable case for getting a precise
(small error) answer for the source contributions occurs when the difference between the two sources is large, and when there is little variability
associated with the average δ values of sources and samples. When you
average fewer source or sample δ values, variability around these averages
increases and so does the error in the final answer. And when the isotope
differences between sources decrease, the errors in the final answer also
increase. You can think about this as signal and noise, with the signal being
the difference between the two sources and noise being the error terms for
the average values of the sources and the sample. Maximize the signal and
minimize the noise, and if the signal shrinks, then do more sampling to
shrink the noise along with it. That is the advice. (And, it is hoped, the
increased sampling will not reveal new populations that actually increase
the error terms; the point is to decrease the error!)

But almost no one propagates errors or shows that they do in the isotope
literature, perhaps because it would be embarrassing. Let's check. Suppose
we had two sources and a sample in the middle, so that we are dealing with
a 50/50 mixture. And further suppose that we make isotope measurements
on the sources and sample, and find that (conveniently and for simplicity)
they all have about the same error term, measured in 95% confidence limits.

Box 5.2. Propagation of Errors in Mixing Models

Here is the problem. For the mixing model results when there are two sources that contribute to the sample, and the contribution of source 1 is f_1 where

$$f_1 = (\delta_{SAMPLE} - \delta_{SOURCE2})/(\delta_{SOURCE1} - \delta_{SOURCE2}),$$

what is the final (propagated) error we should use for f_1? The following shows how to calculate this final propagated error, using a set of example data.

Do this problem in three steps:

1. Measure values and errors for each term in the equation, for example,

    Value ± Error (95% confidence limit) for $\delta_{SAMPLE} = 5 \pm 0.9‰$.

    Value ± Error (95% confidence limit) for $\delta_{SOURCE1} = 0 \pm 1.1‰$.

    Value ± Error (95% confidence limit) for $\delta_{SOURCE2} = 10 \pm 1.0‰$.

    $f_1 = (5-10)/(0-10) = 0.5$, but what is the error for f_1?

2. Propagate errors in the numerator and denominator terms, using the general propagation formula for subtractions: If $A = B - C$, and there are errors in both B and C (error for $B = e_B$, error for $C = e_C$), the propagated error for A (error for A or e_A) is:

$$e_A = \left(e_B^2 + e_C^2\right)^{0.5}.$$

    With this equation, solve for the propagated errors in the numerator and denominator of the first f equation.

$$\text{Error in numerator} = \left(0.9^2 + 1^2\right)^{0.5} = 1.345.$$

$$\text{Error in denominator} = \left(1.1^2 + 1^2\right)^{0.5} = 1.487.$$

3. For the final result, use the "propagation of errors" formula for divisions: If $f = N/D = $ numerator/denominator, then the propagated error for $f(e_f)$ can be calculated from the equation

$$(e_f/f)^2 = (e_A/N)^2 + (e_B/D)^2.$$

Solving with the values given above, $(e_f/0.5)^2 = (1.345/-5)^2 + (1.487/-10)^2 = 0.0945$, and the final propagated error $= e_f = 0.154$ for this particular example.

Conclusion

Note that the propagated error 0.154 is larger than you might estimate from the individual error terms; that is, $0.154 > 0.1$, where 0.1 is the individual noise estimate (95% confidence limit of about 1‰) divided by the signal (difference between the sources $= 10‰$). Also note that when f_1 and its final error are expressed in percentages, $50 \pm 15.4\%$, the propagated error is fairly large, about one third of the estimated 50% value.

Finally, let's say that we want an error of less than ±10% around our 50% average, then fiddle with the signal-to-noise ratio to see what this means for our field program. For a 10‰ signal or difference between sources, we need to determine errors in all three terms (source 1, source 2, and sample) to 0.66‰ noise or less (Figure 5.16). If we halve the signal to a 5‰ difference between sources, then we need to up our sampling effort to maintain a ±10% error around the 50% average, and find we need to get down to a 0.33‰ noise level. This is actually pretty challenging in many studies, and really errors of ±20 to ±40% are not uncommon. At the very least, your sampling strategies should include enough samples so that you can estimate variability and propagate errors in your final estimates of source contributions.

The audience's hands immediately go up in protest, and shouts are heard echoing off the ceiling, "If you know something at 50 ± 40%, do you really know the answer very well. Do you really know anything, Mr. Polychaete?"

This is hardball and doesn't look too good for isotope sourcery. But it is real. Mr. Polychaete replies, "The interesting part is that ecologists routinely deal with this kind of data, which is ultimately 'qualitative' when errors get too large. Errors of ±50% are unfortunately not uncommon in ecology, so ecologists know to get several types of measures as they tackle problems, looking for concordance of results from different approaches, rather than standing too heavily on a single plank that might break. Ahem, well, now you know to propagate errors and think about signal-to-noise in your isotope mixing models." That is the first part of Mr. Polychaete's answer.

Mr. Polychaete pauses here for a drink of water, then continues. "There is more regarding these signal-to-noise ratios that is worth thinking about.

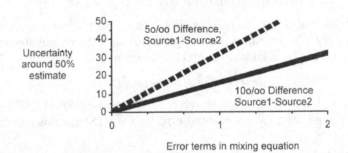

FIGURE 5.16. Propagated errors (uncertainty) for a sample that is a 50/50 mix of two sources, when sources differ by 5 or 10‰ (difference = 5‰ or 10‰). At an equal level of error in the replicates (i.e., at any fixed value along the x-axis) uncertainty is higher when the sources are closer in isotope values. For example, when the difference between sources is 10‰ (bottom line), to achieve a level of 10% uncertainty (y value), the replication errors (x value) must be about 0.66‰. When the difference between sources is 5‰ (top line), replication errors must shrink to 0.33‰ to achieve the same 10% level of uncertainty.

And here it is: when the isotope differences between sources collapse too much, you can lose your signal completely. Sorry, but it happens, and more frequently than you might suppose from the literature that tends to feature reports of successes, not the cases that failed. Those failed cases, the mixing muddles where the source signal is small relative to errors, might be, in reality, half the cases. What to do? My advice is to measure a few samples before you leap into all this, to make sure you are working with a good signal. Some sites have much stronger signals than others. (Here Mr. Poly-chaete shows a slide of Figure 5.9 from Section 5.1, where it is evident that at some sites seagrasses have similar $\delta^{13}C$ values to epiphytes but at other sites, the difference is much larger and more favorable for food web studies). In fact it is often good to check a few systems to see where you should concentrate your isotope studies. The few initial measurements are the best way to see if your system is well-poised with a good signal-to-noise ratio for your research. To make sure you hear this clearly, I repeat that my estimate is that a favorable signal-to-noise ratio occurs only about half the time, so buyer beware!"

Well, at this point, about half the audience leaves, and the rest look really queasy. But Mr. Polychaete decides to discuss two other sources of error, for completeness, even if this is frankly quite boring to most folks. Lectur-ers are used to this, so he drones on, "Let me finish this discussion of errors by mentioning two other common places where errors creep into your results."

"First, it turns out that source values always vary somewhat from place to place and time to time. You often need to measure local isotope values for sources and test for seasonal and spatial variation so that when your field program concludes, you can make source error estimates in your mixing models. If the sources vary strongly in isotope values and are 'noisy,' you will need more sampling to compensate for the increased noise. Also you will need more local sampling where the sources have isotope values that are close together, in areas where the source signal-to-noise ratio is smaller. Overall, this means more analyses for both sources and samples in 'noisy' systems."

"Second, in the mixing diagrams, there are 'fractionation corrections' that need to be made, and these corrections increase the propagated error. For example, in food web diagrams with consumers and potential plant food resources, the ecologist needs to bring the isotope measurements to a common scale of either all plants or all animals, using fractionation rules to achieve this. So, if $\delta^{13}C$ values increase 0.5‰ from a plant diet to animal tissue, then in a food web diagram with mostly plants, one should subtract 0.5‰ from the value of a plant-eating animal to get the true value of the plants consumed by the animal, and then calculate which mixture of sources would give this 'inferred plant' or 'plant diet' value. Making such correc-tions to a common denominator level is often necessary to avoid mixing muddles, but inevitably increases the propagated error."

Mr. Polychaete now summarizes in mid-lecture, "Regarding errors, many things contribute to an aggregated or propagated error that often gives the final result a rather large uncertainty. You can work to minimize all these errors, but you rapidly reach a point of diminishing returns; the most important thing working in your favor in these mixing models is the resolution or ‰ separation between sources. Determine that early on, and if it is small, perhaps isotopes won't work for you."

Students are listening to this advice, which is sounding practical, if somewhat downbeat. "But," he continues, "I have saved the worst for last! The largest problem for the isotope mixing is this: too many sources and not enough tracers. The two-source mixing model completely dissolves into pure muddle when multiple sources start dropping in. And in reality, the two-source business has inevitably involved a lot of lumping (aggregating) of sources already. Full of errors, and now lumpy too! Yes, that is how it is, the isotope sourcery dark side."

Here the audience is definitely gone, but the lights are dim, and Mr. Polychaete fortunately can't see that he has lost his listeners. This is a good thing, because actually a few students are still there, just slumped way down in their seats. They are being polite by not walking away, and they still have some hope from all the papers they have read that Mr. Polychaete will finally reverse and tell them something positive. And sure enough, Mr. Polychaete does start to change the tone of the lecture at this point.

"What to do about mixing muddles? There are four approaches, none of which is really perfect, but any one of which can be helpful. These approaches are (1) working at multiple sites or times, (2) statistical modeling for minmax solutions, (3) using multiple tracers, and (4) adding tracers. Let's look at these in turn."

1. *Multiple sites and times.* In many cases, particular sources fade in and out of the picture along landscape transects. Trees are important in forests, but become unimportant as one walks out of a forest into a grassland—that sort of thing. So you can sample in a forest versus in a grassland to gain insight about the importance of tree inputs, comparing mixing diagrams from two sites that each have multiple sources. Tree inputs will be stronger in the forest isotope diagrams; that is the effect you are looking for. This approach is often termed the transect approach and is essentially comparative. You can also take a transect in time, sampling seasonally in a forest,

for example, to see if normal leaf inputs to soils in the fall create expected isotope signals that are different than in the spring when leaf inputs don't normally occur. In the end, this approach means generating not just one mixing diagram, but making several mixing diagrams for comparative purposes. Don't get stuck in a narrow single mixing diagram, or muddles likely will ensue!

2. *Statistical Modeling*. There is also a statistical approach to dealing with mixing muddles caused by too many sources, an approach that gives a range of possible or feasible solutions (see Phillips and Gregg, 2003, and also the end of Section 5.4). Box 5.3 considers the proper way to deal with these statistical models, which is to use them carefully and report the range of minimum and maximum values they calculate for each source. Workbook 5.5 in the Chapter 5 folder on the accompanying CD gives a spreadsheet guide to calculating minmax solutions for the common case of three sources but just one tracer.

Box 5.3. Minmax "Solutions" to Mixing Muddles

A common problem in mixing models is too many sources and not enough tracers. For example, if you have a sample that measures 5‰, and there are three sources at 0, 5, and 10‰, you may come to the frustrated conclusion that you can't solve for the source contributions. In this example, it is equally possible that the sample represents a 50/50 mix of the two end-member sources, 0‰ and 10‰, or alternatively, that neither of those two sources contributes at all, but that the sample is formed exclusively from the 5‰ source. You grimace, frustrated, and realize that the system is underdetermined, and there is no truly unique solution.

However, in such cases the isotopes still can help provide some useful constraints, especially estimates of the minimum and maximum possible contributions from each source. Not a unique solution, but information that narrows down the range of possibilities. In the too-many-sources cases, there are many alternative source mixes that can produce the same result (there are actually an infinite number of solutions to the simple problem outlined above, such as the following splits among the three sources: 50/0/50, 0/100/0, 1/98/1, 10/80/10, 33.01/33.98/33.01, etc.), but we don't know which one is the correct one. So, we punt, and just report the minimum and maximum (or minmax) values possible for each source, hoping that the next generation of scientists will work on further narrowing this minmax range towards the true correct answer.

The minmax solutions for the problem above are: 0–50% for the 0‰ source, 0–100% for the 5‰ source, and 0–50% for the 10‰ source. We learn from these ranges that contributions from the two end-member 0‰ and 10‰ sources cannot exceed 50%, perhaps useful information, but we actually learn nothing useful about the middle 5‰ source, where contributions are just somewhere in the total range of possibilities, 0–100%. Working with these minmax solutions generally shows that as in the example presented here, sources in the middle are least constrained, whereas constraints are much narrower and often quite informative for the end-member sources.

Workbook 5.5 in the Chapter 5 folder on the accompanying CD lets you explore variations on the three-source, one-tracer problem outlined above: type in numbers for the three sources and your sample, and the spreadsheet will calculate the minmax contributions for each source. There is also a Web site (www.epa.gov/wed/pages/models.html) that helps you calculate minmax distributions for multiple sources and multiple tracers. A caution as you use that programming is to report the actual minmax range and not the mean of the calculated solutions. The real answer is somewhere in the minmax range, and may not be at all close to a mean value derived from statistical models used in that Web site.

Note: In the attached worksheet for three sources, the minmax solutions derive from two mixing equations:

$$f_1 + f_2 + f_3 = 1 \quad \text{and}$$
$$\delta_1 f * f_1 + \delta_2 * f_2 + \delta_3 * f_3 = \delta S * 1,$$

where f gives the fractional contribution of each of the three sources denoted by subscripts 1 to 3, and the second equation gives the weighted average mixing for the three sources with isotope values of δ_1, δ_2, and δ_3, and δS is the isotope value of the sample. With two equations and three unknown fractional contributions, it would seem there is no solution. But you can sequentially set each of the fractions equal to zero, and then you have two simpler equations that can be solved. For example, if $f_1 = 0$, then the equations are $f_2 + f_3 = 1$, $\delta_2 * f_2 + \delta_3 * f_3 = \delta S * 1$, and the solutions for f_2 and f_3 are $f_2 = (\delta S - \delta_3)/(\delta_2 - \delta_3)$ and $f_3 = 1 - f_2$. Doing the parallel exercises for $f_2 = 0$ and then $f_3 = 0$ yields three sets of solutions for f_1, f_2, and f_3, and minmax solutions can be obtained from examining the solution sets in which all three fractions are in the 0 to 1 range. Although this sounds a bit technical, it is easy. Have a look at workbook 5.5 in the Chapter 5 folder on the accompanying CD.

3. *Multiple tracers.* An often successful way to resolve mixing muddles stemming from too many sources is to measure more tracers to resolve more sources. This follows the logic, if one tracer separates two sources, two tracers separate three sources, three tracers separate four sources, and so on. Logic seems to have less to do with following this approach than price tag, with more analyses costing more money. But costs aside, resolution generally does improve with multiple tracer measurements from the same sample.

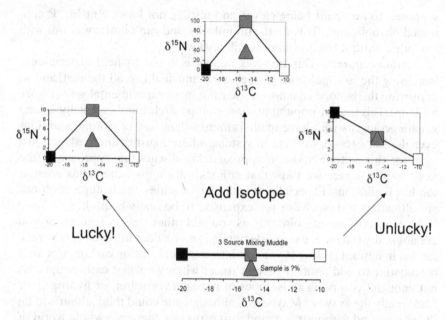

FIGURE 5.17. Mixing models and muddles. Bottom graph shows mixing muddle where there are three sources and no unique solution for source contributions to the sample, which is shown as a filled triangle and sources are depicted as squares. To resolve the muddle, one can measure another tracer, gaining resolution if lucky (left middle) or not gaining resolution if unlucky (right middle). A surer way to gain resolution is to add isotope artificially to one source (top). All sources contribute equally to the sample in these examples.

Let's illustrate these problems. Imagine a case where you start with using just one tracer, say the carbon isotope measurements. You start with two distinct sources at –20 and –10‰ and a sample that is in the middle at –15‰. The solution is easy, a 50/50 split. But then add in a third source with a value of –15‰. Suddenly, you don't know the answer any more (Figure 5.17 bottom). The answer could still be the same 50/50 split, or it could be 100% the –15‰ source, or it could be 1/3 contributed from each of the three sources; you just don't know what is right. This is definitely a mixed up muddle!

But what about measuring a second tracer to resolve this? Well, it depends if you are lucky or unlucky. If you are lucky, the three sources now form a mixing triangle with the sample in the middle (Figure 5.17 left) and it is easy to see the right answer: each source contributes about 1/3 of the total. But if you are unlucky, the sources again fall on a straight line and you still don't know the answer (Figure 5.17 right). So, roll the dice and see what Lady Luck brings!" (The audience hated this part because science is supposed to be about being clever and testing, not luck). But Mr. P. continued on, oblivious. "If you are still unlucky, and can't find a way out with any other natural tracers, there is still a way."

4. *Adding tracers.* This involves adding isotopes to field experiments, increasing the isotope value of one source much above all the rest, and so improving the isotope signal-to-noise ratio in an experimental way (Figure 5.17 top). This is quite popular in some isotope circles, and especially among stream ecologists who have made rather brilliant use of isotope additions over the last decade, working in systems where logistics and costs are just right for this isotope addition approach. We discuss this approach in the next chapter, where we show that although this approach sounds great, it too has limitations. Especially at some larger scales, the isotope additions are ultimately too patchy or too expensive to be really helpful.

Also, a purist might point out, as you add other tracers, sometimes you are not really tracing the same element any more: does nitrogen really track carbon in natural cycles? If you are interested in tracking carbon, wouldn't it be better to add another carbon tracer when your first carbon tracer is not enough? Is turning to isotopes of nitrogen (or sulfur, or hydrogen, or lead) really the answer? Maybe not, although one could think about adding ^{14}C as a second carbon tracer, and as it turns out, there is a whole world of natural abundance ^{14}C just waiting to be explored (Schell 1983), once prices come down.

"But for now, most ecologists are slipping by, measuring those extra tracers, then hoping those extra tracers also track the carbon. This can be a slippery slope." Mr. Polychaete is wheezing here, showing his age and tiredness at this litany of woes. He tries to sum up, but it isn't easy, for he is a rambling sort of lecturer. He puts up this quote from Peterson (1999) for the audience to think about while he collects his thoughts for a final conclusion.

It is important to appreciate that stable isotope techniques are most effectively used (in benthic food web studies) in combination with other techniques such as feeding experiments, gut content analysis, behavioral studies and process rate measurements. This is worth noting because there is a tendency to believe that isotopic tracer techniques are sufficiently powerful that they can, by themselves, provide relatively quick and reliable information on benthic community trophic relationships. *In my experience, this belief is not valid.* Isotopes can address some aspects of these issues, but not others, and normally a combination of techniques appropriate to a particular question will ensure the most robust conclusions.

Conclusion

In the end, there is some art and wisdom in all this business with mixing models. The art is keeping track of the signal-to-noise ratios, and being clever enough to find and recognize situations where this ratio is high, then dig in at that point and do a solid study. If you are forced into a corner and stuck with a system with little signal, then reinvent yourself as an isotope artist, expanding the palette of approaches, or manipulating and directing the action. The newborn artist starts using multiple tracers, or paints with added isotopes to lend color and context to an otherwise boring field study. And, as pointed out twice, both at the beginning and again at the end of this lecture, don't rely on isotopes alone, but see how the isotope results fit into a broader approach that has multiple lines of inquiry.

Mr. Polychaete concludes in a low mumble. "For those who are interested in more about how to really apply these mixing models, there is a practical handout about food webs and four I Chi model handouts on the table at the front of the classroom." (These handouts are presented as the next four sections of this chapter, Sections 5.6 to 5.9, and the last handout is extended problem 10 for Chapter 5 on the accompanying CD). The I Chi examples deal with rivers, mud, marshes, and tuna. These are opportunities to practice your I Chi mixing skills. A little practice and I Chi modeling in advance will make you think about where you might have problems and errors in your own field projects. Anticipating problems is usually more valuable than focusing in depth on the somewhat prosaic errors discussed in this lecture. "Will you be stumped by all the problems, or learn to take them in stride and become a Mixmaster Extraordinaire?" With that question left hanging in the air, he turns and shuffles off, saved by the bell that sounds for the next class.

5.6 River Sulfate and Mass-Weighted Mixing

Here is a story that features an isotope mixing mistake. Getting caught in a mistake is more common than you might think. Be patient as you read and see if you can help resolve the situation.

Suppose we were tracking sulfate pollution of rivers. We become especially interested in how the various sewage treatment facilities all along the Mississippi River contribute to a growing sulfate load as the Mighty Muddy Mississippi winds its way down the mid-section of North America. Could isotopes help? You aren't sure. You start to read and think more about this. Well, there is not much sulfate removal going on in the main part of the river, so we are dealing mostly with tracking new inputs, such as an input from a sewage (wastewater) treatment facility. It sounds like we are adding two kinds of sulfate, source 1 natural sulfate and source 2 wastewater sulfate. And each of those sources is bound to have an isotope value. This sounds good so far. So, the river might start upstream with a low sulfate concentration that has some unknown isotope value. How would concentrations (masses) and isotopes change as the river moves past a wastewater facility? This is clearly a mixing problem where both masses and isotopes are on the table.

To tackle this problem, we remember that the concentration in a river does not really give the total amount that passes a treatment plant each day. You have to multiply the concentration by the flux of water to get what is called the daily load, or the total amount per day. Let's say we did that, and found that 10 tonnes (10,000 kg) of sulfate is passing our sewage facility each day. How does this compare with the daily discharge of our large municipal wastewater treatment facility? Here, let's choose a number for daily load (amount) added to the river, say 0.25 tonnes (250 kg). These kinds of numbers are usually known, as you find out by asking the engineers working on the river, but you find also that no one knows the isotope numbers for sulfate. So, donning your isotope hat, you go down to the river, and take three samples of river water: a sample from just upstream of the wastewater discharge, a sample of the discharge effluent, and a sample of river water about 2 km downstream of the discharge point, the 2 km site being a point where you estimate the wastewater has been well mixed into the huge river. Just for fun, back in the lab, you also make a 10:1 mixture of river water:effluent water, to check that these mixing models really work. You pack up the samples and send them off for analysis.

Results come back from the lab in two pages. On page 1, you find sulfur isotope values of the sulfate in the upriver and wastewater samples listed as −10 and +5‰. You know the loads and amounts involved in kg sulfate per day, so you can substitute into the mass-weighted mixing equation to make a prediction about what the mixed downriver sample should be:

$$\delta_{SAMPLE} = (m_1 * \delta_{SOURCE1} + m_2 * \delta_{SOURCE2})/(m_1 + m_2)$$
$$\delta_{SAMPLE} = [(10{,}000)*(-10)+(250)*(5)]/(10000+250) = -9.63‰$$

and doing a similar substitution for the 10:1 mixture you made in the lab

$$\delta_{SAMPLE} = [(10)*(-10)+(1)*(5)]/(10+1) = -8.63‰.$$

Now that you have your prediction, you turn to page 2 of the lab results, and find δ = −9.5‰ for the downriver samples (a close and good agreement with your predicted value of −9.63‰, great!), but −2.5‰ for the 10:1 laboratory mixture you made, a value that doesn't agree well with your −8.6‰ prediction or with your stomach!

What did you do wrong in the lab? You don't know right away, so you puzzle along for a while. Isotopes seem right in the natural system, but if they are so great, shouldn't they work in the lab too? You ponder this, but can't crack the mystery over the next two weeks, and finally decide at an instinctive level that you don't trust isotopes. Being the practical person that you are, you move on to other ways to track wastewater effluents. Time marches on.

A year later, you are at a national-level science meeting and see a session full of isotope talks. In a fit of curiosity and bravery, you decide to introduce yourself to one of the more harmless-looking of these folks, and outline your river puzzle. You are shy at first, but become more comfortable as you begin talking about something you love, sulfate. You say that this isotope approach looked really promising for tracking wastewater sulfates in the Mississippi, but, you add a little sheepishly, the lab experiments didn't work out and so really you haven't done any more work with isotopes. "What do you think?" you hesitatingly ask the expert. "Is there something that I missed?"

The isotope expert (isotope guru or "isogoo") nods sympathetically and tries a few questions, then you both write down the isotope values and amounts on a piece of napkin there at the coffee break bar. Suddenly a light goes on for the isogoo expert. "The concentration," she says, "the concentration is important in the lab!" Yes, in the field, the amounts are given by the load (concentrations times the volume of water in the river), but in the lab, the amounts are a little different (concentrations times the volume of water in the mixtures you made). You need the concentrations to calculate the amounts of sulfate present in the lab experiment. She looks up at you with building triumph in her eyes, and asks, "What were the actual concentrations in the river water and in the wastewater effluent?" You can barely remember those old numbers, but they are somehow there, etched in your memory, and they inch forward into your consciousness and then parachute out of your mouth. The concentrations were $400\,mmol\,m^{-3}$ in the river and $4000\,mmol\,m^{-3}$ in the wastewater. "Aha!" says the isogoo, "let's write the mixing equation again, adding W terms to denote the concentration weightings for the sources:

$$\delta_{SAMPLE} = (m_1 * W_1 * \delta_{SOURCE1} + m_2 * W_2 * \delta_{SOURCE2})/(m_1 * W_1 * + m_2 * W_2).$$

Now put in the numerical values, remembering that you used 10 parts river water at $400\,mmol\,m^{-3}$ (4000 total units of sulfate) and 1 part wastewater at $4000\,mmol\,m^{-3}$ (4000 units again):

$$\delta_{SAMPLE} = [(10)*(400)*(-10)+(1)*(4000)*(5)]/[(10*400)+(1*4000)]$$
$$= -2.5‰.$$

Your predicted δ value should have been $-2.5‰$, not $-8.63‰$, and what did you actually measure?" she asks. Unfortunately, at this climactic moment, you find you can't remember that isotope value from page 2 of the results, but you promise to let this helpful isogoo person know, once you get home. You shake hands, and wander your separate ways, wondering will the veils of uncertainty descend again? Once you reopen those old lab reports, will the answer still be wrong? Is isotope doom imminent?

So, a few days later, back home, you do find that old data sheet mouldering in the files, and there it is, on page 2, $\delta = -2.5‰$! Just as predicted. Your first reaction is actually this: this isotope stuff is really incomprehensible. It didn't work, and now it does! Who can trust something like that? But then you start thinking that it is definitely cool that there is an agreement here, so you decide to go home and let this settle overnight.

By the next day, you are clear-headed, and just realize that you made a mistake. It wasn't the first time, after all, and it is nice to know that isotopes were working well all along. So you get in touch with the kindly isogoo, apologetically at first because of the mistake, but you find a flood of nurturing comments on the other end of the phone, to the effect that you had a new and interesting idea, that no one ever thought of trying before, wonderful creative innovative explorer of the universe you, and you shouldn't let a little mistake bog you down. (All this bubbly support comes from one of those really satisfying experiences in science, the experience of really solving something so that things that were once murky are now clear.)

The following day, you find that this conversation has settled extremely well with you, and you sit down to try to puzzle out some predictions for a stretch of the river you know fairly well, the middle river. You use a spreadsheet and try to use the mixing formulas to predict sulfate dynamics in the river. Keeping it simple, you assume 10 large municipalities located like beads along the river, one every 100 km, with each municipality having a wastewater facility that adds a relatively small amount of water with high sulfate concentration and high isotope value. In detail, the assumptions you make are that the wastewater additions are the same for each municipality, that each adds 100 units ($m^3 s^{-1}$) of water, with the water having a sulfate concentration of 4000 mmol m^{-3} and an isotope value of +5‰.

What you find out when you put this in a spreadsheet and use the mass-weighted isotope formula is a simple and straightforward pattern (Figure 5.18). The relatively small wastewater inputs have fairly minor effects on the total amount of water because this is a big river that already had substantial flows. Water fluxes change <10%, from 15,000 to 16,000 m^3/s (Figure 5.18, top). But because the sulfate concentration is fairly low in the river and very high in the wastewater, there is a much stronger effect on sulfate concentrations, with sulfate concentrations nearly doubling over the 1000 km (Figure 5.18, middle). Finally, as the concentrations change, so do the isotopes, on their way up towards the +5‰ value of the wastewater source (Figure 5.18, bottom). Yep, add those chocolates (oops, sulfates) and watch the isotope color change, just like Mr. Polychaete said. You can see that the isotopes are providing additional information about the mixing dynamics, and that they will help you check your sulfate balances. And overall, you are pleased as a scientist because you have some predictions that you can test with field work.

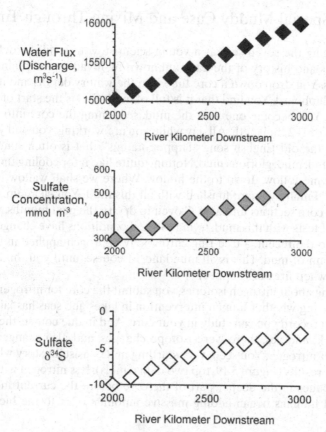

FIGURE 5.18. Hypothetical water flux, sulfate concentrations, and sulfate isotopes along a stretch of the Mississippi River.

So, within a few days time, you decide to set out next summer to visit the 10 wastewater treatment facilities, collecting data on water discharges, sulfate concentrations, and sulfate isotopes. Later that winter, the data come back, and you see that this is really working well. You have to adjust the predictions for the actual input values, and after doing this, the data fit pretty well, except in some interesting spots where you suspect sulfate removal in wetlands or extra sulfate additions from nonpoint source agricultural inputs. You eventually expand your sampling program to include more detailed sampling in these areas, and also in pristine and polluted tributaries. It works very well overall, and you conclude by publishing a nice scientific article about sources and accumulation of sulfate downstream in the Mississippi River. Late at night, you dream this paper brings you a big raise, with fame reaching out to all the galaxies of the known and unknown universe. After all, what is life without dreams?

5.7 A Special Muddy Case and Mixing Through Time

While sailing the seven seas as a young scientist, you decide to delve into the depths and history of the dark unknown by invading the realm of the sediments. You drop down a core tube into the watery depths and it comes back up shipboard. You find that it is full of mud. This is the start of a mud-filled day. You become one with the mud, sectioning the core into histori-cal tidbits of 1–2 cm length. All day while you are working, you find yourself humming the old English song "Hippopotamus" that is often sung in the round: "Mud, mud, glorious mud, Nothing quite like it for cooling the blood, So follow me follow, down to the hollow, Where we shall wallow in glori-ous mud." Finally you are finished with all this mud. You clean up and put the sliced core sections into a 60°C oven to dry. In the next months, you run all sorts of tests with this mud, trying to see if conditions have changed from the past to the recent, a tree-ring chronology approach applied in the ver-tical column of mud. This is all mundane, of course, until you think about isotopes, when fireworks appear.

Thinking about nitrogen isotopes, you submit the core for nitrogen analy-sis, wondering whether human intervention in lakes and seas has laid down a nitrogen record you can study in your core. And in due course, the results come back, and yes, there were isotope changes, and also changes in the amount of nitrogen. You see a very startling and consistent story when you graph the results (Figure 5.19, top two data panels): less nitrogen with lower isotope values in the deeper part of the core before the current industrial age when humans began adding massive amounts of N to the biosphere,

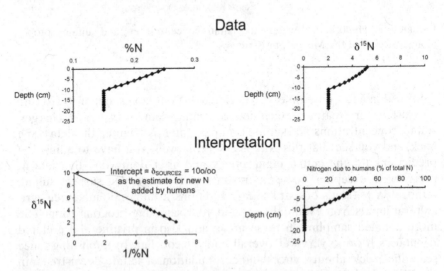

FIGURE 5.19. Nitrogen concentrations and isotopes in an hypothetical lake core (top panels), and interpretation of the source contributions from mixing model equa-tions (bottom panels).

and in the more recent top sections, lots of nitrogen with higher isotope values. This definitely seems like an isotope mixing problem, but where are source 1, source 2, and sample in all this N data? Can we use the basics of isotope sourcery to help sort this out?

Let's begin. Source 1, well, that would be the background conditions. No problem, we say, lots of deep core with no changes anywhere in amounts of nitrogen or isotopes of nitrogen. That nails source 1 for masses and isotopes; source 1 values will be the average of all those preindustrial values. What about source 2? That is harder, because the rest of the core, the upper core affected by humans, is actually the sample, and not source 2. So, now this is just confusing, until you finally look again at those source mixing equations really carefully:

$$\delta_{SAMPLE} = [(m_1)*(\delta_{SOURCE1})+(m_2)*(\delta_{SOURCE2})]/(m_1 + m_2).$$

Doing this, you count five items in this equation, and realize (finally) that you know four items, so should be able to solve for item 5, which is the isotope value of good old source 2. You also realize there is a convenient second equation, $m_T = m_1 + m_2$: just the total mass is the sum of the two masses contributed by sources 1 and 2. You rearrange this to $m_2 = m_T - m_1$ and combine equations:

$$\delta_{SAMPLE} = [(m_1)*(\delta_{SOURCE1})+(m_T - m_1)*(\delta_{SOURCE2})]/(m_T)$$

then rearrange so that you can solve for $\delta_{SOURCE2}$ at each point in the core:

$$\delta_{SOURCE2} = [(m_T)*\delta_{SAMPLE} - (m_1)*(\delta_{SOURCE1})]/(m_T - m_1).$$

This is a good result, and you can use it to solve your problem. But, in a flash of insight while working with all these equations, you find that you can solve across all the points with a different rearrangement of the same equation:

$$\delta_{SAMPLE} = \delta_{SOURCE2} + (\delta_{SOURCE1} - \delta_{SOURCE2})*(m_1/m_T).$$

The only variables in this last equation are actually δ_{SAMPLE} and m_T, so that this equation surprisingly has the form of a straight line, when (x, y) values are plotted as $(1/x, y)$. Most importantly, the intercept of this line is $\delta_{SOURCE2}$. So, now you can solve for $\delta_{SOURCE2}$ on a point-by-point basis, or by combining all data in a regression model.

You decide to use the regression model approach and find the isotope value for the second source is 10‰ (the intercept of line in bottom left panel of Figure 5.19). You also check this with the point-by-point calculations, and are relieved and comforted that you are on the right track when you find the same answer of 10‰. Then you go get a cup of coffee, and while standing in line, start thinking a subversive thought, which is this: so what did we learn that was extra from all this isotope work?

You sit down and slowly sip your coffee, thinking this through. Well, what you learned was that source 2 had an isotope value of 10‰, a value which you can check against other isotope values. Some time spent in the library and on the Web shows that 10‰ is a relatively rare and high value, but it is about what other scientists are finding in other systems affected by human pollution. It's the final nail in the coffin, the smoking gun. Really, by looking at the mass increases over background, you already suspected this core was in the human impact category, but having the high isotope value confirms human agents are responsible.

Later, it also occurs to you that you can use the isotopes to estimate the human contributions to the N in the core: now that we know the 10‰ source value, we can calculate the history of anthropogenic N deposition in this core (Figure 5.19, bottom right), and in other cores where the % N trends might not be so clear-cut. The next step would be to date sections of the core, so that we could assign a chronology to these interesting increases in N pollution.

As the years go by, you find that many scientists have used this kind of mixing analysis, wherever one source is fixed in isotopes and mass. It is used in analysis of blanks in lab situations, of CO_2 dynamics in forests and grasslands, in modeling nitrate dilution dynamics, and on and on. But learning this also makes you wonder about whether source 1 was really fixed in your mud work; that is, did the historical background really stay constant and not change its contribution through the more recent portions of the core? What if we took away all the human impact today, would you bet that the background would still be the same? "Perhaps, but perhaps not," you think to yourself, and realize that there is something slippery in these models, isn't there? Actually measuring the $\delta^{15}N$ value of source 2 would have been the key to really interpreting this example, to make sure it was 10‰ and not 6‰. If it were 6‰, then the human contribution would be much larger than estimated in Figure 5.19, so this is definitely worth checking in future work, you decide. You also realize that some of these isotope equations have assumptions in them, and you have to watch for errors creeping in from poor assumptions. Error creep: this sounds bad and is bad. You resolve to be more mentally alert and on your guard the next time.

But at the moment, we are just happy to have solved one of those rather difficult mixing problems where sources and amounts are involved. Other thoughts come to us later over beer, when we start thinking about those geochemical footprints down there in the mud. The results shown in Figure 5.19 are isotope (and mass) traces or tracks that we are leaving for future generations to know us by, something like the dinosaur tracks the paleontologists have discovered from past ages of the earth. We sigh and hope that in the future, we can reverse these pollution trends, so that the isotope legacy will also document our attempts to clean up the biosphere.

Note for Workbook 5.7 on Accompanying CD,
Chapter 5 Folder

This workbook implements a spatial I Chi approach for the mud core, and
the spatial approach used here (and in the next section) is a little different
from the time-based I Chi models used otherwise in this book. The difference is that in these spatial models, there is no time-based wraparound from
one section of the core to the next. (The time-based wraparound step is
described in Section 4.4, step 6). Instead, there is a different initial condition for each section of the core. In this case, the different initial condition
in each row is a different amount of human nitrogen that can be determined
by the reader. The subsequent equations of the row use this initial condition for that row to calculate the total amount and isotope value of the
nitrogen, but these results do not wrap around and influence the calculations of the next row. Thus, the model simulates N deposited sequentially,
without disturbance from processes such as bioturbation or diagenesis that
would operate across core sections.

5.8 The Qualquan Chronicles and Mixing
Across Landscapes

A practical concern for coastal managers concerns protecting salt marshes.
These beautiful systems are composed of salt-tolerant grasses that flourish
at the tide's edge. Unfortunately, this is also a lovely place for building vaca-
tion houses and marinas, so, as you could imagine, the natural marshes are
rarer each year. The save-the-marsh conservationists want to prove these
marshes are valuable, and seemingly have lots of evidence. They see lots of
big plants growing fast in luxuriant meadows, and these natural systems are
in fact tremendously productive, on a par with fertilized agricultural
systems. What these ecologists don't see is that big plants are there because
the plants are not good to eat. The quality is poor, even though the quantity is high, a quality–quantity or "qualquan" puzzle for grazers. And in
fact, plants are involved in coevolution with animal grazers, trying to
achieve impalatibility while a few hardcore grazer species evolve to defeat
each new antigrazer defense posed by the plants. At the time of this writing,
it seems that crab predators control some of the strongest snail grazing,
keeping grazing pressure low and allowing luxuriant development of plant
communities.

But other questions remain for managers, whether the marsh plants are
really necessary for coastal systems, or whether the marshes can be replaced
by marinas and housing developments. Over the last decades, marsh ecologists have begun using isotopes to test linkages between marsh plants,
grazers, and the wider food web of estuarine bays. The record is mixed, with
the isotope evidence tilting away from a strong food web connection

between marsh plants and most coastal consumers. This is unpopular with save-the-marsh conservationists, but some business interests are happy with these development-friendly isotope answers. However, our conservationist buddies are definitely thinking bad karma thoughts our way, especially, "Why can't those isotopes get the right answer anyway? The wrong answers give us the salt marsh blues."

Can an isotope mixing model help us solve the qualquan puzzle, and show important ecological linkages that exist in these tidal marsh systems? Here we consider a qualquan mixing model that mixes two kinds of plant foods to support coastal food webs and fisheries.

But first, let's find out more about the coastal marsh environment. The main plant food source in the sea is phytoplankton, small floating algae that are abundant in coastal waters. Consumers usually rely on phytoplankton, but in coastal areas, consumers can also rely on organic matter produced by the marsh grasses, which we consider as *Spartina* (cordgrass) from temperate salt marshes. Living *Spartina* is not grazed heavily (so that is why we see it waving in beautiful meadows in the breeze), but enters food webs after it dies as dead plant material (detritus). Microorganisms such as fungi and bacteria are active decomposers of *Spartina* in the detrital food webs. As these microbes colonize and feed on the detritus, they blow off much of the energy in the food as respiration, a kind of ecological tax. This tax leads to inefficient transfer of the *Spartina* plant energy into food webs, so we can think of the *Spartina* as contributing higher quantities of food, but food that has lower quality or lower energy yield. A unit of *Spartina* production might be good for feeding bacteria, but there is only a small amount of energy left over after the eco-taxes from the microbes. In this case, very little energy would be passed on to fish that many humans care about most. Further coastal nuances are that some *Spartina* passes into food webs in dissolved form (DOM, dissolved organic matter or DOC, dissolved organic carbon), and some in particulate form (POM particulate organic matter or POC, particulate organic carbon), with the marsh contributions strongest in the marshes, but declining seawards outside the marshes.

Now that you are an expert in coastal marshes, let's consider the weighted average mixing model, and do the qualquan (quality–quantity) dance, the qualquan cancan. The qualquan model is just a weighted average played out across a coastal landscape. How do you do this? It's a three-step dance. Step 1. First, set up some typical quantities and qualities for the phytoplankton and *Spartina* food sources. For example, let's assume that typical productivities for *Spartina* export in terms of POC and DOC are respectively 50 and $50 \, g \, C \, m^{-2} \, yr^{-1}$, the respective phytoplankton values are 250 and 25, and relative qualities for supporting the larger food web leading to fish are 0.1 and 1. Step 2. Then assign a virtual isotope value (a qualquan isotope value that ranges from 0 to 100) to each of these food sources, say 100‰ for *Spartina* and 0‰ for phytoplankton. Step 3. Finally, do the weighted averages for landscapes that have different mixtures of marsh and open water, using W to denote the various relative weightings:

Weighted qualquan δ value =
$$(W_1 * W_2 * W_3 * \delta_1 + W_4 * W_5 * W_6 * \delta_2)/(W_1 * W_2 * W_3 + W_4 * W_5 * W_6).$$

Let's remind ourselves of how this fits together. Consider the POC pools for a piece of estuary with 1 unit of marsh and 1 unit of open water. There will be $50 * 0.1 * 1$ units (productivity * quality * area units $= W_1 * W_2 * W_3$) of Spartina and $250 * 1 * 1$ units (productivity * quality * area units $= W_4 * W_5 * W_6$) of phytoplankton available to the food web. We multiply these amounts by their qualquan isotope values to obtain $(50 * 0.1 * 1) * (100$ isotope units) of Spartina and $(250 * 1 * 1) * (0$ isotope units) of phytoplankton, then divide by the sum of the quantities for the weighted average. For the qualquan cancan numbers used in this paragraph, the value of the weighted qualquan δ value is $(500)/(5 + 250) = 1.96‰$ or 2‰. This 2‰ value is very close to the assigned 0‰ value of phytoplankton, and means a near-zero contribution for Spartina when there is an equal mix of marsh and open water area.

This rather dramatic result encourages us make some further calculations with the qualquan model. We begin investigating some thought experiments that develop spatially across the landscape, rather than sequentially through time. First we can consider the simpler case where quality does not come into play, and we are looking only at the total quantities of POC and DOC. When we don't consider quality, and just think about quantities in the POC and DOC pools, then Spartina does contribute to local offshore carbon stocks, although more for DOC than POC (Figure 5.20, upper). The higher DOC contributions are due to higher DOC production by marshes (50 instead of 25 production units). Said another way, the marsh export of 50 DOCs goes a lot farther mixing against a phytoplankton pool of 25 DOCs than does marsh export of 50 POCs mixing against a phytoplankton pool of 250 POCs. As you move away from shore, the amount of open water increases and the amount of marsh declines, so that the open water:marsh ratio increases. Judged on this landscape scale, marsh importance for DOC or POC pools is already only moderate by the time there is twice as much open water as marsh (Figure 5.20, upper).

Now that we understand half the picture depicted in the upper part of Figure 5.20, let's add in some quality thinking, actually some thinking about poor food quality. If we assign phytoplankton a food quality of 100% and a lower food quality of 10% for detrital Spartina, we find that the food web based on particulates (and on particulates + DOC) does not favor Spartina, except in the marshiest, most nearshore areas (Figure 5.20, bottom). The low quality of the Spartina POC detritus rapidly diminishes the overall importance of the detrital Spartina as a food resource. DOC also suffers the

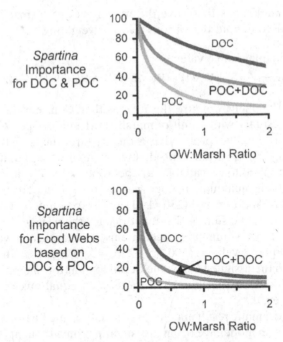

FIGURE 5.20. Hypothetical mixing results for two sources across an estuarine land-scape composed of open water (ow) and marsh, with the ow:marsh ratio varying from pure marsh (ratio = 0) towards higher values characteristic of mostly open water. The two sources creating the mixing dynamic are phytoplankton in open water and *Spartina* plant carbon exported from marshes. The *y*-axis scales the rela-tive contributions of the two sources. The numerical *y*-values show an index that varies from 0 to 100, and could be % contribution of *Spartina*, or also the ‰ value of samples in the hypothetical case that Spartina has a δ value of 100‰ and phyto-plankton a value of 0‰. Both phytoplankton and *Spartina* sources contribute to DOC (dissolved organic carbon) and POC (particulate organic carbon) pools. Top graph: phytoplankton and *Spartina* have equal quality for DOC and POC pools. Bottom graph: In food webs, *Spartina* is much less important because it has a ten times lower quality than phytoplankton.

ecological tax of bacterial processing, and so including DOC in the food web qualquan models does not greatly increase the calculated importance of marsh outwelling. Marsh outwelling is pretty much a bust by these calculations, calculations that perhaps we should have kept to ourselves. Well, sometimes you just keep getting the politically unpopular result, don't you?

But then again, no one has tested these models, so perhaps it is too early to really put our faith in these results. To work further on these matters, you can try out qualquan models in the I Chi spreadsheets of Workbook 5.8 on the accompanying CD and put in your own values; see what you get with isotope index values (*Spartina* = 100, phytoplankton = 0) or realistic $\delta^{13}C$ values (*Spartina* = –13‰, phytoplankton = –25‰).

And actually, there is an argument that if marshes are not important for carbon, then they still may be important for the overall nitrogen (N) cycle of estuaries. Tidal export or outwelling from marshes may supply inorganic and organic nitrogen compounds that are used by phytoplankton. The nitrogen supply often determines and fertilizes the overall rate of plant productivity in estuaries, so marshes supplying N could be important, especially in estuaries with low N inputs from rivers and other external N sources. The marsh outwelling connection to fisheries would be indirect, via phytoplankton stimulation, rather than via a direct subsidy of marsh organic matter ingested by fish, shrimp, and clams.

Not surprisingly, we can use the qualquan model to also investigate these potential N linkages. Using liberal (high) estimates of marsh N supply, there does seem to be a possible case that marshes are important, especially via the DOM export (Figure 5.21, top). However, in most estuaries, rivers currently supply the vast majority of new N entering the systems, so that over time, the marsh contributions to the N cycle would be miniscule (Figure 5.21, bottom).

So, what is the real answer? Are marshes important for estuarine food webs or aren't they? The models would suggest no, and that is the hard-to-take answer. But even so, we contact our conservationist friends with the results, and ask them to think carefully about other values of the marsh. For example, marsh detritus may not be important fish food, but marshes still may be important for sustainable fisheries because these shallow marsh areas are nursery refuges for small fisheries species seeking to escape larger predators. Although we are using this idea to blunt the bad news from the models, we find unexpected resistance from our conservationist colleagues who still are not convinced that the models are right. And they make an interesting point. They listen carefully, but then point out that perhaps we still do not have the right measurements to support the model conclusions. One important unknown concerns the amount of new nitrogen that gets added to estuaries. How much is really added via N fixation, and of the added N, how much is exported from the marsh? These are difficult questions to answer at the aggregated ecosystem level, but important for future management of the marshes. The I Chi advantage is that now we have an isotope mixing model that generates predictions about how C and N dynamics are linked between marshes and open water, predictions that can be tested during the next years in landscape-level studies. Slow though it is, this is progress.

Overall, this example has dealt with estuaries and marshes where there is a long history of using isotopes to investigate landscape-level mixing (Fry and Sherr 1984; Peterson et al. 1985; Gearing 1988; Peterson 1999; Currin et al. 2004). But more generally, quality and quantity of organic matter inter-

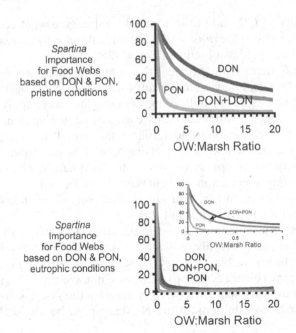

FIGURE 5.21. As previous figure, but for nitrogen stocks, not carbon. DON = dissolved organic nitrogen, PON = particulate organic nitrogen. See Figure 5.19 for explanation of the *x* and *y* axes. Top: maximum contribution scenario for *Spartina* exported from marshes, assuming high inputs of labile DON from marshes and low N inputs from other sources. Bottom: more likely scenario that shows marsh inputs are quite minor for food webs in most current estuaries that have high anthropogenic N loading, with high inputs of labile DON from marshes, but inputs from other sources that are 50 times higher than marsh inputs. The inset panel shows an expanded view of the results for Open Water:Marsh ratios of 1 or less.

act in many parts of the organic matter cycle, controlling dynamics in sediments and soils as well as in food webs. In this regard, qualquan modeling should help ecologists investigate the spatial dynamics of many other transitional regions such as ecotones between forests and grasslands or the nutrient cycling interface between canopy and understory plants.

Note for Workbook 5.8 in the Chapter 5 Folder on the Accompanying CD

This workbook implements a spatial I Chi approach for the qualquan landscape, and the spatial approach used here (and in the previous section) is a little different from the time-based I Chi models used otherwise in this book. The difference is that in these spatial I Chi models, there is no time-based wraparound from one row of the model to the next. (The time-based wraparound step is described in Section 4.4, step 6). Instead, there is a different initial condition for each row, that is, a different value for the open water:marsh ratio. The subsequent equations of the row use this initial condition to calculate qualquan results, but these results do not wrap around and influence the calculations of the next row. The consequence is that Figures 5.20 and 5.21 compile qualquan results into smooth trends using many individual landscapes that differ in their open water:marsh ratios. These figures do not portray results for a single landscape with linked production and consumption processes.

5.9 Dietary Mixing, Turnover, and a Stable Isotope Clock

Nature has written ecological histories with the invisible ink of stable isotopes. Once made visible, this isotope record can be found in many hard materials such as rocks, claws, scales, hair, and whale baleen. Ecologists are beginning to use such materials to assess the historical development of populations and ecosystems, and to establish historical reference baselines useful in restoration ecology.

But sometimes there are no hard materials available for historical study. So here we turn our attention towards soft tissues that via their different metabolic activities record the visceral dynamics of changing lifestyles and individual histories. The difference in rates of tissue turnover leads to an isotope clock, and this section considers an example of how this works. Mind you, this is a virtual experiment, never validated in the real world, but that does not stop Mr. Polychaete.

In this thought experiment, a small tuna gets bored one day and swims to a new ocean home where the diet is different and better for fast growth. With this new diet, the weight goes up and up over the next year (Figure 5.22, top panel). We have X-ray vision and keep track of the weight gain at different levels, including the muscle, liver, and blood plasma. Weight goes up over the year, increasing 3.5 times in all tissues that have very different starting points. Initial weights are large, medium, and small for the muscle, liver, and plasma, respectively, and turnover rates for these tissues are reversed; slow, medium, and fast. This means that the muscle accumulates most of the new mass over time with relatively little loss, blood plasma is a low-mass, fast-cycling pool that rapidly reflects the new diet, and liver is in the middle.

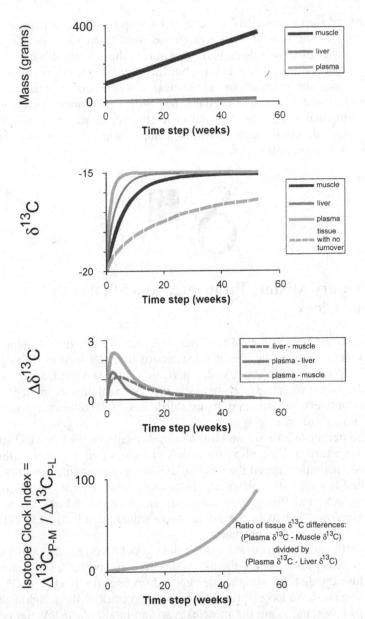

FIGURE 5.22. Hypothetical growth and isotope changes for a tuna that is switching diets. Muscle, liver, and blood plasma are three tissues that turn over at, respectively, slow, medium, and fast rates as the tuna grows. The new diet has a different isotope label than the starting diet, and differential turnover in the three tissues leads to different isotope changes with time. Isotope changes can be used to construct a clock that times progression of the turnover (bottom).

So now picture the accompanying isotope parade, that secret procession normally writ invisible. Thanks to virtual technology, we have an "isoscope" instrument that resolves all. Isotopes can run, but they can't hide from our prying isoscope detector. We find that the muscle is slow to equilibrate with the new diet, the blood plasma is very quick to equilibrate, and the liver is in the middle (Figure 5.22, second panel). The three tissues are humming along at different rates. Our virtual case has liver metabolizing two times faster than muscle, and plasma turning over faster still, two times faster than liver. These metabolic turnover rates really accelerate the isotope dynamics compared to what occurs when no turnover occurs (Figure 5.22, second panel).

These isotope turnover dynamics have been known for over 20 years (Fry and Arnold 1982; Tieszen et al. 1983), but now let's use this information as a clock. Let's subtract isotope values to magnify the relative metabolic differences among the tissues, and sure enough, weird curves result: the isotope disequilibrium curves (Figure 5.22, third panel). These curves are near zero, the equilibrium setpoint, at the start of growth when tissues are in equilibrium with the old diet, and again at the end, after the year of new growth when they are in equilibrium with the new diet. In between, there is a time when fast turnover tissues are mostly equilibrated with the new diet and slow turnover tissues are still mostly equilibrated with the old diet, and this leads to the isotope disequilibrium curves.

The final step is using the isotope disequilibria to clock the diet change. Measuring only one tissue will yield a clock estimate if you know the starting and final diet isotope values, but usually you don't know these diet isotope values for field situations. In these cases, isotope comparisons between tissues, the disequilibrium comparisons, allow an estimate of time elapsed since a diet switch began. Dividing one disequilibrium value by another, we find a nicely behaved curve relating time to isotope dynamics (Figure 5.22, bottom). This result also allows the reverse, using isotope information (y values, Figure 5.22 bottom) to estimate time elapsed (x values, Figure 5.22 bottom), or the isotope clock. Isotope sourcery has struck here, and because the ups and downs of isotope disequilibria have cancelled out; we have a clock—amazing, isn't it? We might not have predicted the complex variables that give the clock (y axis in Figure 5.22, bottom), but it is fundamentally the different tissue turnover rates that propel the regular motion of this clock.

We come back to this perhaps puzzling result in a moment, but first, let's review this sequence of events. The dietary changes set up a cascade of isotope dynamos running at different rates in the different tissues. The isotope differences among tissues clock the turnover dynamics, starting with no change and ending with no change. All's well that ends well. In the middle, isotope disequilibria get big, then small, and the way these disequilibria play out gives a clock. Having multiple isotope measurements from a single individual is key to making these clock estimates.

Today's technology will allow you to use a gas chromatograph system to measure isotopes in 20+ different biochemicals (lipids) from an animal within a single two-hour analysis. With 20+ numbers to work with instead of just three from plasma, liver, and muscle, the number of potential isotope clocks becomes enormous. With this big improvement, we should be able to create order out of all the seeming chaos of isotope disequilibria, and hindcast ecohistory with considerable refinement. What was eaten and when it was eaten should emerge from this approach, a way to conduct detailed dietary interviews for animals that are normally silent about their pasts. Animals spend much of their time finding food and eating, so using an isotope clock interview technique could tell us a lot about how animals ov (including humans) behave.

Stable isotope clocks based on tissue turnover rates are being used to help estimate animal migration routes and strategies, and which foods support consumers as they switch diets. An overall equation for isotope change in a rapidly growing animal that switches from an initial (I) diet to a final (F) diet is

$$\delta_{ANIMAL} = \delta_F + (\delta_I - \delta_F) * (M_T / M_I)^c,$$

where the size of the animal is given as a ratio of mass at time T (M_T) relative to an initial mass (M_I) measured when the diet change started. The exponent c is a metabolic turnover factor, with $c = -1$ indicating no turnover of the initial mass (Fry and Arnold 1982). More complex isotope turnover equations have been developed (Hesslein et al. 1993), based on bioenergetics (Harvey et al. 2002), and for those cases when multiple nutrient pools contribute to turnover (Ayliffe et al. 2004; West et al. 2004).

Problems 8 to 10 in the Chapter 5 folder on the accompanying CD ask you to develop an I Chi model to further investigate these isotope turnover dynamics associated with diet changes.

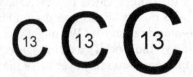

5.10 Chapter Summary

Chapter 5 deals with mixing, a seemingly simple subject. The magic of isotopes usually resides in mixing, where sources combine or mix to form a sample, and isotopes record the fractional contributions of the sources. The isotopes index and record the sources, so that mixing is isotope sourcery in action.

As you might expect, learning sourcery is not always fast or easy, and Section 5.1 considers a long-term estuarine seagrass project where sourcery worked sometimes and failed other times. The research investigated the importance of detrital seagrasses in the diets of small consumers, part of a conservation-minded effort to help preserve the beautiful coastal seagrass meadows. These underwater grasslands, like the salt marshes of Section 5.8, are often threatened by coastal housing developments and marinas. The isotope sourcery failures were rather typical, occurring when there were too many sources and not enough tracers, and when extra tracers still did not resolve important sources. The successes were also fairly typical, with comparative survey work yielding the most reliable results. Algal foods rather than detrital seagrasses were often the most important foods in seagrass systems. The isotope information was not obtainable by other methods, but still was not complete when considered in isolation. Simple additional experiments that involved watching what consumers actually ate proved invaluable in helping decipher the isotope information. The isotope information was most useful when coupled with other approaches.

With this real-life story of isotope sourcery and seagrasses as background, Sections 5.2 to 5.5 consider technical details of isotope mixing, the errors and problems as well as the successes. Section 5.2 gives an illustrated introduction to isotope mixing for all the HCNOS elements, and Section 5.3 gives the simple equations for weighted averages that underlie all isotope mixing. Section 5.4 considers more complex mixing dynamics associated with various kinds of weighted averages, especially those associated with concentrations. Keeping track of the mass-related mixing with various weighting factors is key to predicting correct isotope answers in more complex mixing situations.

Section 5.5 provides a short lecture on the pitfalls of the isotope approaches, introducing the idea that scientists can add isotopes to improve resolution and better solve ecological problems, an approach explored in detail in the next chapter. Propagating errors in mixing models emerges as an important constraint in Section 5.5, a constraint that makes many of the isotope mixing interpretations seem more qualitative than quantitative, or "semi-quantitative." Still, it is possible to use normal statistical and multivariate approaches to be quantitative with isotope mixing, and to check the reasonableness of isotope mixing results.

With this technical background of Sections 5.2 to 5.5 completed, the last sections of Chapter 5 consider four examples where isotope mixing is important for tracing and solving ecological problems. These four examples all have attached spreadsheets (see Chapter 5 folder on accompanying CD) for exploring different ways isotopes mix and circulate. Working through these examples helps practice using isotope sourcery to your advantage. The examples emphasize complex mixing related to concentrations of sulfate (Section 5.6) and marsh organic matter (Section 5.8), with isotopes showing ecological regions and boundaries. These examples show that isotopes

provide a novel nonintuitive perspective for studying and evaluating ecological landscapes. Mixing also occurs over time, and concluding examples
of this chapter concern the history of nitrogen deposited in sediments over
the last century (Section 5.7) and the timing of tissue turnover in animals
switching diets (Section 5.9). These examples show that stable isotope
studies can help distinguish the history and general timing of ecological
events, even though clocks and rates are usually considered the province of
radioisotopes rather than stable isotopes.

An important summary message of this chapter is that isotope sourcery
does not always work. The key requirement for isotope tracing is that
sources must have strong isotope differences, so that it is easy to sort out
which sources are most important in mixtures. If preliminary sampling
shows that sources are not different in a system you are interested in, you
may want to add isotopes to create differences among sources, a strategy
discussed in detail in the next chapter.

Further Reading

Section 5.1

Anderson, W.T. and J.W. Fourqurean. 2003. Intra- and interannual variability in seagrass carbon
and nitrogen stable isotopes from south Florida, a preliminary study. *Organic Geochemistry*
34:185–194.
Currin, C.A., S.C. Wainright, K.W. Able, M.P. Weinstein, and C.M. Fuller. 2003. Determination
of food web support and trophic position of the mummichog, *Fundulus heteroclitus*, in New
Jersey smooth cordgrass (*Spartina alterniflora*), common reed (*Phragmites australis*), and
restored salt marshes. *Estuaries* 26:495–510.
Day, J.H. 1967. The biology of Knysna estuary, South Africa. In G.H. Lauff (ed.), *Estuaries*.
American Association for the Advancement of Science (AAAS), pp. 397–407. Washington,
D.C.
Estep, M.F. and H. Dabrowski. 1980. Tracing food webs with stable hydrogen isotopes. *Science*
209:1537–1538.
Finlay, J.C., S. Khandwala, and M.E. Power. 2002. Spatial scales of carbon flow in a river food
web. *Ecology* 83:1845–1859.
Fourqurean, J.W., S.P. Escorcia, W.T. Anderson, and J.C. Zieman. 2005. Spatial and seasonal
variability in elemental content, $\delta^{13}C$ and $\delta^{15}N$ of *Thalassia testudinum* from South Florida
and its implications for ecosystem studies. *Estuaries* 28:447–461.
Fry, B. 1981. Tracing shrimp migrations and diets using natural variations in stable isotopes.
Ph.D. dissertation, University of Texas.
Fry, B. 1984. $^{13}C/^{12}C$ ratios and the trophic importance of algae in Florida *Syringodium* seagrass meadows. *Marine Biology* 75:11–19.
Fry, B. 1988. Food web structure on Georges Bank from stable C, N and S isotopic composi-
tions. *Limnology and Oceanography* 33:1182–1190.
Fry, B. and P.L. Parker. 1979. Animal diet in Texas seagrass meadows: $\delta^{13}C$ evidence for the
importance of benthic plants. *Estuarine and Coastal Marine Science* 8:499–509.
Fry, B. and E. Sherr. 1984. $\delta^{13}C$ measurements as indicators of carbon flow in marine and freshwater ecosystems. *Contributions in Marine Science* 27:13–47.

Fry, B., S.A. Macko, and J.C. Zieman. 1987. Review of stable isotopic investigations of food webs in seagrass meadows. *Florida Marine Research Publications* 42:189–209.

Fry, B., R.S. Scalan, and P.L. Parker. 1977. Stable carbon isotope evidence for two sources of organic matter in coastal sediments: Seagrasses and plankton. *Geochimica et Cosmochimica Acta* 41:1875–1877.

Gearing, J.N. 1991. The study of diet and trophic relationships through natural abundance ^{13}C. In D.C. Coleman and B. Fry (eds.), *Carbon Isotope Techniques*. Academic, San Diego, CA, pp. 201–218.

Grey, J., S.J. Thackeray, R.I. Jones, and A. Shine. 2002. Ferox trout (*Salmo trutta*) as "Russian dolls": Complementary gut content and stable isotope analyses of the Loch Ness foodweb. *Freshwater Biology* 47:1235–1243.

Harrigan, P., J.C. Zieman, and S.A. Macko. 1989. The base of nutritional support for the gray snapper (*Lutjanus griseus*): an evaluation based on a combined stomach content and stable isotope analysis. *Bulletin of Marine Science* 44:65–77.

Hyndes, G.A. and P.S. Lavery. 2003. Seagrass litter: Trash or treasure. *Gulf of Mexico Science* 21:111.

Kitting, C.L., B. Fry, and M.L. Morgan. 1984. Detection of inconspicuous epiphytic algae supporting food webs in seagrass meadows. *Oecologia (Berlin)* 62:145–149.

Lepoint, G., P. Dauby, and S. Gobert. 2004. Applications of C and N stable isotopes to ecological and environmental studies in seagrass ecosystems. *Marine Pollution Bulletin* 49:887–891.

Loneragan, N.R., S.E. Bunn, and D.M. Kellaway. 1997. Are mangroves and seagrasses sources of organic carbon for penaeid prawns in a tropical Australian estuary? A multiple stable-isotope study. *Marine Biology* 130:289–300.

McCutchan, J.H. Jr. and W.M. Lewis Jr. 2002. Relative importance of carbon sources for macroinvertebrates in a Rocky Mountain stream. *Limnology and Oceanography* 47: 742–752.

Melville, A.J. and R.M. Connolly. 2003. Spatial analysis of stable isotope data to determine primary sources of nutrition for fish. *Oecologia* 136:499–507.

Moncreiff, C.A. and M.J. Sullivan. 2001. The trophic importance of epiphytic algae in subtropical seagrass beds: Evidence from multiple stable isotope analyses. *Marine Ecology Progress Series* 215:93–106.

Moore, J.C., E.L. Berlow, D.C. Coleman, P.C. de Ruiter, Q. Dong, A. Hastings, N.C. Johnson, M.S. McCann, K. Melville, P.J. Morin, K. Nadelhoffer, A.D. Rosemond, D.M. Post, J.L. Sabo, K.M. Scow, M.J. Vanni, and D.H. Wall. 2004. Detritus, trophic dynamics and biodiversity. *Ecology Letters* 7:584–600.

Mumford, P.L. 1999. The effects of environmental stress and primary productivity on food chain length in Florida Bay. Master of Science thesis, Florida International University.

Mutchler, T., M.J. Sullivan, and B. Fry. 2004. Potential of ^{14}N isotope enrichment to resolve ambiguities in coastal trophic relationships. *Marine Ecology Progress Series* 266:27–33.

Parker, P.L. 1964. The biogeochemistry of the stable isotopes of carbon in a marine bay. *Geochimica et Cosmochimica Acta* 28:1155–1164.

Phillips, D.L. 2001. Mixing models in analyses of diet using multiple stable isotopes: A critique. *Oecologia* 127:166–170.

Phillips, D.L. and J.W. Gregg. 2001. Uncertainty in source partitioning using stable isotopes. *Oecologia* 127:171–179 (see also erratum, *Oecologia* 128: 204).

Phillips, D.L. and J.W. Gregg. 2003. Source partitioning using stable isotopes: Coping with too many sources. *Oecologia* 136:261–269.

Phillips, D.L. and P.L. Koch. 2002. Incorporating concentration dependence in stable isotope mixing models. *Oecologia* 130:114–125.

Yamamuro, M, H. Kayanne, and H. Yamano. 2003. δ^{15}N of seagrass leaves for monitoring anthropogenic nutrient increases in coral reef ecosystems. *Marine Pollution Bulletin* 46: 452–458.

Section 5.4

Case, J.W. and H.R. Krouse. 1980. Variations in sulphur content and stable sulphur isotope composition of vegetation near a SO_2 source at Fox Creek, Alberta, Canada. *Oecologia* 44:248–257.

Fry, B. 2002. Conservative mixing of stable isotopes across estuarine salinity gradients: A conceptual framework for monitoring watershed influences on downstream fisheries production. *Estuaries* 25:264–271.

Harrigan et al. 1989. Listed above; see Section 5.1 readings.

Krouse, H.R. 1980. Sulphur isotopes in our environment. In P. Fritz and J. Ch. Fontes (eds.), *Handbook of Environmental Isotope Geochemistry, Vol. 1, The Terrestrial Environment, A.* Elsevier, Amsterdam, pp. 435–471.

Krouse, H.R., A.H. Legge, and H.M. Brown. 1984. Sulphur gas emissions in the boreal forest: The West Whitecourt case study. V. Stable sulphur isotopes. *Water, Air and Soil Pollution* 22:321–347.

Peterson, B.J. 1999. Stable isotopes as tracers of organic matter input and transfer in benthic food webs: A review. *Acta Oecologica* 20:479–487.

Phillips and Gregg. 2003. Listed above; see Section 5.1 readings.

Phillips and Koch. 2002. Listed above; see Section 5.1 readings.

Winner, W.E., J.D. Bewley, H.R. Krouse, and H.M. Brown. 1978. Stable sulfur isotope analysis of SO_2 pollution impact on vegetation. *Oecologia* 36:351–361.

Section 5.5

Kendall, C. and J.J. McDonnell. 1998. *Isotope Tracers in Catchment Hydrology.* Elsevier Health Sciences, Amsterdam.

McCutchan, J.H. Jr., W.M. Lewis Jr., C. Kendall, and C.C. McGrath. 2003. Variation in trophic shift for stable isotope ratios of carbon, nitrogen, and sulfur. *Oikos* 102:378–390.

Peterson. 1999. Listed above; see Section 5.4 readings.

Phillips and Gregg. 2001. Listed above; see Section 5.1 readings.

Phillips and Gregg. 2003. Listed above, see Section 5.1 readings.

Post, D.M. 2002. Using stable isotope methods to estimate trophic position: Models, methods, and assumptions. *Ecology* 83:703–718.

Schell, D.M. 1983. Carbon-13 and carbon-14 in Alaskan aquatic organisms: Delayed production from peat in Arctic food webs. *Science* 219:1968–1071.

Vander Zanden, J.M. and J.B. Rasmussen. 2001. Variation in $\delta^{15}N$ and $\delta^{13}C$ trophic fractionation: Implications for aquatic food web studies. *Limnology and Oceanography* 46:2061–2066.

Section 5.7

Fry, B., W. Brand, F.J. Mersch, K. Tholke, and R. Garritt. 1992. Automated analysis system for coupled $\delta^{13}C$ and $\delta^{15}N$ measurements. *Analytical Chemistry* 64:288–291.

Mortazavi, B. and J.P. Chanton. 2004. Use of Keeling plots to determine sources of dissolved organic carbon in nearshore and open ocean systems. *Limnology and Oceanography* 49:102–108.

Voss, M., B. Larsen, M. Leivuori, and H. Vallius. 2000. Eutrophication signals in coastal Baltic Sea sediments. *Journal of Marine Systems (Special Issue)* 25:287–298.

Section 5.8

Boesch, D.F. and R.E. Turner. 1984. Dependence of fishery species on salt marshes: The role of food and refuge. *Estuaries* 7:460–468.

Currin, C.A., et al. 2003. Listed above; see Section 5.1 readings.

DeLaune, R.D. and W.H. Patrick Jr. 1990. Nitrogen cycling in Louisiana Gulf Coast brackish marshes. *Hydrobiologia* 199:73–79.

Eldridge, P.M. and L.A. Cifuentes. 2001. A stable isotope model approach to estimating the contribution of organic matter from marshes to estuaries. In M.P. Weinstein (ed.), *Concepts and Controversies in Marsh Ecology*. Kluwer Academic, Hingham, MA, pp. 495–513.

Fry and Sherr. 1984. Listed above; see Section 5.1 readings.

Gearing, J.N. 1988. The use of stable isotope ratios for tracing the nearshore-offshore exchange of organic matter. In B.-O. Jansson (ed.), *Lecture Notes on Coastal and Estuarine Studies*, Vol. 22. Springer-Verlag, New York, pp. 69–101.

Haines, E.B. 1976. Relation between the stable carbon isotope composition of fiddler crabs, plants, and soils in a salt marsh. *Limnology and Oceanography* 21:880–883.

Haines, E.B. 1977. The origins of detritus in Georgia salt marsh estuaries. *Oikos* 29:254–260.

Haines, E.B. 1979. Interactions between Georgia salt marshes and coastal waters: A changing paradigm. In R.I. Livingston (ed.), *Ecological Processes in Coastal and Marine Systems*. Plenum, New York, pp. 35–46.

Haines, E.B. and C.L. Montague. 1979. Food sources of estuarine invertebrates analyzed using $^{13}C/^{12}C$ ratios. *Ecology* 60:48–56.

Neff, J.C., F.S. Chapin III, and P.M. Vitousek. 2003. Breaks in the cycle: Dissolved organic nitrogen in terrestrial ecosystems. *Frontiers in Ecology and Environment* 1:205–211.

Nixon, S.W. 1980. Between coastal marshes and coastal wates—A review of twenty years of speculation and research on the role of salt marshes in estuarine productivity and water chemistry. In P. Hamilton and K.B. McDonald (eds.), *Estuarine and Wetland Processes*. Plenum, New York, pp. 437–521.

Page, H.M. 1997. Importance of vascular plant and algal production to macro-invertebrate consumers in a southern California salt marsh. *Estuarine, Coastal and Shelf Science* 45:823–834.

Peterson, B.J. 1999. Stable isotopes as tracers of organic matter input and transfer in benthic food webs: A review. *Acta Oecologica* 20:479–487.

Peterson, B.J., R.W. Howarth, and R.H. Garritt. 1985. Multiple stable isotopes used to trace the flow of organic matter in estuarine food webs. *Science* 227:1361–1363.

Sherr, E.B. 1982. Carbon isotope composition of organic seston and sediments in a Georgia salt marsh estuary. *Geochimica et Cosmochimica Acta* 46:1227–1232.

Silliman, B.R. and J.C. Zieman. 2001. Top-down control of *Spartina alteniflora* production by periwinkle grazing in a Virginia salt marsh. *Ecology* 82:2830–2845.

Silliman, B.R., C.A. Layman, K. Geyer, and J.C. Zieman. 2004. Predation by the black-clawed mud crab, *Panopeus herbstii*, in Mid-Atlantic salt marshes: Further evidence for top-down control of marsh grass production. *Estuaries* 27:188–196.

Tenore, K.R. 1977. Growth of *Capitella capitata* cultured on various levels of detritus derived from different sources. *Limnology and Oceanography* 22:936–941.

Tenore, K.R. 1981. Organic nitrogen and caloric content of detritus. I. Utilization by the deposit-feeding polychaete, *Capitella capitata*. *Estuarine and Coastal Marine Science* 12:39–47.

Tenore, K.R., R.B. Hanson, B.E. Dornseif, and C.N. Wiederhold. 1979. The effect of organic nitrogen supplement on the utilization of different sources of detritus. *Limnology and Oceanography* 24:350–355.

Tenore, K.R., R.B. Hanson, J. McClain, A.E. Maccubbin, and R.E. Hodson. 1984. Changes in composition and nutritional value to a benthic deposit feeder of decomposing detritus pools. *Bulletin of Marine Science* 35:299–311.

Turner, R.E. 1978. Community plankton respiration in a salt marsh estuary and the importance of macrophytic leachates. *Limnology and Oceanography* 23:442–451.

Turner, R.E. 1993. Carbon, nitrogen and phosphorus leaching rates from *Spartina alterniflora* salt marshes. *Marine Ecology Progress Series* 92:135–140.

Section 5.9

Ayliffe, L.K., T.E. Cerling, T. Robinson, A.G. West, M. Sponheimer, B.H. Passey, J. Hammer, B. Roeder, M.D. Dearing, and J.R. Ehleringer. 2004. Turnover of carbon isotopes in tail hair and breath CO_2 of horses fed an isotopically varied diet. *Oecologia* 139:11–22.

Fry, B. 1981. Natural stable carbon isotope tag traces Texas shrimp migrations. *Fishery Bulletin* 79:337–345.

Fry, B. and C.K. Arnold. 1982. Rapid $^{13}C/^{12}C$ turnover during growth of brown shrimp (*Penaeus aztecus*). *Oecologia (Berlin)* 54:200–204.

Fry et al. 1987. Listed above; see Section 5.1 readings.

Harvey, C.J., P.C. Hanson, T.E. Essington, P.B. Brown, and J.F. Kitchell. 2002. Using bioenergetics models to predict stable isotope ratios in fishes. *Canadian Journal of Fisheries and Aquatic Sciences* 59:115–124.

Herzka, S.Z. and G.J. Holt. 2000. Changes in isotopic composition of red drum (*Sciaenops ocellatus*) larvae in response to dietary shifts: Potential applications to settlement studies. *Canadian Journal of Fisheries and Aquatic Sciences* 57:137–147.

Herzka, S.Z., S.A. Holt, and G.J. Holt. 2001. Documenting the settlement history of individual fish larvae using stable isotope ratios: Model development and validation. *Journal of Experimental Marine Biology and Ecology* 265:49–74.

Herzka, S.Z., S.A. Holt, and G.J. Holt. 2002. Toward the characterization of settlement patterns of red drum (*Sciaenops ocellatus*) larvae to estuarine nursery habitat: A stable isotope approach. *Marine Ecology Progress Series.* 226:143–156.

Hesslein R.H., K.A. Hallard, and P. Ramlal. 1993. Relpacement of sulfur, carbon and nitrogen intissue of growing broad whitefish (*Coregonous nasus*) in response to a change in diet traced by $\delta^{34}S$, $\delta^{13}C$, and $\delta^{15}N$. *Canadian Journal of Fisheries and Aquatic Sciences* 50: 2071–2076.

Olive, P.J.W., J.K. Pinnegar, N.V.C. Polunin, G. Richards, and R. Welch. 2003. Isotope trophic-step fractionation: A dynamic equilibrium model. *Journal of Animal Ecology* 72:608–617.

Podlesak, D.W., S.R. McWilliams, and K.A. Hatch. 2005. Stable isotopes in breath, blood, feces and feathers can indicate intra-individual changes in the diet of migratory songbirds. *Oecologia* 142:501–510.

Suzuki, I.W., A. Kasai, K. Nakayama, and M. Tanaka. 2005. Differential isotopic enrichment and half-life among tissues in Japanese temperate bass (*Lateolabrax japonicus*) juveniles: implications for analyzing migration. *Canadian Journal of Fisheries and Aquatic Sciences* 62:671–678.

Tieszen, L.L., T.W. Boutton, K.G. Tesdahl, and N.A. Slade. 1983. Fractionation and turnover of stable carbon isotopes in animal tissues: Implications for $\delta^{13}C$ analysis of diet. *Oecologia* 57:32–37.

West, A.G., L.K. Ayliffe, T.E. Cerling, T.F. Robinson, B. Karren, M.D. Dearing, and J.R. Ehleringer. 2004. Short-term diet changes revealed using stable carbon isotopes in horse tail-hair. *Functional Ecology* 18:616–624.

6
Isotope Additions

Overview

Ecologists can add isotopes to field and laboratory experiments. This chapter considers what is good about these addition experiments, and what might go wrong. For those who actually perform one of these addition experiments, two technical supplements are included on the accompanying CD to help you in your detailed planning.

6.1. *Addition Addiction.* You don't have to be satisfied with nature's isotope handiwork. You can purchase isotopes and add them to your own field experiment. Experimenters do this in streams, lakes, estuaries, and in many terrestrial settings including fields and even whole forests and watersheds. In fact, adding isotopes can be addictive because you control the progress of the experiment, and because you don't have to worry about fractionation. Although this all sounds good, isotope additions usually require good modeling to really make sense out of what happened.

6.2. *The Golden Spike Award for Isotopes.* The Golden Spike is a virtual award to those who can bring together two isotope approaches in a fruitful marriage. The two approaches are survey work of natural, pre-existing isotope distributions, and the more laborious, but more controlled isotope addition work.

Technical Supplement 6A. How Much Isotope Should I Add? (See accompanying CD, Chapter 6 folder.) This example goes through a typical planning sequence for an isotope addition experiment. After a little practice, the calculations are straightforward.

Technical Supplement 6B. Noisy Data and Data Analysis with Enriched Samples. (See accompanying CD, Chapter 6 folder.) This section highlights technical issues that become important only when working with data from enriched samples. Working with these details can improve accuracy in data analysis with highly enriched samples.

Main points to learn. Isotope additions are important experimental tools that are profitably used together with surveys of natural isotope distributions. The big advantage of addition experiments is that they can really quantify the importance of a source you label, solving many a mixing puzzle. The main disadvantages are that you usually need a good knowledge of turnover rates and some good modeling to correctly interpret the isotope enrichment data.

6.1 Addition Addiction

A colleague once compared natural isotope distributions to subtle pastel paintings, things of beauty for the connoisseur. But for every art form, there is a counterview. The next artist comes along and spills a big bucket of isotope red all over the pastel landscape, and the subtle natural isotope variations with all their encoded information is all gone, covered in shades of isotope red. Although this portrayal sounds bad, the isotope addition approach has many advantages. First, it puts you, the experimenter, in control so that you can really work on the problem that most interests you. The other major benefit is that you won't have to worry about isotope fractionation. Or, if you drown the system in tracer, then the subtleties of slight isotope variations (due to fractionation) are just plain irrelevant. Because fractionation is complex, as we show in Chapter 7, it is appealing to think we don't have to bother with it. So, a new program would be to focus on controlled isotope additions, and forget fractionation.

Let's look at how one of these addition experiments might work, in a simplified lake ecosystem. You start at a baseline (natural abundance) level, add isotope typically via ^{15}N-labeled nutrients such as nitrate or ammonium, and watch isotope values of plants, herbivores, and carnivores rise to great heights. You are careful to add only a small amount of tracer so that this is not a fertilization experiment, but even a small amount of highly enriched tracer is enough to energize the isotope dynamics as the tracer starts to spread through the ecosystem (Figure 6.1).

In the food web of Figure 6.1, the plants grow over 90 days, fueling subsequent growth of herbivores that then support growth of carnivores. The left panel of Figure 6.1 shows growth of plants, herbivores, and carnivores, with growth expressed as relative increase versus 100 initial units, so that a value of 200 indicates a doubling of biomass. The right panel of Figure 6.1 shows the isotope expectations where herbivores depend solely on the labeled plants and the carnivores depend solely on the labeled herbivores. With isotope added continuously to the nutrient supply to a target label of 1000‰, the food web approaches this 1000‰ value over time. We see that the plants rapidly equilibrate to the 1000‰ nutrient δ value, the herbivores lag because of slower growth and slower turnover than the plants, and this slowdown is even more exaggerated in the carnivores.

Overall, the first part of the isotope curves in the right panel of Figure 6.1 mainly reflect the uptake and turnover dynamics, whereas the latter

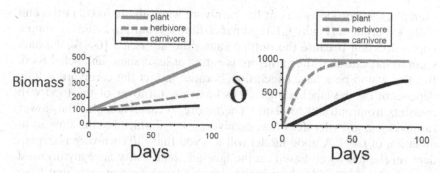

FIGURE 6.1. Time course of changes in a food web involving plants, herbivores that eat the plants, and carnivores that eat the herbivores. Plants grow rapidly during the 90-day experiment, supporting the slower growth of the animal consumers. Plant isotopes change quickly towards the value of the added N nutrient that in this case has an isotope label of 1000‰. Animal isotope values approach the 1000‰ value more slowly due to lags as label is transferred from one trophic level to the next, (a) from plants to herbivores, and (b) from herbivores to carnivores, plus (c) slower growth of animals, and (d) slower tissue turnover rates in animals.

parts of the experiment reflect which sources are important, the labeled sources or the unlabeled sources. In principle, one can estimate both turnover dynamics and differential use of sources from these addition experiments, very nice indeed. And in the I Chi workbook for this section (see workbook 6.1 in the Chapter 6 folder on the accompanying CD) you can add in fractionation effects of a few permil, and find, *voilà*, it doesn't make a difference one way or the other when you are dealing with labels of 500‰ or more. So, solve the system and banish fractionation: who would not be addicted to these addition experiments?

However, these experiments are not so simple. If you could add isotopes for years, you would see the complete isotope turnover dynamic in all the trophic levels, even in long-turnover massive pools such as soils and sediments. The final long-term isotope results after complete turnover and equilibration would give you the source dependencies. But unfortunately, the practical experiment is all-too-often a truncated experiment lasting the one to three months of a field season. With short-term data, you need a good model to interpret the data, a model that requires, guess what: good knowledge of turnover times, trophic levels, and sources. Why, this is just the information you were trying to solve for in the first place! Sounds suspicious, doesn't it? It is. In the final balance, although you think you are on a superior winning path with the isotope addition approach where especially you do not have to worry about fractionation, in the end, you may fail if your modeling of turnover dynamics is flawed.

Careful consideration of Figure 6.1 shows that without a good model, you really can reach a wrong conclusion in these label addition studies. For example, the isotope values of the carnivore were about 700‰ at the end of the experiment (Figure 6.1, right panel), and it is hard to avoid the simple impression that because the isotope values did not reach 1000‰, the data seem to indicate that the carnivore is eating at least some unlabeled food. But this would be a false conclusion, because in fact the carnivore is fully dependent on the labeled algae. It is just slow turnover of the food web transfers from nutrient to plant to herbivore to carnivore, plus the slower turnover and growth rates of the carnivore itself, that lead to the slow accumulation of label. A good model will tell you that the carnivore is dependent on the food web based on the labeled algae, but you really do need very good models indeed for higher trophic levels or materials that have long turnover times.

The art of the addition experiment is to continue long enough and sample intensively enough to be able to estimate the final asymptotic isotope values. Longer experiments help gather results needed for better extrapolation to asymptotic values, but more field work will not always help because many organisms will not equilibrate in even one full field season. For these reasons, good modeling becomes important for whole ecosystem labeling experiments. Labeling studies in smaller, fast-turnover systems that closely approach equilibrium in a few days will be much less dependent on complex modeling.

To better understand these points, find workbook 6.1 in the Chapter 6 folder on the accompanying CD, and try out a tracer addition experiment in a virtual food web, to see what you get for a simple plant–herbivore–carnivore system. Mr. Polychaete thinks that it is possible to make good sense of the results, but it usually depends on the adequacy of the understanding of bioenergetics and turnover, and on the food web modeling. Do you agree?

In many labeling experiments, isotope additions are not made uniformly across a landscape so that spatial dynamics as well as temporal dynamics become important. In these cases, sampling should be done generally in transects to characterize the spread of the spatial signal, as well as over time. Models used to evaluate the resulting data will need to consider spatial as well as temporal dynamics.

Practically, these addition experiments are excellent for bringing together many scientists in ecosystem-level efforts, including scientists interested in nutrients, plants, and animals. But the experiments are also a lot of work, and can have the difficulties highlighted above. For these and other reasons, the addition addiction stands as a complement, not a substitute for natural abundance studies where systems are often near steady state. The next section examines this theme more closely, why it makes sense to do natural abundance studies alongside the tracer addition studies.

Problems 1 to 5 in the Chapter 6 folder on the accompanying CD ask you to calculate isotope enrichments for tracer addition experiments.

Problems 6 and 7 consider whether you can truly ignore fractionation in these tracer addition experiments.

6.2 The Golden Spike Award for Isotopes

The completion of the U.S. transcontinental railroad in 1869 was honored by driving a golden spike into the earth, an award that marked the completion of a long venture, and the opening of a new era. A new era in isotope mixing is opening up as biologists are beginning to use isotope additions more and more to clarify details of their mixing models. And so we end our thoughts on mixing with a challenge to compete for the golden (isotope) spike award.

But first, let's look at a short snapshot of a label addition experiment, to get a feel for what might be involved. Here was the problem: a team of arctic river ecologists could not easily decipher from the natural isotopes which plants were the most important food resources for stream food webs. There were two major possible sources, in-stream algae and detrital terrestrial plants that washed into streams. The isotope difference between these two food resources was small in the naturally occurring, natural abundance isotopes, so ecologists were facing a mixing muddle. What to do? Perhaps wisely, they gave up on the natural isotopes, and started spiking for fun and fame. Natural isotopes would not answer their research question, so the creative researchers turned to an alternative isotope approach, the isotope addition experiment.

Their spike was a stream "dripper", like the intravenous or IV dripper in a hospital, but located outdoors to add nitrogen isotope label slowly but surely, day after day, to a small stream ecosystem. They added small amounts of ^{15}N-labeled ammonium, tracer amounts that were <5% of natural background amounts and small enough to avoid fertilizing the stream. This left the stream unchanged except in its (invisible) isotope color.

The stream naturally mixed up the labeled ammonium and distributed it downstream where plants took it up to satisfy their nitrogen demands. With the label addition, isotope signals in the stream algae climbed to nearly 1000‰, well outside the range of usual values observed in nature, the natural abundance values that varied between about −10 and +10‰ $\delta^{15}N$. The added label created a giant signal, so large that any fractionation reactions that could change isotope values by 5 to 10‰ were just minor noise. With fractionation unimportant, all the isotope action could be followed as simple mixing. Extensive sampling over time and space showed the label took a few weeks to reach constant values in most small stream consumers, and label uptake was strongest nearest the dripper. The results led to nice models of stream N cycling through 15 food web compartments, with an overall conclusion that it was in-stream algal foods, rather than detrital foods washed in from the surrounding land, that were most important for fish production.

The stream ecologists went home happy campers, ready to spike further anytime, anywhere, a 24/7/365 kind of attitude. However, this spiking proved quite intensive in demands of time and money. So there was also a continuing love affair with the natural abundance isotopes that could produce less-precise answers, but at a large number of sites and with much less overall effort. Also the natural abundance measurements can be used to address somewhat different questions involving slow-turnover pools that are hard to label. An evolving view might be that a happy marriage of both approaches, spiking plus surveying the natural abundances, would make for a wise and fruitful research strategy.

Here are some specifics about why natural abundance studies should not be neglected. The natural abundance studies generally help distinguish trophic levels and sources. They are usually superior in studies of slow-turnover pools. Added tracer does not readily infiltrate slow-turnover pools in short-term experiments, so for these pools, the natural abundance distributions often provide better constraints on pathways and sources. Also, ecologists are beginning to recognize and document natural seasonal and spatial isotope variations that propagate bottom-up through ecosystems, using these natural tracer perturbation results to advantage in their studies of ecosystem dynamics.

Overall, there are complementary strengths and weaknesses in the natural abundance surveys and added tracer approaches. The survey approach really is strongest at excluding sources that might be important. But it is much weaker at identifying the importance of sources in a positive sense. The spike experiments are strong at this weak point of the natural abundance approach, because adding label to a source will show if it is important. However, the spike results are weak in that the results only apply really at one particular time and place, and generalizing from these specifics may be difficult. So, because the survey and spike approaches have weaknesses as well as different strengths, it may be wise to do both. It is generally beneficial to think about the precise scientific question before proceeding with either approach or both, but here we consider the advantages of a dual approach, surveying and spiking.

Who among you will really marry the advantages of the surveys of natural isotopes with the experimental isotope additions? Many isotope-interested scientists today are at the edge of this marriage, not sure whether to embark, for it will need a very even-handed, compromise-friendly approach. Perhaps we can give some early marriage guidance to get going in this new arena. Here is Mr. Polychaete's attempt at counseling this isotope addition scene, where advice comes to Nat and Addy, two sides of a scientist interested in the golden spike award, but split between the Natural approach and the Addition approach.

First of all, you two should recognize that a rich life awaits you if you combine, commingle, and conjugate. There are those stakeholders out there, folks with fixed interests who review your grant proposals and say it cannot be done. They hold stakes ready to stab you if you pay them too much attention, or they will feed you too many steaks so that you become oblivious. Those stakeholders and steakholders who oppose successful unions are always diverse in their tactics. But foremost, it is not those outside forces, it is your internal commitment that is important. A fertile life beckons, if you only pursue it in good faith.

Second, there is no right or wrong in these relationships. You each should recognize that you have special talents and opportunities. Is it better to survey then spike, or should we make a straight line of spikes through the career landscape? Neither path is right or wrong, and you should recognize you are halves of a whole. Be pliant and observant, seek joint paths that multiply the fruits of your endeavors. Avoid the pit of mixing muddle misery and the barbs of spikey righteousness, work judiciously together. Travel extensively with Nat and see the wonders of nature's isotope distribution patterns, but then settle down from time to time, and set some of Addy's spikes in the landscape to establish what really happened and force nature's hand where the answers are otherwise muddled.

Lastly, have patience with the ways of each other. Addy needs careful attention to repetitive sampling to chase the fate of the spike. Nat needs a free-roaming scheme to maximize the information present in the signals from natural systems. There is room for both sedentary detail and wandering wonder in the right marriage.

In any case, we are looking forward to giving the golden spike award to scientists who make the Nat–Addy marriage work in a routine and fruitful way. Consulting some of the pioneering studies in the Further Reading section for this chapter, studies that date back to the 1950s, is a good way to start. Also, Technical Supplements 6A and 6B on the accompanying CD give practical examples about isotope additions for those of you who really want to do some of these interesting experiments.

P.S. Some scientists who work with truly large systems, such as the global inventories of methane and CO_2 in the atmosphere, are also thinking about spike experiments. At the global scale, human activities are inadvertently altering the methane and CO_2 atmospheric pools, putting in the isotope spike. But the isotope addition is low-level, so that fractionation as well as mixing is important for isotopes in methane and CO_2. The global modelers are busy making budgets with sources and sinks (like income and expenditures, respectively, in a household budget) to better understand climate change from these methane and CO_2 measurements. Isotopes help constrain these budgets; the isotopes as well as the amounts have to balance. The thinking is this: no balance in the budget indicates that something is missing or wrong. But because the isotope spike is fairly small at this large scale, these scientists have to consider not only mixing, but also open the door to fractionation as a control on isotope values. So, the moral is that as spikers think about scaling up their efforts to larger and larger systems, they will need to know about isotope fractionation, the subject of Chapter 7. Even spikers acknowledge the underlying fractionation patterns by normalizing for the natural abundance background.

6.3 Chapter Summary

Chapter 6 considers a special kind of isotope mixing, one in which the scientist adds isotopes to ecological systems to create strong isotope differences among sources. These isotope additions are possible because commercial companies can routinely boost heavy isotope contents from <5% to >99%, so that δ values rise from values near 0‰ to >1,000,000‰ in products such as ^{15}N-labeled ammonium sulfate and ^{13}C-labeled bicarbonate. A small amount of these added tracers usually will provide very strong signals for ecological experiments done in mesocosms and whole ecosystems such as smaller streams. This seems like a dream come true, because you can create designer labels that enter systems in just the places and times you want. And in theory you can label multiple sources simultaneously for a rather complete and simultaneous evaluation of competing ecological processes. The combination of large source signals and good source targeting ensures that isotope addition experiments have a great appeal for ecological researchers. A final bonus is that by adding isotopes, source signals are so large that effects of fractionation can be ignored in a simple mixing approach. With all these good points, it is no wonder the mere thought of isotope addition experiments can be addictive.

However, there is a flaw in many of these isotope addition experiments. The flaw is that there is not enough time to bring all components of the systems to equilibrium where isotope values are steady and not changing, and reflect only source mixing. If experiments are too short, as they often are due to practical considerations, one has to correctly model other factors that also influence isotope dynamics, before one can calculate the source contributions. These other factors are the turnover rates of the pools within a system. If pools turn over quickly, they equilibrate and source mixing dominates the isotope pattern. But if pools turn over slowly relative to the length of the addition experiment, then the resulting isotope pattern reflects a combination of source mixing and turnover. In this typical case, good models are needed to sort out what is mixing and what is turnover.

The good news in this is that the isotope addition experiments provide an abundance of data to constrain models quickly towards correct answers. But it is also very important to find ways to estimate turnover dynamics. In the end, good modeling is a necessary part of most label addition experiments.

Section 6.1 shows a food web experiment where the investigator adds label continuously for 90 days, but still must use a dynamic simulation model to interpret the results. An attached spreadsheet gives a generic model for the label addition experiment, so you can try out some of your own ideas about what would happen if you started your own label addition experiment. Section 6.2 recommends a joint strategy for tracing source contributions, using both the natural abundance isotope distributions together with label additions. This is more sophisticated and more effective. Technical Supplement 6A on the accompanying CD leads you through the calculations you will need to make when actually starting a label experiment. Technical Supplement 6B gives technical details about measuring and calculating isotope values associated with label additions.

In the end, you will probably need some practice, in virtual reality spreadsheets and in the field with a pilot project, before you master how to add tracer and how to evaluate the outcomes. Tracer additions are a very powerful approach for estimating rates and source contributions at specific times and places, an approach that is well complemented by natural abundance surveys that apply widely in different systems and through history. The challenge continues to be how to wisely balance the isotope addition and isotope survey approaches to obtain the best overall results. The place to start is simply to use both approaches in tandem, rather than relying solely on one or the other.

Further Reading

Section 6.1

Dyckmans, J., C.M. Scrimgeour, and O. Schmidt. 2005. A simple and rapid method for labeling earthworms with ^{15}N and ^{13}C. *Soil Biology and Biochemistry* 37:989–993.

Mulholland, P.J., J.L. Tank, D.M. Sanzone, W.M. Wollheim, B.J. Peterson, J.R. Webster, and J.L. Meyer. 2000. Food resources of stream macroinvertebrates determined by natural-abundance stable C and N isotope and a ^{15}N tracer addition. *Journal of the North American Benthological Society* 19:145–157.

Sharp, Z.D., V. Atudorei, H.O. Panarello, J. Fernandez, and C. Douthitt. 2003. Hydrogen isotope systematics of hair: archeological and forensic applications. *Journal of Archaerological Science* 30:1709–1716.

Van den Meersche, K., J.J. Middelburg, K. Soetaret, P. Van Rijswifk, H.T.S. Boschker, and C.H.R. Heip. 2004. Carbon-nitrogen coupling and algal-bacterial interactions during an experimental bloom: modeling a ^{13}C tracer experiment. *Limnology and Oceanography* 49: 862–878.

Wolfe, R.R. and D.L. Chinkes. 2004. *Isotope Tracers in Metabolic Research: Principles and Practice of Kinetic Analysis*. Wiley-Liss, New York.

Wollheim, W.M., B.J. Peterson, L.A. Deegan, M. Bahr, J.E. Hobbie, D. Jones, W.B. Bowden, A.E. Hershey, G.W. Kling, and M.C. Miller. 1999. A coupled field and modeling approach for the analysis of nitrogen cycling in streams. *Journal of the North American Benthological Society* 18:199–221.

Section 6.2

Blair, N.E., L.A. Levin, D.J. DeMaster, and G. Plaia. 1996. The short-term fate of fresh algal carbon in continental slope sediments. *Limnology and Oceanography* 41:1208–1219.

Chae, Y.M. and H.R. Krouse. 1986. Alteration of S^{34} natural abundance in soil by application of feedlot manure. *Soil Science Society of America Journal* 50:1425–1430.

Coleman, D., S. Fu, P. Hendrix, and D. Crossley Jr. 2002. Soil foodwebs in agroecosystems: Impacts of herbivory and tillage management. *European Journal of Soil Biology* 38:21–28.

DeMaster, D.J., C.J. Thomas, N.E. Blair, W.L. Fornes, G. Plaia, and L.A. Levin. 2002. Deposition of bomb ^{14}C in continental slope sediments of the Mid-Atlantic Bight: Assessing organic matter sources and burial rates. *Deep-Sea Research II.* 49:4667–4685.

Dornblaser, M., A.E. Giblin, B. Fry, and B.J. Peterson. 1994. Effects of sulfate concentration in the overlying water on sulfate reduction and sulfur storage in lake sediments. *Biogeochemistry* 24:129–144.

Friedli, H., H. Loetscher, H. Oeschger, U. Siegenthaler, and B. Stauffer. 1986. Ice core record of the ^{13}C/^{12}C ratio of atmospheric CO_2 in the past to centuries. *Nature* 324:237–238.

Fry, B., D.E. Jones, G.W. Kling, R.B. McKane, K.J. Nadelhoffer, and B.J. Peterson. 1995. Adding ^{15}N tracers to ecosystem experiments. In E. Wada, T. Yoneyama, M. Minagawa, T. Ando, and B. Fry (eds.), *Stable Isotopes in the Biosphere*. Kyoto University Press, Japan, pp. 171–192.

Hall, R.O. and J.L. Meyer. 1998. The trophic significance of bacteria in a detritus-based stream food web. *Ecology* 79:1995–2012.

Hall, R.O., B.J. Peterson, and J.L. Meyer. 1998. Testing a nitrogen cycling model of a forest stream by using a nitrogen-15 tracer addition. *Ecosystems* 1:283–298.

Hart, E.A. and J.R. Lovvorn. 2002. Interpreting stable isotopes from macroinvertebrate foodwebs in saline wetlands. *Limnology and Oceanography* 47:580–584.

Holmes, R.M., B.J. Peterson, L.A. Deegan, J.E. Hughes, and B. Fry. 2000. Nitrogen biogeochemistry in the oligohaline zone of a New England estuary. *Ecology* 81:416–432.

Hughes, J.E., L.A. Deegan, B.J. Peterson, R.M. Holmes, and B. Fry. 2000. Nitrogen flow through the food web in the oligohaline zone of a New England estuary. *Ecology* 81:433–452.

Kling, G. 1994. Ecosystem-scale experiments in freshwaters: The use of stable isotopes. In L.A. Baker (ed.), *Environmental Chemistry of Lakes and Reservoirs*, Advances in Chemistry Series 237. American Chemical Society, Washington D.C. pp. 91–120.

Marino, B.D., M.B. McElroy, R.J. Salawitch, and W.G. Spaulding. 1992. Glacial-to-interglacial variations in the carbon isotopic composition of atmospheric CO_2. Nature 357:461–466.

Mayer, B. and H.R. Krouse. 1996. Prospects and limitations of an isotope tracer technique for understanding sulfur cycling in forested and agro-ecosystems. *Isotopes in Environmental and Health Studies* 32:191–201.

McCarthy, M.C., K.A. Boering, A. Rice, S. Tyler, P. Connell, and E. Atlas. 2003. Carbon and hydrogen isotopic compositions of stratospheric methane: 2. Two-dimensional model results and implications for kinetic isotope effects. *Journal of Geophysical Research* 88: ACH 12–1 to ACH 12–18, doi:10.1029/2002JD003183.

Middleburg, J.J., C. Barranguet, H.T.S. Boschkler, P.M.J. Herman, T. Moens, and C.H.R. Heip. 2000. The fate of intertidal microphytobenthos carbon: An *in situ* ^{13}C labeling study. *Limnology and Oceanography* 45:1224–1234.

Odum, E.P. 1971. *Fundamentals of Ecology*, 3rd edition, W.B. Saunders, New York, especially pp. 459–464 and references cited there.

O'Reilly, C.M., R.E. Hecky, A.S. Cohen, and P.D. Plisnier. 2002. Interpreting stable isotopes in food webs: Recognizing the role of time averaging at different trophic levels. *Limnology and Oceanography* 47:306–309.

Pendleton, R.C. and A.W. Grundman. 1954. Use of P^{32} in tracing some insect-plant relationships of the thistle, *Cirsium undulatum*. *Ecology* 35:187–191.

Peterson, B.J., M. Bahr, and G.W. Kling. 1997. A tracer investigation of nitrogen cycling in a pristine tundra river. *Canadian Journal of Fisheries and Aquatic Science* 54:2361–2367.

Schimel, D.S. 1993. *Theory and Application of Tracers*. Academic, San Diego, CA.

Shearer, G. and D.H. Kohl. 1989. Estimates of N_2 fixation in ecosystems: The need for and basis of the ^{15}N natural abundance method. In P.W. Rundel, J.R. Ehleringer, and K.A. Nagy (eds.), *Stable Isotopes in Ecological Research*. Springer-Verlag, New York, pp. 342–374.

Snover, A.K., P.D. Quay, and W.M. Hao. 2000. The D/H content of methane emitted from biomass burning. *Global Biogeochemical Cycles* 14:11–24.

Tyler, S.C. 1986. Stable carbon isotope ratios in atmospheric methane and some of its sources. *Journal of Geophysical Research* 91:13232–13238.

7
Fractionation

Overview

This chapter starts with atoms and ends with the whole biosphere, showing how isotope fractionation works in theory and practice. Fractionation starts with atomic-level considerations, but usually starts to make sense in larger ecological contexts only when you grasp the idea of mass balance. Mass balance is an accounting idea that masses and isotopes entering a reaction must equal masses and isotopes exiting the same reaction. This sounds simple, but it forces us to budget several things at once, masses and isotopes, in a kind of multitasking consciousness. This demands a juggling skill that takes practice to learn, so be patient and take time to practice, especially using workbook 7.2 of in the Chapter 7 folder on the accompanying CD.

Sections 7.1, 7.2, 7.6, 7.7, and 7.10 contain the more theoretical sections, and may need rereading several times for full comprehension. Sections 7.3 to 7.5 and 7.8 and 7.9 provide examples. Technical Supplements 7A and 7B on the accompanying CD are reference sections for advanced and interested readers.

7.1. *Fractionation Fundamentals*. Physical and chemical reactions fractionate isotopes at the atomic level, creating a rich treasure trove of isotope information at the atomic level within molecules and also within the biosphere as a whole. Is fractionation an inherent property of the universe, or just the work of a bothersome quantum genie, Fractionation Frank, described in Box 7.1? (Hint: chemists use quantum mechanics to calculate potential fractionations, though not always with success; for some quantum details, see Technical Supplement 7A in the Chapter 7 folder on the accompanying CD).

7.2. *Isotopium and Fractionation in Closed Systems*. Isotopium is an imaginary, equal-opportunity element with 50% heavy isotopes and 50% light isotopes. Open workbook 7.2 in the Chapter 7 folder on the CD that accompanies this book and follow the lead of the quantum isotope genie, Fractionation Frank, as you make fractionation happen in a virtual closed system, a sealed laboratory jar. If you want to know more about the equations for the closed systems, consult Technical Supplement 7B in the Chapter 7 folder on the accompanying CD.

7.3. *A Strange and Routine Case.* This essay examines isotope fractionation that occurs routinely during isotope measurements made in modern laboratories.

7.4. *A Genuine Puzzle—Fractionation or Mixing?* Sometimes it is easy to confuse effects of fractionation and mixing, as this example shows. Gathering more information beyond the isotopes is usually key to resolving which is active, fractionation, mixing, or both.

7.5. *Cracking the Closed Systems.* Isotopes circulate in well-described, predictable ways in closed flasks and bottles in the laboratory, with fractionation driving this circulation. However, two field examples show that these laboratory models are often too simple, and that mixing occurs in unappreciated ways in the field, offsetting the effects of fractionation.

7.6. *Equilibrium Fractionation, Subtle Drama in the Cold.* Fractionation occurs in some interesting exchange reactions, and can provide a thermometer for measuring temperatures of ancient seas. This section also gives some of the elegant algebra for equilibrium reactions.

7.7. *A Supply/Demand Model for Open System Fractionation.* Most I Chi modeling in this book is based on open systems such as flow-through continuous cultures. The open system isotope equations are based on mass balance and are derived in this section. These equations lead to a strong prediction that fractionation should be related to supply/demand relationships in many ecological settings.

7.8. *Open System Fractionation and Evolution of the Earth's Sulfur Cycle.* Our planet has changed a great deal over its 4.5 billion year lifetime, and one transition was to the oxidized biosphere we enjoy today. This essay concerns one of the stages in this transition, studied with sulfur isotopes that are stable and thus survive in the geological record as indicators of past isotope ecology. The supply/demand isotope fractionation model derived in Section 7.7 is important in interpreting these ancient ecologies.

7.9. *Open System Legacies.* Isotope ecology rests on the efforts of chemists, geologists, and physicists who first worked on isotope circulation in the biosphere. This section briefly considers four geochemist pioneers and their work with isotope fractionation in open systems.

7.10. *Conducting Fractionation Experiments.* Fractionation is less mysterious than you might think, and can be measured easily in laboratory settings. Here are some practical tips on how to measure fractionation in the lab.

Technical Supplement 7A. A Chemist's View of Isotope Effects. (See accompanying CD, Chapter 7 folder.) This reference section shows how chemists use a formula from quantum mechanics, the Bigeleisen equation, to estimate fractionation effects.

Technical Supplement 7B. Closed System Derivations. (See accompanying CD, Chapter 7 folder.) This is a technical reference section that gives the mathematical derivation of equations for fractionation in closed systems.

Main points to learn. Fractionation is the most complex topic in the isotope lexicon, and almost always lurks behind the simpler mixing dynamics found in earlier parts of this book. Mass balance and supply/demand models are important concepts related to fractionation. You should take the time to practice with fractionation in many contexts, getting to know it well. Fractionation is the heartbeat of isotope circulation in the biosphere, setting the stage for isotope mixing. It will always be worth your while to learn more about fractionation, even if you have to reread this chapter several times.

7.1 Fractionation Fundamentals

Isotope fractionation occurs in many reactions, and at two different levels. First, fractionation occurs at atomic and subatomic levels where chemistry and physics are the best guides. At this basement level of the universe, subtle isotope differences in bond strengths can be evaluated to estimate maximum possible or potential fractionations. But these maximum fractionations are not always realized, with isotope fractionation dampened at a second, macroscopic level. Here reactions occur in more recognizable ecological settings involving closed systems and open systems. A closed system can be a sealed laboratory vessel containing a reaction that proceeds forward to form product from substrate, but with no new additions of substrate once the experiment starts. An open system can be a flow-through reactor or continuous culture at steady state, where inputs are continual and balanced by outputs. The chemistries contained in the closed and open systems are often familiar ecological processes such as photosynthesis and respiration, and examples in this chapter illustrate how fractionation works in ecological settings.

This section introduces isotope fractionation at both the atomic and system levels. Later sections in this chapter present more details about fractionation in closed and potentially closed systems (Sections 7.2 to 7.5), in equilibrium systems (Section 7.6), and in open systems (Sections 7.7 to 7.9). Fractionation Frank, a quantum genie who may be behind all this fractionation commotion (Box 7.1), will be your guide in parts of this chapter.

Atomic Level—Physical Fractionation

At the atomic level, fractionation occurs in both physical and chemical reactions. One important physical reaction is diffusion, where the extra neutron difference in the nucleus means a slower speed. Calculation shows that dif-

Box 7.1. Wanted—Fractionation Frank

Crime: Disturbs the unified peaceful nature of the universe, a separatist.

Modus operandi: Uses atoms of the light elements, picks and chooses their variant isotopes like colors on an artist's palette, and overlays all nature with a (to him) "glorious" tapestry of isotope colors.

Name: Answers to the name of Frank, but is extremely slippery otherwise.

Description: Take a fistful of air then open your hand and see what is left—that is this character.

Special superpower: Creates inequality in everything he touches.

Venue: Works behind closed doors but also in the open.

Hangout: In the middle of the activated complex and underneath bonds, both places very hard to locate precisely within the virtual universe of quantum mechanics.

Tactics: Always picks away at the margins where the percentages are small.

Alert level: Harmless until it is too late.

Suspected Employers: Internal Revenue Service, Mr. Polychaete.

Apprehension: Generally impossible to catch in the act.

Favorite Haunts: An invisible middle zone between an easy-to-see before and an evidently different afterwards.

Favorite Game: Quantum mechanics roulette, trying to be two places at once and cause inverse fractionation (an arcane pastime if ever there was one, but sometimes successfully played with hydrogen isotopes while tunneling in reaction basements, at the bottom of zero point energy wells).

Age: Ancient as the universe.

Appearance: Unknown, but we suspect he laughs a lot and enjoys artistic license.

Status: Like the tax man and death, a certainty in this life.

Reward for information leading to capture: A good cup of coffee over which you can tell us the lie you have concocted to make us think we could capture this amazing fellow.

fusion isotope effects are largest in a vacuum, smaller in air, and negligible in water. Frequent collisions in air and water attenuate the maximum isotope effects that are expressed in a vacuum setting where few collisions occur (Box 7.2). Physical effects such as those in diffusion lead to a partial separation of the isotopes, and more complete separations are possible when these reactions occur in repeated cycles.

Harold Urey and co-workers began investigating physical and chemical means for separating isotopes during the 1930s, and this work continues

Box 7.2. Isotope Fractionation During Diffusion

Diffusion in a Vacuum

Most of the time, we think of chemical processes fractionating isotopes. But some physical processes also fractionate isotopes. Here is an example. Imagine two stable isotope twins, gases in a vacuum, like bowling balls. You give each a push with the same force. What will happen?

Let's think of CO_2 gas with twins $^{12}CO_2$ and $^{13}CO_2$ of masses 44 and 45 that are, respectively, "light" and "heavy." If we push with equal force, $F = \frac{1}{2}m * v^2$ where v is the velocity, then

$$F = \frac{1}{2}(44)*(v_{LIGHT})^2 = \frac{1}{2}(45)*(v_{HEAVY})^2,$$

and we can calcate the ratio of the velocities (although we don't calculate the actual velocities in this example), v_{LIGHT}/v_{HEAVY} = square root of $(45/44) = 1.0113$.

This result is that the light CO_2 will travel 1.13% (11.3‰) faster than the heavy CO_2 molecule. This is fractionation in action.

If you do this calculation for hydrogen that has two stable isotopes (^1H = protium and ^2H = deuterium, or D), you find that H_2 (mass 2) gas travels 22.47% (224.7‰) faster than HD (mass 3) gas. Because the difference between protium and deuterium is very large (2×), most hydrogen isotope fractionations are larger and more obvious.

You might think that physical separations of isotopes are rare, but you would be wrong. They are the central principle used in mass spectrometers, our main instruments for measuring isotope values (see Figure 2.2 in Chapter 2).

Diffusion in Air and Water

Collisions affect the diffusional separation of isotopes. If there are many collisions, then the isotope differences are dampened. In a vacuum, there are minimal collisions, so the physical fractionations are maximal. At the other extreme, in liquids, molecules collide so frequently with water that the mass differences of the isotopes don't make a difference. Solvation quenches the physical isotope effects. But what about diffusion in air?

The effect in air is given by a formidable equation involving "reduced mass ratios" (the square root term):

$$D_1/D_2 = [(C_2 + C')/(C_1 + C')]*\text{square root of}$$
$$\{[(M_1 + M')/(M_1 * M')]*[(M_2 * M')/(M_2 + M')]\},$$

where D_1/D_2 is the ratio of diffusion coefficients, C and M refer to the concentrations and molecular weights of the two isotope molecules 1 & 2, and the primed (') terms refer to C and M in air (*Geochimica et*

Cosmochimica Acta 3, p. 73, 1953). Solving this equation for diffusion of light ($^{12}CO_2$) and heavy ($^{13}CO_2$) carbon dioxide in air, one obtains 4.4‰ faster diffusion of light CO_2 than heavy CO_2 (*Phytochemistry* 20:553–567; 1981), a much reduced effect versus the 11.3‰ effect calculated above for diffusion in a vacuum. Collisions make the difference.

In most biological systems, the presence of abundant water ensures many collisions. This makes physical fractionation effects during diffusion small and near zero. But when gases diffuse in air, such as entry of CO_2 into plant stomata during photosynthesis, diffusion effects can be important. These diffusion effects are included in photosynthesis models described in Section 7.7.

today, with commercial companies separating isotopes to high purity. The physical process of distillation is often part of these separation schemes that can enrich heavy isotopes from natural levels of <1% abundance to> 99% purity (Urey 1939; Bigeleisen 1969). Centrifugation is also used to separate fluorinated uranium isotopes for commercial and military uses.

Enrichment makes possible the isotope addition experiments described in the previous chapter, and can also lead to interesting laboratory experiments. One such experiment is to purchase D_2O (^2H_2O), the heavy isotope form of water, or "deuterated water," and weigh it in a side-by-side test with normal water, H_2O (^1H_2O). The predicted weight or mass difference for equal volumes of D_2O and H_2O can be calculated from the formula weights as approximately $D_2O/H_2O = 20/18 = 1.11$. This difference can be readily measured with a laboratory balance, in one of the few tangible demonstrations of the existence of isotopes. Mass spectrometers also make use of physical principles for separating isotopes, especially the principle of inertia as explained in Section 2.1.

Atomic Level—Chemical Fractionation

Chemical isotope fractionation occurs at the atomic level during the making and breaking of bonds. Heavy isotope atoms with extra neutrons make bonds that are harder to make and harder to break. Chemists have had considerable success describing bonds and calculating bond strengths with quantum theory (Bigeleisen and Mayer 1947; Urey 1947; Bigeleisen and Wolfsberg 1958). Classical calculations about fractionation were made by assuming two substances were in equilibrium with each other, then comparing vibrational energies of their bonds (Urey 1947). These calculations led to early estimates of the maximum fractionation effects likely for different elements (Table 7.1). The estimates have proven robust and reinforce the concept that the common site of fractionation for chemical reactions lies in bond formation and rupture.

TABLE 7.1. Calculated Maximum Possible Isotope Fractionation or Δ.[a]

Light Isotope	Heavy Isotope	Maximum Theoretical Fractionation (‰)	Observed Natural δ Range (‰)
^1H	^2H	18,000	700
^1H	^3H	60,000	—
^{12}C	^{13}C	250	110
^{12}C	^{14}C	500	—
^{14}N	^{15}N	140	90
^{16}O	^{18}O	190	100
^{32}S	^{34}S	—	150
^{32}S	^{35}S	50	—
^{35}Cl	^{37}Cl	>30	1

[a] Δ in ‰ $= (k_1/k_2 - 1) * 1000$ for selected elements at 25°C, based on the ratio of rate constants, k_1/k_2, for light (k_1) and heavy (k_2) isotope reactions. Theoretical fractionations give a guide to ranges expected for isotope variations. The natural range of δ values actually observed on earth is also given, as listed by Anderson and Arthur (1983). In the wider cosmos, much larger variations in δ values are observed than on earth, and the study of meteorites in the field of cosmochemistry often shows these larger isotope variations.
Source: Adapted from Bigeleisen, J. 1949a. The validity of the use of tracers to follow chemical reactions. *Science* 110:14–16.

At this point, we slow down and take some time to expand on the points just made in the previous paragraph, to make sure we understand these fundamentals of fractionation. These ideas about chemical bonds are not everyday territory for ecologists, but they are not really difficult if we focus in, thinking carefully with Figure 7.1 as our guide. There you can find heavy and light isotope bonds depicted as springs in a simple analogy. The light isotope bond is depicted as a lightweight bouncy spring, and the heavy isotope bond is depicted as a stiffer, more massive spring that is harder to perturb. Isotope chemists use springs to illustrate bond differences with isotopes because the isotopes affect the vibrational or springlike components of bonds. In quantum mechanics, there are two other important components of bonds besides the vibrational components, the rotational and translational components. These rotational and translational components are coupled to larger molecular-level movements and are generally unaffected by isotope substitution (with some exceptions for deuterium). Because of this relative insensitivity, the isotope focus centers on the springy vibrational component of bonds formed between atoms within molecules. And here the idea is that with the same amount of energy applied, the light isotope bonds spring apart and break, whereas the heavy isotope bonds break less frequently. This leads to a faster reaction of the light isotope atoms that more easily escape bonds, and this faster or differential reaction is fractionation in action.

Using these background ideas, chemists also diagram energy relationships in bonds in a more formal way. The x-axis in Figure 7.1 is really a side view of two bonds stretched out in front of us inside an energy well. The x-axis is also an imagined distance between the centers of two atoms. Atoms are imagined to be far apart and not connected at the right of the diagram, or smashed together and fused in the leftmost part of the diagram. In the middle is an energy well, a region of stability in quantum mechanics where atoms are connected by chemical bonds. The horizontal lines in the energy well are the bonds, and you are left to imagine the atoms at the ends of these bonds.

The bonds of Figure 7.1 sit towards the bottom of the potential energy well, where they have more stability and less potential energy. Adding vastly more energy could possibly fuse the atoms, pushing the atoms together and to the left in this diagram. But more commonly, added energy will simply break the bonds, with interatomic distance increasing to the right as bonds break. When a bond elongates far enough to the right in Figure 7.1, it is no longer there and atoms are no longer connected. The bond is broken.

But there is an isotope difference in the rates at which the bonds break, and this difference causes isotope fractionation. The light isotope bonds are "springier;" that is, they are wider and they have more potential energy, sitting higher up in the potential energy well in the middle of Figure 7.1. These light isotope bonds are wider because they are vibrating more. They need to gain only a relatively small amount of energy before leaving the potential energy well, where atoms move apart and bonds break.

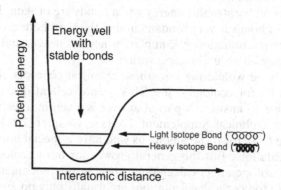

FIGURE 7.1. A chemical diagram showing why fractionation occurs when bonds are broken in kinetic reactions. Bonds are often compared to springs, with light isotope bonds depicted as the less massive, easier-to-break springs. Light isotope bonds are slightly wider and have more potential energy than heavy isotope bonds. Adding equal energy to both kinds of bonds results in more rapid bond breaking for the light isotope bonds that need less energy to climb out of the energy well as they elongate and break. When bonds are broken, atoms move apart to the right in this diagram, and interatomic distance increases. Bonds are only stable within the energy well. See text for further explanation. (Reprinted with permission from Bigeleisen, J. 1965. Chemistry of isotopes. *Science* 147:463–471. Copyright 1965, AAAS.)

Bond characteristics are different when heavy isotopes are involved. These heavy isotope bonds vibrate less and must gain more energy to move up out of the energy well (Figure 7.1), before atoms can move apart and bonds break. So for a given amount of energy added, relatively more light isotope bonds will break than heavy isotope bonds, and the difference in reaction rates gives the isotope fractionation.

Maybe now you can see also why once a bond is formed with heavy isotopes, it is harder to break this more stable and less energetic bond. This idea accounts for the fact that in equilibrium exchange reactions between molecules, heavy isotopes concentrate in the molecules where bonding is strongest. Those bonds will be most difficult to break and effectively trap or concentrate the atoms containing heavy isotopes.

But what about bond formation? As it turns out, these processes of bond rupture that we have been examining do not simply reverse themselves for bond formation. When bonds are formed, the more massive heavy-isotope atoms require more activation energy, so that bond formation is slower for heavy isotope reactions. Figure 1.6 shows this in cartoon form in the Introduction. Overall, for either bond formation or rupture, less energy input is needed for the light isotopes, accounting for the important fractionation rule that lighter isotopes react faster.

As you think about these issues in a more general way, you may realize that fractionation should be common. After all, metabolism makes and breaks bonds all the time. Photosynthesis captures the energy of sunlight in the chemical bonding of CO_2 to make plant sugar, and our metabolism of these sugars reliberates this energy when bonds are broken. The making and breaking of bonds is very fundamental in the biosphere ecology of our planet, so that fractionation is completely normal, usual, and routine. It occurs every day, all around us and within us.

But you may be wondering how these chemical concepts about bonds can be scaled up for ecological use. This is an excellent question for the isotope chemists to answer. A partial answer is given in a technical letter reproduced in Technical Supplement 7A (see Chapter 7 folder on the accompanying CD), for those among us who have a special interest in fundamental calculations. But the general answer for most ecologists is this. Although the isotope effect calculations made at the fundamental level of bonds continue today, these calculations are usually only possible for fairly simple reactions and molecules, and are of limited use for most biological reactions. Nonetheless, the outcomes of these calculations are important in a larger sense, because they constrain our ecological thinking about fractionation. The calculations set some overall limits, so that fractionations of 1‰ are routinely possible, 10‰ are common, but fractionations greater than 100‰ are very uncommon, with the exception that fractionations associated with hydrogen can be much greater (Table 7.1; Bigeleisen 1969). Fractionation effects for hydrogen isotopes are larger because the mass ratio for ^2H versus ^1H is very large at 2:1 compared to mass ratios near 1 for other elements (see Table 1.1). Fractionation differences can be so large for hydrogen isotopes that chemists are tempted to think of the ^2H and ^1H isotopes as separate elements, rather than as chemical twins.

Although fractionation in biological systems may be difficult for chemists to calculate from considerations of chemical bonds, ecologists still can estimate fractionation in careful laboratory studies. Biological uptake and loss reactions involved with nutrients and CO_2 are especially amenable to study, and show fractionations that are typically variable, from no fractionation to maximal values in the range of 10–100‰. The lower values often represent diffusion or advection limitation of the overall reaction, so an important part of the experimental design is to ensure a large excess in supplies of substrate entering the reaction to avoid this somewhat artificial result of little or no fractionation. The larger fractionations expressed when substrate is abundant generally reflect bond cleavage or bond formation when nutrients are incorporated or lost from organic matter. These reactions are kinetic isotope effects (KIE), and a well-known example is uptake of CO_2 during photosynthesis (Box 7.3). Equilibrium isotope effects (EIE) are also known for many exchange reactions important in global element cycles (Box 7.3). These equilibrium effects can also be thought of as arising from two opposing kinetic isotope effects in the forward and reverse reactions of the equilibrium.

There is also a newer recognition of unusual fractionations involving atmospheric ozone and SO_2, and some of these reactions are important for budgeting global sulfur and oxygen dynamics (Luz et al. 1999; Luz and Barkan 2000; Farquhar et al. 2000, 2002; Clark et al. 2004, reviewed in Hoefs 2004). These unusual fractionations are departures from the usual rule that the amount of fractionation should increase proportionately with increasing numbers of extra neutrons (Fry and Calvin 1952; Young et al. 2002). The usual proportional rule is termed "mass-dependent fractionation" (MDF), and the departures arise because of "mass-independent fractionation" (MIF). [Sidenote. The Fry in the 1952 Fry and Calvin reference above is the author's father, Arthur Fry. Lois M. Fry, the author's mother, was also an isotope chemist; one of her papers is also cited at the end of this section. This book is written by a second-generation isotope scientist.]

An important consequence of chemical fractionation at the atomic level is that the basic isotope information in nature exists on an atom-by-atom basis or "position-specific" basis within molecules (Box 7.4). Biosynthesis involves many fractionations that accumulate as the differential labeling of the individual atoms within molecules, so that this atomic level is the fundamental level of isotope information available to us. However, as of the time of this writing, only a very few studies have been able to access this most detailed atom-specific level of isotopic information. Interesting exceptions that have focused on the atomic-level isotope information include nuclear magnetic resonance studies of compounds important in wine and beer making (Martin 1995) and in maple sugar (Martin et al. 1996). Recent

Box 7.3. Kinetic Isotope Effects (KIE) and Equilibrium Isotope Effects (EIE), Carbon Isotope Examples

For most chemical reactions, the light isotope molecules react faster than heavy isotope molecules. If the rates of reaction are kinetic k rates, then the rates for molecules with light and heavy carbon isotopes, ^{12}C and ^{13}C, can be abbreviated as ^{12}k and ^{13}k. The kinetic isotope effect (KIE) is the contrast in the rate constants α, where $\alpha = {}^{12}k/{}^{13}k$ and Δ is the permil fractionation factor derived from α: $\Delta = (\alpha - 1)*1000$. Most biological reactions involve KIEs and show no or weak dependence on temperature.

An equilibrium isotope effect (EIE) is the net sum of two opposing kinetic isotope effects that apply in an exchange reaction. In these two-way reactions, the heavy isotope concentrates where it is bound most strongly. In these reactions, this equilibrium fractionation is: $\alpha_{EQ} = R_{\text{HEAVY MOLECULE}}/R_{\text{LIGHT MOLECULE}}$ where R is the $^{13}C/^{12}C$ ratio. EIEs often change predictably with temperature.

A Carbon KIE for Photosynthesis

The Rubisco enzyme fixes carbon in plant photosynthesis, adding CO_2 to a five-carbon compound to form a six-carbon sugar. The lighter isotope reacts faster in this kinetic reaction with bond formation, and the KIE is $\alpha = 1.029$, or $\Delta = 29‰$. So, if CO_2 in air currently has a carbon isotope value of $-8‰$, and conditions allow full expression of the fractionation, the KIE lowers this value to $-37‰$ for added photosynthetic carbon. However, other reactions often partly control the overall kinetics of photosynthesis, so that the final fractionation is usually reduced from $29‰$, to, for example, about $20‰$ for common terrestrial C_3 plants.

A Carbon EIE for Atmospheric CO_2

Carbon dioxide gas dissolves in water where it can further react with water. Dissolved CO_2 hydrates to form carbonic acid that then dissociates to bicarbonate, all in reversible exchange reactions:

$$CO_2 + H_2O = H_2CO_3 = H^+ + HCO_3^-.$$

In EIE exchange reactions, the rule is that the heavy isotope concentrates where it is most strongly bonded, and because dissolved CO_2 is not bound up with water (and is termed "free" CO_2), you might correctly guess that the heavy isotope of carbon concentrates in HCO_3^-. For this particular exchange at 15°C, $\alpha = 1.009‰$, or $\Delta = 9‰$, with bicarbonate about $9‰$ heavier than dissolved CO_2. The CO_2–bicarbonate exchange reaction plays out constantly on a global scale between the atmosphere and the ocean, where the ocean has an average bicarbonate value near $+1‰$. Further equilibria between bicarbonate and carbonate occur in seawater, but do little to change this overall fractionation relationship. Atmospheric CO_2 has a $\delta^{13}C$ value of $-8‰$, $9‰$ lower than that of bicarbonate due to equilibrium exchange. This CO_2–bicarbonate exchange thus largely controls the $-8‰$ isotope value of atmospheric CO_2 used by plants in the KIE above.

Box 7.4. Atomic-Level Isotope Compositions for Six Carbon Atoms in a Glucose Molecule

We know that DNA has four bases that spell out a language of genetic information. Isotope information also exists at a very fundamental level, the atomic level. Take the six-carbon molecule, sweet glucose. You might think all six carbons have the same isotope compositions. Wrong! If you extract glucose from corn it has a normal C_4-type carbon isotope value of −10.3‰. But if you break apart this sugar molecule, you find different carbon isotope values for each of the six component atoms (the diagram shows the numbers of the six carbon atoms in glucose that includes an oxygen atom in its ring form),

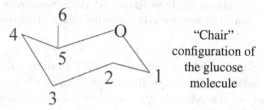

"Chair" configuration of the glucose molecule

measured $\delta^{13}C$

Carbon 1: −9.4‰
Carbon 2: −10.4‰
Carbon 3: −11.0‰
Carbon 4: −5.1‰
Carbon 5: −10.4‰
Carbon 6: −15.1‰

(from fermentation results for glucose from corn, Table 1 of Rossman et al. 1991).

You can see that atoms 1, 2, 3, and 5 have about the same isotope values, close to the −10.3‰ average value for the whole molecule. Why are the numbers for carbons 4 and 6 so different? It has to do with isotope cycling (fractionations) in carbon metabolism. Most interesting, however, is that nature is recording isotope information at this atom-by-atom or position-specific basis. Few studies currently work at this most detailed level because of technical problems, but we know that the information is rich at this fundamental atomic level, a gold mine for future isotope tracing.

studies of nitrogen in amino acids also show position-specific isotope information, providing a nuanced view of trophic dynamics in marine systems (McClelland et al. 2003). The last ten years have produced new technologies that allow isotope studies of individual compounds, and the future will likely bring atom-specific isotope studies being made on a routine basis (Brenna 2001).

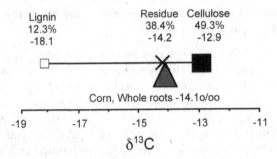

FIGURE. 7.2. Carbon isotope compositions of different biochemical fractions of corn (from Fernandez et al. 2003). The percentages of the three fractions lignin, cellulose, and the residue add up nicely to 100%, and the isotope mass balance is also very good, better than 0.1‰ agreement between the total and the summed fractions.

Against this detailed background, ecological studies usually focus at a much more general level, the so-called "bulk" level where whole tissues, organisms, or soils are analyzed as a composite or aggregated sample. The fractionation expressed at this bulk level really reflects the averaging of many effects that can be traced to isotope differences at the level of compounds (Benner et al. 1987; Figure 7.2) and atoms within compounds (Box 7.4). Ecologists working with bulk organic materials have worked to establish empirical fractionation rules at this bulk or aggregated level, rules such as, "You are what you eat" for carbon isotopes (DeNiro and Epstein 1976, 1978). A modern approach is to analyze these bulk materials and track isotopes with empirical fractionation rules, but to supplement this approach with molecular-level approaches known as compound-specific isotope analysis, or CSIA (e.g., Hayes 2001).

System Level Fractionation

Although fractionation is a process that occurs at the atomic level, scientists use a budget approach to keep track of the balanced effects of fractionation in reactions, following the isotope action in both substrates and products. Reactions convert substrates to products over time, so that if a system initially contains only substrate, it may contain only product at the end of a reaction. During intermediate stages of the reaction, both substrate and product will be present, but the sum of the two should add up always to the initial amount of material or substrate. This is the accounting principle of mass balance: that no matter how the total mass is split or divided between substrates and products, it all still sums up to equal the input. Said a different way, mass balance dictates that if the input material is not present in the product pool, then it must be in the pool of unreacted residual substrate. This subject of mass balance appears again and again in the isotope literature, and we revisit this topic in three more sections of this chapter, Sections 7.2, 7.6, and 7.7 that, respectively, give more detail on mass balance in closed systems, equilibrium systems, and open systems.

Similar to the overall mass balance, there is also an isotope mass balance that applies to substrates and products. The essence of the isotope balance is this: if light isotopes are accumulating faster in the product, then the slower-reacting heavy isotopes must be accumulating in the residual substrate. The details of how this isotope balance occurs varies according to inputs and exchanges in different systems, but there are two general frameworks or systems for understanding isotope balance during biological reactions (Figure 7.3). The closed system is usually distinguished by the lack of new inputs, and reactions progress in a sequential fashion over time, consuming substrate that was present at the beginning. Examples are closed-bottle batch cultures in the laboratory or plankton blooms in the sea. Plankton blooms that use up nutrients in a few days approximate closed systems where the nutrient uptake greatly exceeds new inputs from vertical mixing. The contrasting system is an open system in which substrate is continually supplied to the point of reaction, and products and residual substrate exit the system in a flow-through manner. Examples of open systems include continuous cultures with balanced inputs and outputs, and also our

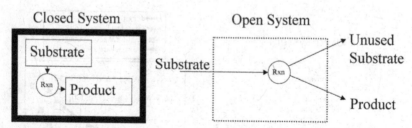

Closed System

Open System

- Examples: batch culture, bank account with only withdrawals allowed from initial capital, plug reactor.
- Substrate is added once, then *used up over time*, sequentially declining in concentration.
- Product accumulates over time (although some product can be exported; important point is that the system is closed and non-renewing for the substrate).
- Non-sustainable system, non-renewing, concentrations change over time.
- Reaction goes only forward and substrate pool is not renewed.
- Reaction proceeds over time as progressive consumption of substrate. Reaction monitored as cumulative "**fraction reacted**" of substrate.

- Examples: continuous culture, balanced bank account with deposits = withdrawals, flow-through reactor.
- Substrate enters and *exits continuously*, maintaining a steady state concentration at the "split" or "branch" point of reaction.
- Product exits the system.
- Sustainable steady state system, renewable system, concentrations remain the same over time.
- Unused substrate may return to larger pool of input substrate.
- Odd factoid: a sequence of linked open systems that consume substrate over time act as a closed system.
- Reaction proceeds in a steady state balance. Reaction monitored as steady state "**fraction reacted**" of substrate.

FIGURE 7.3. Comparison of closed and open systems that are important settings for isotope fractionation. Rxn = reaction.

global atmosphere where many atmospheric gases share or nearly share these types of balanced inputs and outputs. Some fine points of the distinctions between open and closed systems are elaborated at the end of this section.

The equations governing the isotope dynamics in closed and open systems are well-known, and in both systems, the extent of reaction or fraction reacted f is key to understanding isotope compositions. In the closed system, this fractional reaction also represents time, and isotope values for substrates follow exponential trajectories over time (Figure 7.4). There are two products in the closed system, the long-term product that accumulates and the short-term instantaneous product that continually forms instant by instant only to become part of the accumulating product pool. Section 7.2 considers closed systems in detail. In the open systems, there are simpler overall isotope dynamics (Figure 7.5), with linear changes in isotope composition. Only one product forms from substrate at an instant in time, then both product and residual substrate are exported. Overall, although there are differences between the two systems, especially difference in the equations, there is actually a good deal of similarity between the two systems of predictions for isotope changes accompanying f, the progression of the reaction (Figure 7.6).

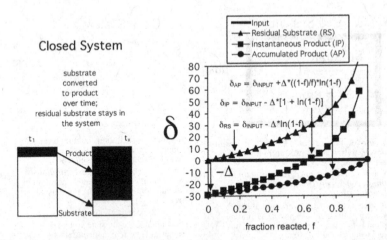

FIGURE 7.4. Isotope dynamics in closed systems. Equations are those derived in Technical Supplement 7B on the accompanying CD. Isotope fractionation is 30‰ in the illustrated example, and leads to lighter products (lower δ) and by difference, heavier substrates (higher δ). There are two products, one that accumulates over time (accumulated product) and one that is transient, forming instant by instant in time (instantaneous product). Isotopes in the instantaneous product track the substrate isotopes offset by the fractionation factor Δ, and mass balance imparts a more gradual trend to the isotope values of the accumulated product. See text and Section 7.2 for further details.

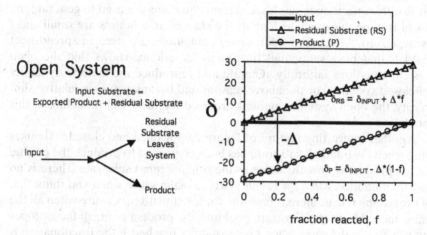

FIGURE 7.5. Isotope dynamics in open, flow-through systems where reactions are split or branched. Isotope fractionation is 30‰ in the illustrated example, and leads to lighter products (lower δ) and by difference, heavier substrates (higher δ). See text and Section 7.8 for further details.

The open and closed systems are most similar in isotope values of products and substrates when *f* is near zero, and both systems form products with the same low δ values (Figure 7.6). With the nearly identical loss of product when *f* is near zero, changes in δ values of residual substrates for both systems are also essentially identical. This is important for the I Chi modeling of this book. This near-identity for isotopes in products and a similar near-identity for isotopes in substrates is the reason why it makes

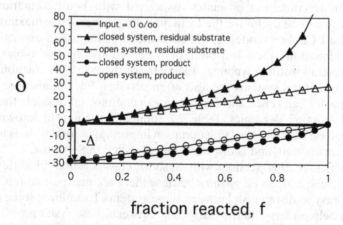

FIGURE 7.6. Comparison of isotope dynamics in closed and open systems. For simplicity, the instantaneous product of the closed system (see Figure 7.4) has been omitted.

little difference if open or closed system equations are used to generate the
I Chi isotope dynamics, as long as the changes in amounts are small and f
is near zero. Also, when open system equations are used in spreadsheet
models that have sequential time steps, experience shows that the open
system equations faithfully generate and reproduce the isotope dynamics
of closed systems. For the above reasons and because of their relative simplicity, the open system equations are used for the I Chi modeling in this
book.

Another interesting feature of Figure 7.6 for open and closed systems is
that when $f = 1$ and all substrate has been converted to product, the δ value
of the product is now the same as the original input substrate. There is no
net fractionation when $f = 1$. This starts to make sense when you think that
if everything reacts, there is no room for fractionation, because when all the
mass moves from the substrate pool into the product pool, all the isotopes
go too. But in the cases when f has not quite reached 1, the fractionation is
still there, with product isotope values less than those of input substrate.
Near 1, the effect of the fractionation has been split mostly into the substrate, rather than being expressed mostly in the product (Figure 7.6). We
considered this effect in Section 4.3 and consider this split fractionation
again in more detail in Section 7.2, using an I Chi model to help investigate
the split effects for both mass and isotopes together.

Overall, the effects of fractionation are often complex and split between
product and substrates, but this complexity can yield valuable insight into
how processes are split and balanced in nature. This is perhaps the good
news about fractionation. The bad news is that fractionation can upset
simple mixing models if it is not carefully measured and accounted for.

As you accumulate information about isotope dynamics in systems you
are particularly interested in, pay attention to fractionation estimates. The
maximum fractionation estimates associated with bond formation and
breaking are quite useful for the I Chi modeling. Given these maximal estimates, the I Chi box modeling will calculate any diminished expression of
this fractionation when diffusion or other competing reactions partly
control overall isotope dynamics. Estimates of these smaller fractionations
realized in natural systems are also often very helpful, and also are worth
tabulating for reference and for use in calculations and models that lack
detailed reaction dynamics. Table 7.2 summarizes some of known facts
about isotope fractionation. Fractionation important in food webs is termed
trophic enrichment, and is discussed and modeled in Section 4.7.

We conclude this section with a short consideration of differences
between open and closed systems because there are many situations where
it is not easy to distinguish between these systems. Examining some examples can help understand the distinction between these systems.

The most important distinction between open and closed and systems lies
with the substrate. Substrate is used up progressively over time in closed
systems without renewal or export, whereas in open systems, new substrate

TABLE 7.2. Ten Fractionation Facts.

1. Fractionation is an important counterpoint to mixing, with fractionation separating isotopes whereas mixing combines isotopes. Without fractionation and isotope separation, there would be no meaningful mixing or isotope tracing.
2. Fractionation is a difference in reaction rates for isotopologues, molecules that are twins differing by an isotope substitution at one atom.
3. Reaction rates for heavy and light isotopologues are much more similar than different, with differences usually less than 2% (20‰). Fractionation is generally subtle.
4. The lighter isotope reacts faster than the heavy isotope in "normal" kinetic isotope effects (KIEs). Faster reaction of heavy isotopes can occur in some relatively rare cases, yielding an "inverse" isotope effect.
5. At the level of simple compounds, chemical and physical theory can predict maximum possible fractionation by considering aspects of molecular bond strengths.
6. At the level of complex ecosystems, fractionations are usually the observed net results of several competing processes, many of which are poorly characterized. There is a strong empirical aspect to characterizing fractionations important in ecological settings.
7. Fractionation leads to predictable isotope changes in three different types of well-characterized systems: equilibrium systems, closed systems, and open systems.
8. Equilibrium isotope distributions are particularly important in geology where larger isotopic differences reflect lower temperatures, and provide a way to take the temperature of ancient rocks and reactions (see Section 7.6).
9. Closed systems were characterized over 100 years ago by Lord Rayleigh, and we still use his mathematics for distillations to guide many careful laboratory studies that establish ground-truth isotope fractionation factors.
10. Open, flow-through systems are common for the many elements whose cycling can be represented in box-and-arrow diagrams, and it is possible to build elegant I Chi models of isotope cycling for microcosms or the whole globe using only the simple algebra of open systems.

is added continually and unused substrate exits continually. As it turns out, the concept of an open system is not restricted to isotopes, but is more general and concerns balancing fluxes in and out of a reaction chamber (Figure 7.7). You might think that substrate builds up over time in the internal reservoir and can be used sequentially over time, but the flow-through dynamics provide a strong constraint of balanced inputs and outputs. On balance, substrate is used once to form product, with remaining unused substrate exported.

Now consider two examples that look similar, yet represent open versus closed systems (Figure 7.8). The top panel of Figure 7.8 represents a continuous culture, one type of open system. Substrate is used once or exported. But the bottom panel also represents a kind of continuous culture, where substrate is passed through a sediment plug containing bacteria. The successive use of substrate in this flow-through or plug-flow system will yield closed system dynamics. Do the differences make sense to you? The closed system dynamics arise when several seemingly "open systems" are linked. The system becomes closed because there is multiple (successive) use of same substrate and repeated product formation from the dwindling substrate pool.

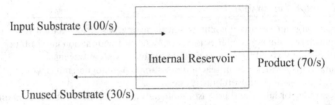

FIGURE 7.7. Example of an open system where inputs equal outputs. Because of this balance, open systems are steady-state systems at (or near) equilibrium. Reaction rates are given in arbitrary units of amount reacted/second. Note that the input flow pressurizes output flows, especially so that substrate enters but also exits. The internal reservoir can be small or large. Substrate flowing past a reaction site and exiting is the characteristic feature of open systems.

Here is another example. Consider methane gas diffusing from a landfill, with bacteria oxidizing the methane in both open and closed system settings (Figure 7.9). The open system (Figure 7.9, left panel) corresponds to methane passing a thin layer of bacteria poised at an aerobic–anaerobic interface, so that the bacterial consumption and methane loss proceeds all at once. But a closed system could also occur around some soil particles in gas pockets where methane arrives in bubbles (Figure 7.9, right panel), then is used up partially over time. Overall, this leads to a combination of open system dynamics and closed system dynamics in the same ecosystem (Figure 7.9).

In many systems such as this methane example, it seems possible that reactions proceed in a mixed way, with a part of the overall reaction occurring in open systems and part in closed systems. It is possible to allow for this mixture of effects in a simple way, by partitioning one fraction of the

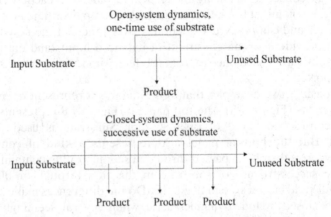

FIGURE 7.8. More examples of open versus closed system dynamics. Top panel: open system dynamics with a one-time use of substrate. Bottom panel: closed system dynamics develop in a plug-flow reactor when there is multiple successive use of substrate.

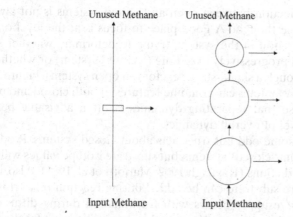

FIGURE 7.9. Both open and closed system dynamics can occur in some instances. In this example for methane diffusing upwards at a landfill site, methanotrophs (methane-consuming bacteria) at an aerobic–anaerobic interface can utilize methane essentially once in a flow-through, open system (left panel), or sequentially if methane bubbles linger near bacterial cells and are slowly consumed (right panel). Arrows to the right represent methane consumption by bacteria.

reaction to open system dynamics and the remaining fraction to closed system dynamics. For example, the isotope compositions of substrate exiting these systems can be readily predicted from well-known equations for closed systems (Figure 7.4) and open systems (Figure 7.5), leading to two types of isotope change for residual substrate, an exponential increase in δ values for the closed system and a linear increase in the open system (Figure 7.10). The result for a 50/50 mixed open plus closed system is intermediate between these two extremes (Figure 7.10).

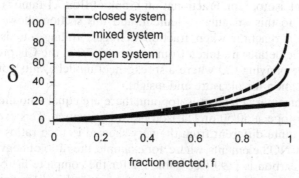

FIGURE 7.10. δ values expected for residual substrate in a closed system, an open system, and in a 50/50 mixed system that combines both closed and open system components. The fractionation factor is assumed to be 20‰ for substrate consumption in all cases.

Overall, the distinction between open and closed systems is not always easy, and takes some thought. A good place to focus is at the level of the most important reaction in the system, trying to determine whether substrate is being used progressively over time (a closed system) or whether it is being flushed through a single-stage reactor (an open system). Figure 7.10 illustrates that many systems can combine features of both closed and open system dynamics, so that separating dynamics is often advisable before summing to the level of overall dynamics.

Lastly, here are some odd but true facts about closed systems. Products can be removed from "closed" systems, but substrate isotope values will still follow classical predictions (Rayleigh 1902; Mariotti et al. 1981). Also, in at least one case, more substrate can be added during reaction in a "closed" system, and if this resupply occurs with fractionation during diffusional addition, the same "closed system" isotope dynamics will persist for the substrate (see CD, Chapter 7 folder, Problem 12). But generally, closed system dynamics depend on substrate remaining in the system, without resupply and without export.

7.2 Isotopium and Fractionation in Closed Systems

To help you sharpen up and practice your skills with fractionation, this section considers isotope fractionation in a closed system using an imaginary element, isotopium. Fractionation Frank of Box 7.1 fame is our imaginary guide to this imaginary element, and this section shows just what happens in a reaction when fractionation gets rolling and is taken to extremes. The details are taken from workbook 7.2 in the Chapter 7 folder on the accompanying CD where a spreadsheet model awaits you, to practice fractionation for pleasure and insight.

For the imaginary element isotopium, there are equal amounts of heavy and light isotope, a 50/50 or 1:1 isotope ratio situation. This even-handed 1:1 ratio is quite different from the very skewed isotope ratios that apply to all the HCNOS elements, where, for example, the ratio of heavy-to-light-isotope for carbon is 1:99 (see Table 2.1 for the complete list of HCNOS ratios). So here and now, ladies and gentlemen, and just for your viewing pleasure, Fractionation Frank has provided the equal opportunity element, isotopium.

The main thing about fractionation is that the light isotope reacts faster than the heavy isotope. Ok, that wasn't so hard was it? But there is more, and it is a little subtle, as follows. The light isotope is going to react faster, yes, but the heavy isotope is also reacting too. That is the important point. Both isotopes are busily reacting, it is just that one is reacting a little faster than the other. So, after a while, there is very little left of either the heavy isotope or the light isotope. Both reacted, poof, and are gone, all gone!

Chemists have a jargon phrase for how these reactions occur, the "first-order" reaction. Let's back up and see if we can understand this. Say you have 100 units of stuff, and every time step you take away one unit, leaving 99, then 98, then 97, and so on. This subtraction of equal amounts is called a "zero-order reaction;" that is, the amount subtracted does not depend on the concentration present. It is a flat fee at each step with no % discounts allowed. Moving on quickly (Fractionation Frank stands still for no one), now we come to the first-order reaction where the amount subtracted is no longer a fixed amount, but a percentage, like the tax you pay at the store. If you spend less, the tax amount is less. As a reaction goes along, and there is less and less of the starting stuff, the amount reacted is less and less as well, but remains a constant percentage. The fixed percentage always applies, and leads to a reaction that seems to be slowing down, approaching a final asymptotic value of nothing (zero).

So, here is the picture for isotopium, where our first-order reaction starts the race with 50 units of heavy isotope and an equal 50 units of light isotope, then starts losing both, reacting and disappearing away. The first-order reactions are gentle curves that we examine in detail, whereas the zero-order picture is simpler, a line that goes quickly to zero (Figure 7.11).

Well, now that we have seen the zero-order linear results for reference, we ignore them from here on out, focusing instead on the curves that represent the more realistic first-order results. Looking at this first graph, it is

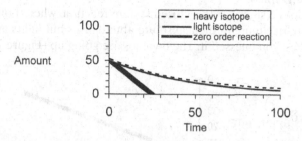

FIGURE 7.11. During a reaction, both heavy and light isotope molecules will disappear from the substrate pool to form product. In a zero-order reaction, the rate of disappearance is linear, a constant amount at each time step. In a first-order reaction, the rate of disappearance is a constant percentage, leading to an exponential decrease in substrate amounts. Differences between reaction rates of the heavy and light isotope are exaggerated in this drawing so that lines are farther apart than in reality.

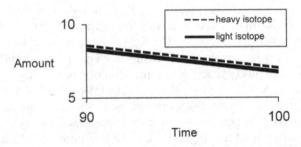

FIGURE 7.12. Magnified view of the first-order reaction dynamics shows the very slightly faster loss of light isotope than heavy isotope towards the end of the reaction, between time steps 90 and 100. Differences between the heavy and light isotope are not exaggerated in this slide that represents the cumulative fractionation effect over 90–100 time steps, with more heavy isotope material surviving than light isotope material.

actually hard to tell the difference in the reaction dynamics of the two isotopes. They look the same until we put the data under the magnifying glass, looking at a more detailed view of the reaction, for example, between 90 and 100 time units (Figure 7.12).

In this expanded view of what is happening between the times 90 and 100, we see that yes, the light isotope is indeed disappearing faster. The difference is small, but there is a little less of the light isotope, and a little more of the heavy isotope is surviving this constant drain to form product. As it turns out, all through the reaction, the isotope composition has been changing, with the heavy isotope left behind more and more. This dynamic is captured by δ, a good index that tracks the relative amount of heavy-to-light isotopes. As the light isotopes react faster, the heavy isotopes become relatively more abundant (although, and this is tricky until you think it through, the heavy isotope is also getting absolutely less abundant as shown in the above graphs).

So what does δ show for this closed system reaction, where both heavies and lights are declining in their absolute abundances but lights are declining faster? Yes, you guessed it. The δ values are going up (Figure 7.13). The

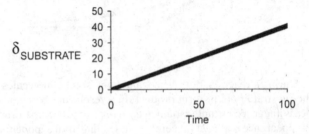

FIGURE 7.13. The faster reaction of light than heavy isotopes shown in Figures 7.11 and 7.12 leaves the shrinking substrate pool relatively enriched in heavy isotopes, and higher δ values (‰) record this enrichment in an easy-to-see fashion.

ratio of heavy-to-light is 1 in the beginning (the initial ratio of 50 units of heavy and 50 of lights = 50/50 = 1), but increases to 1.04 at the end of this example. This is not a big increase, is it? It is, hmm, let's see, a 4.0% difference, or 40‰ difference. Yes, that is what we have got here, a small difference magnified up via the δ definition.

Let's summarize. Reactions consume both heavy and light isotopes, so that both disappear over time. But fractionation magic is that there is a slightly faster disappearance of light than heavy isotope, leading to a relative enrichment in the remaining heavy isotope. That is the secret of fractionation. Also, although the differences start out small, they get larger and larger over time as less and less material remains. The exponential decrease in amounts transforms in the isotope mirror to an exponential increase in the δ values, with highest δ values at the end of the reaction.

Let's take a break for a moment, relax, and consider fractionation again in a chocolate example: You are eating a mixture of white and dark chocolates, but are a little selective for the color (say white) that appeals to you. Pretty soon most of the chocolates have vanished, and you pause to see what's left. You see that the white chocolates have gotten very rare, whereas the dark chocolates are more evident without the whites to fill out the background. There is an increasingly (heavily) dark selection left from which to pick. In analogy to the isotopes, the chocolate amounts are decreasing, but the heavy darks are relatively more and more abundant. As this is sinking in, well, those chocolates wait for no one (Fractionation Frank loves 'em too), so get busy eating! And then poof! All the chocolates are gone (we couldn't stop ourselves, could we?), and it's time to start this example again, isn't it?

You can practice fractionation with isotopium and other elements in a spreadsheet model (see CD, Chapter 7 folder, I Chi Workbook 7.2, and click on a tab in the bottom left "2 graphs" to see the first worksheet model). Change any of these three master variables highlighted in bold in the upper left of that worksheeet, and watch what happens to the isotope dynamics:

Reaction rate (% per time step):   2
Isotopic fractionation, ‰:   20
Abundance of heavy isotope (out of 100):   50

The third line is fun; in the spreadsheet, you will find a lot of choices about which elements with which to work. Try substituting elements such as carbon or nitrogen for isotopium and see if the reactions change. What do you guess or predict? (Note: the ratio-of-ratios is important as you change elements. See Section 2.4 that explains how the δ notation normalizes isotope values across elements, accomplishing this by dividing isotope compositions of samples by the isotope composition of a standard.) Also, if you change the isotope fractionation factor from 20 to 500‰ (see workbook 7.2

on the CD), this makes for a much larger change in reaction rates, 50% instead of 2% faster reaction of the light versus heavy isotope. With this change to 500‰, can you see more easily that the heavy isotope is reacting more slowly?

When you have finished this first part, go on to the next worksheet labeled "3 graphs". There you will find an additional graph that gives the fractionation factor calculated as the slope of a line. The x-axis of this graph involves a logarithmic function, and derives from the exponential reaction rates of the light and heavy isotopes shown in the topmost graph of the worksheet. It turns out that when reaction rates are exponential, logarithmic plots of δ versus the amounts or concentrations measured during a closed system reaction will yield the fractionation factor as the slope of a line. This is very convenient, and explained in several pages of math that the interested reader can consult (or not) in the Technical Supplement 7B on the accompanying CD. In any event, you should be able to see that if you change the isotope fractionation factor, the δ values of the substrate will change, as will the slope of the fractionation line (Figure 7.14).

Moving on to the next worksheet labeled "product", we drop detailed consideration of the substrate for the moment, and include the other side of the reaction, the formation of product (Figure 7.15). Manipulating the reaction rate, we find that more or less product forms from the substrate, in a mirror image. It makes sense—substrate disappears as product forms—but nothing is lost from the system, so the overall total remains the same (= 100 in the example of Figure 7.15). This is the accounting principle of mass balance, and can be written: $mass_{INITIAL} = mass_{SUBSTRATE} + mass_{PRODUCT}$.

We can replot these same dynamics using a different x-axis, the fractional extent of reaction, or fraction reacted, with 0 representing no reaction, and 1 representing complete conversion of substrate to product. This presentation makes straight lines of the exponential reaction rates (Figure 7.16), but

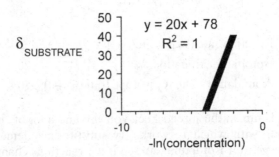

FIGURE 7.14. The isotope fractionation factor can be extracted from the reaction dynamics, when data are plotted on an (x,y) basis with $x = -\ln(\text{concentration or amount})$ and $y =$ the δ value of the remaining substrate. The slope of the line gives the fractionation factor in ‰ units, 20‰ in this case.

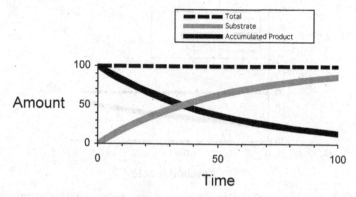

FIGURE 7.15. Mass balance accounting helps us follow reactions in closed systems, with mass balance meaning here that the amount of substrate plus product always adds to a fixed total. Or, said another way, as substrate is converted to product, nothing is lost and the total of substrate plus product always sums to 100%.

the message about the total is the same: substrate plus product always sum to the same value in a closed system, the starting input value.

A further graph from Workbook 7.2 shows the isotopes in the substrate and product, with the total now representing the average of the two isotope pools, weighted by the relative amounts of substrate and product at each time step (Figure 7.17). Mass balance or weighted averages apply for isotopes as well as total amounts, following the mixing equations introduced in Chapter 5,

$$\delta_{INITIAL} * mass_{INITIAL} = \delta_{SUBSTRATE} * mass_{SUBSTRATE} + \delta_{PRODUCT} * mass_{PRODUCT}.$$

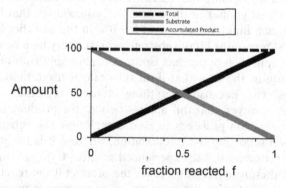

FIGURE 7.16. Mass balance accounting also applies when the x-axis of reaction changes from time (Figure 7.15) to fraction reacted to form product or fractional extent of reaction (this figure). Expressed on this basis, the changes in substrate and product amounts are linear, rather than exponential as in Figure 7.15.

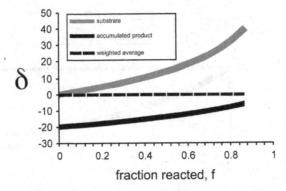

FIGURE 7.17. Isotope changes also conform to mass balance during reactions, with faster segregation of light isotopes into product balanced by increased heavy isotope content in the residual substrate. The isotope balance of the system remains constant at the input value of 0‰, when the mass balance is the mass-weighted average of (amounts × isotopes) or: $\delta_{INPUT} * \text{mass}_{INPUT} = \delta_{SUBSTRATE} * \text{mass}_{SUBSTRATE} + \delta_{PRODUCT} * \text{mass}_{PRODUCT}$.

The total system δ value shows a constant value equal to that of the input, and this is expected when we are just converting substrate to product with no losses or gains.

And as it turns out, the isotope curve for the product is calculated from that of the substrate, explicitly under the assumption that the isotopes accumulated in the product are those lost from the substrate, and that the sum of product plus substrate remains constant. This is also mass balance at work, so that when all material has been converted to product, product isotopes must match isotopes in the original substrate.

There are other interesting features of these seemingly simple graphs. For example, if you manipulate the master reaction rate variable (% reaction) to values of 5% or greater in this worksheet you will see that the product isotopes approach input values (input $\delta = 0$‰ in the workbook example), even as the last remaining bit of substrate achieves very high isotope values. This eventual approach of product isotopes to the input isotope value also finds expression in the idea that if all substrate is used, there is no fractionation observed because everything is used, regardless of isotope content. Note, however, that this applies only to the product, not the substrate. As the reaction proceeds to near completion, the substrate can be highly enriched reflecting the fractionation, whereas it is the product that shows little fractionation. The experienced scientist trying to estimate fractionation will therefore keep an eye on the product if the reaction has not proceeded very far, but on the substrate if the reaction is nearing completion. In the middle, at 20–80% reaction, both the substrate and product isotope values readily indicate the fractionation dynamics, with high isotope values for the substrate and low isotope values for the product.

The final graph in Workbook 7.2 draws together much of the previous three worksheets, with one new addition in the final right-hand graph. This is the "instantaneous product", a product that blinks into existence then joins the accumulating product pool (Figure 7.18). There are thus two product pools, the instantaneous product and the accumulated product. These two pools can be understood by analogy to percolated coffee that drips drop by drop into your coffee pot. The percolating drops are the instantaneous (coffee) product and the coffee you drink from the pot is the accumulated (coffee) product. In another example from the laboratory, bacteria growing in a closed reactor with glucose will respire CO_2 that builds up as an accumulated product, but we could also continually flush out the headspace of the vial and trap the CO_2 given off at any instant in time (the instantaneous product) before it accumulates.

When we collect the instantaneous product from the middle of a reaction, we find it is always offset in isotope composition of the substrate by the fractionation factor, that is, 20‰ in the illustration of Figure 7.18. Indeed, this is the heart of the reaction, with the lighter isotope always bleeding out of the substrate faster than the heavy isotope, and with the fractionation factor giving this difference in loss rates. Note that because this is an instantaneous product, it never accumulates, and has zero mass, so we ignore it in the mass balances of this and the previous worksheet. That is life with the instantaneous product. Now you see it, now you don't.

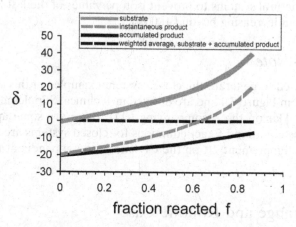

FIGURE 7.18. As Figure 7.17, but with a second product, the "instantaneous" product that forms from substrate but is quickly passed to the total accumulated product. In some reactions, the instantaneous product can be continuously monitored as a gas sparged out of a reaction vessel. In such cases, the system is closed to substrate (no new substrate is added or exported), but still open to product loss.

Overall, the graphs of this section and workbook 7.2 depict the dynamic nature of closed systems, where Fractionation Frank manages to operate with remarkable discrimination. The illustrations show that fractionation creates products with lower isotope values and leaves the residual substrate with higher isotope values. Separating the isotopes into heavier substrate and lighter products via fractionation sets the stage for nature and humans to have mixmaster fun later on, reuniting the separated isotopes into various interesting blends and mixes. Try changing a few of the master variables in workbook 7.2 in the Chapter 7 folder on the accompanying CD. Surprise yourself, amaze your friends, and feel good that you are starting down the trail that Fractionation Frank blazed early on, back when the universe was young.

P.S. One reader wondered if the substrate isotope values really will reach infinity when essentially no substrate is left over. This is indeed the prediction from the equations. However, really high substrate values >1000‰ are seldom seen in nature for two reasons. One reason is that there are usually other sources of substrate lurking about, or being regenerated slowly from the main reaction, and adding in these source contributions will keep isotope values down. The other reason is that as substrate concentrations approach zero, diffusive supply to reaction centers may become slower and slower, so that diffusion starts to limit the reaction rate and all substrate is used upon reaching the main reaction center, regardless of isotope composition. Summarizing, towards the end of a reaction when almost all substrate has been consumed, one might expect to see routinely very high substrate isotope values. But a combination of mixing from low-level sources and reduced fractionation at very low substrate concentrations usually acts together in natural systems to prevent isotope values of the last little bit of substrate from increasing beyond 1000‰.

Technical Note

Equations used to generate the closed system examples of this section are those shown in Figure 7.4 and are derived in Technical Supplement 7B (see Chapter 7 folder on the accompanying CD). There are some approximations in these equations. Exact equations for closed systems are given also in Technical Supplement 7B on the CD and in the appendix at the end of this book.

7.3 A Strange and Routine Case

Every day in isotope laboratories around the world, isotope machines are busy at their routine work, measuring, monitoring, and, if you look closely enough, mystifying! What could be mystifying about routine work? The isotope measurement itself can be quite complex and a little mystifying. Let's look at an example.

We enter an isotope laboratory where machines are taking in samples and producing δ values in routine computerized work. We sit down to watch the action on a computer screen. This is a modern continuous flow system we are watching, where the CO_2 gas has been prepared and purified in an upstream elemental analyzer, then blows by our detectors in an isotope ratio mass spectrometer (IRMS). So now, sportsfans, let's follow the action (Figure 7.19).

We see nice symmetrical peaks rolling by the isotope detectors, beautiful curves that go up to peak values, then back down to the baseline (Figure

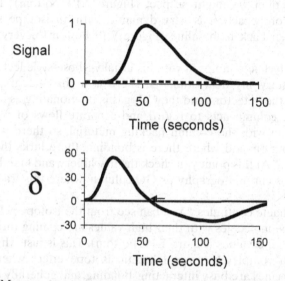

FIGURE 7.19. Mass spectrometer measurements of CO_2 gas in a helium carrier stream. Top panel: the mass spectrometer detectors measure the CO_2 amounts, here normalized so that 1 = the maximum CO_2 signal at mass 44. The chromatogram is from an actual sample analysis with a 3 m GC column. The CO_2 has been generated in an elemental analyzer and passed over a gas separating column, or gas chromatography (GC) column, to purify it from other combustion gases. Computer programming integrates the total detector response and subtracts a baseline value to quantify the amount of CO_2 in the peak. Integration proceeds for three measured isotopologues of CO_2 at masses 44, 45, and 46. Detector responses for only masses 44 (solid line) and 45 (dashed line) are shown here, with no amplification of the small mass 45 signal so that this peak reflects its true natural abundance of about 1.1% versus the mass 44 signal. The mass 46 signal is smaller still. Bottom panel: isotope changes in the CO_2 gas as it passes the mass spectrometer detectors, with heavier components (positive δ values) arriving first, followed by lighter components (negative δ values). Computers integrate the isotope signals for the entire peak using the separate ion beam measurements made at masses 44 and 45, then use the integrated signals to calculate isotope ratios and δ values. The integrated value for the peak shown here is δ = 0.95‰ when the baseline value is used as a reference value of 0‰. The peak maximum at 53 seconds has an associated instantaneous δ value of 5.2‰ indicated by the arrow.

7.19, top). The two curves are the separate kinds of CO_2, "light" CO_2 at mass 44 and "heavy" CO_2 at mass 45. The heavy CO_2 has an extra neutron and is rare, only a minor peak really (Figure 7.19 top). But the machinery normally amplifies up this peak to roughly match the much larger mass 44 signals, and we can imagine integrating these peaks for their total areas. It is the ratio of these areas that gives the measured isotope ratio and δ value. Integrating peaks and calculating ratios and δ values is what the computer does so well and so routinely. If the action stopped there, we could go home happy. But alas, someone has actually calculated the isotope information on a second-by-second basis and the isotope measurements are more complicated than we might suspect (Figure 7.19, bottom). The second-by-second isotope action is very dynamic, with an isotope swing, high, then low, then back to baseline (Figure 7.19 bottom). Everyone puzzles over this.

So, Mr. Polychaete jumps in with his usual verbose-style lecture. "It is all very elementary, my dear students. There is a chromatography column upstream of the detectors, and the CO_2 gas is fractionating as it crosses the column. The column acts to retard and separate flows of passing gases via interaction with the column packing material, so there is a transient bonding going on, and where there is bonding, there lurks the danger of fractionation." At this point you check the machinery, and find the upstream column, a gas chromatography or GC column, where this fractionation is occurring.

Mr. Polychaete continues. "You can see from the isotope action that the heavy isotope molecules with their high values are coming through fastest giving the high δ values (Figure 7.19, bottom). This is just what we might expect for a completely normal kinetic isotope effect where the light isotope molecules are busy interacting, bonding, and generally having a big old time during their visit to the GC column. Some of the heavies get caught up in the fun down on the column, but a greater fraction of the heavies give the column a pass and sail on. The heavies to win the race to the mass spectrometer detectors, so that δ values rise, sailing up the "heavier = higher" highway to heaven. That is the first part of the peak, where the heavies are winning the race to the detectors. The lighter isotopes lag behind and lose the race, showing up in the later lighter part of the overall CO_2 peak."

We ask a scientist to repeat this explanation, to make sure we understand. "There are important isotope kinetics developing across the upstream GC column. The heavy isotope component that is not strongly bound to the GC column arrives as the first part of the peak. The later part of the peak is composed of the lighter isotope component that first accumulated on the column, but finally let go and swept along to the detectors. This is reminiscent of closed system dynamics for isotopes, with progressive formation of light product on the GC column that leaves gas-phase, heavier CO_2 as a 'residual substrate' that is swept on out of the column.

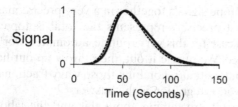

FIGURE 7.20. As Figure 7.19, top panel, but the mass 44 and mass 45 signals have been scaled similarly, so that 1 = maximum peak height in each case. Due to isotope fractionation on the upstream GC column, the mass–45 CO_2 (dashed line) moves more quickly than mass–44 CO_2 (solid line) so that the mass–45 peak leads at the front and back of the mass–44 peak. This isotope-related difference in the timing of the peaks is very slight, and has been magnified severalfold in this figure. The faster movement of the mass–45 peak causes the swings in the instantaneous δ values shown in Figure 7.19 (bottom).

"However, there is a simpler way to understand this, just that the mass 45 peak goes through the GC column faster than the mass 44 peak. The difference is very slight as you might expect of isotope fractionation kinetics, but if you look carefully at Figure 7.20, you can see that the mass 45 peak leads slightly in the beginning and at the end, consistent with a faster overall transit rate across the GC column."

The scientist pauses, then adds a little extra information. "An important detail is that not all gases show this heavy-to-light isotope swing seen in Figure 7.19 (bottom). This heavy-to-light swing seems characteristic for gases such as CO_2 and SO_2 that bind strongly to GC columns. You can routinely observe the heavy-to-light isotope swings for CO_2 and SO_2 eluting from GC columns in elemental analyzers. But these GC interactions with other gases such as N_2 are much weaker, and correspondingly the isotope swings are also much weaker for N_2. Also, and here is an interesting part, the nature of the gas interaction with the column can change. The light isotope gas can come through first with N_2, a result not expected for normal kinetic isotope effects, but a result that is consistent with equilibrium isotope effects. In these reactions, heavy isotope components are left on the GC column, so that light isotope N_2 comes out first. Overall, the sample preparation devices produce many different types of fractionations in slower versus faster peaks. But in the end, it is rather simple for the computer. The task is just to integrate the total areas under the peaks and divide these areas to get the isotope ratios and δ values. So it is only the final peak areas that really count, regardless of fractionation patterns, and peak areas are (relatively) easy and routine to measure sample after sample after sample."

We understand this explanation about peaks, but decide it is also a minor miracle that the machinery works with all this isotope separation and fractionation. It still seems amazing that the machines can integrate and combine those isotope signals together in a very precise manner to get one single number that precisely represents the total isotope content of the sample. A strange case for this very routine, automated work, a strange and routine case indeed. We puzzle it out, then throw up our hands and thank heavens for computers that recombine those wavy fractionated isotopes to get the right answer. Integration is the answer.

Later over beer we are talking about this and the laboratory scientist drifts by. She adds, "You know, it is worse than you think. Ok, the computer is reuniting those GC-fractionated isotopes, but there is usually additional fractionation in various dilution and splitting devices in the sample preparation systems. Many seemingly small details in these gas handling systems upset and fractionate the isotopes. But modern computers handle it all, especially by running reference samples through the same steps as the samples in these multifractionating sample preparation devices. It all boils down to simple peak detection programs that are very good these days. In the old days, we prepared samples by hand, and were careful to get 100% yields and introduce sample and standard gases to the mass spectrometers in nonfractionating ways. But those days of careful manual work have yielded to automation, with only minor losses in precision, and even those "minor" losses get "more minor" each year. It is getting to be beyond me. These days, you only have to add your reference samples, tweak your baseline, and make sure your peak identification programs are good. Then you can ignore all those fractionations, and leave it to the computers." The scientist sighs and wanders off.

We sit a while longer, thinking that maybe in a few more years, the computers will move beyond data generation, and take over data interpretation as well. Where will that leave people? We smile as we think about this welcome "problem," then all drink to the day when we will have more free time.

7.4 A Genuine Puzzle—Fractionation or Mixing?

Nitrogen isotopes (nitrotopes) are proving useful tracers of human pollution in many settings around the world. Nitrates can build up to high levels in some groundwaters with strong inputs from agricultural runoff, and nitrate concentrations greater than about 10 mg/l can cause human health problems. A classic study examined high nitrate levels in groundwaters of France, and measured the data shown in Figure 7.21. The high levels of nitrate at the right of the graph were undoubtedly due to pollution, but what about the low levels of nitrate at the left of the graph, the points with the high $\delta^{15}N$ values? Investigation showed that the data could be understood from two different viewpoints. In the first scenario, microbial respiration of nitrate, or "denitrification," a process that converts nitrate to mostly N_2 gas, was removing nitrate while fractionating in a closed system manner. Deni-

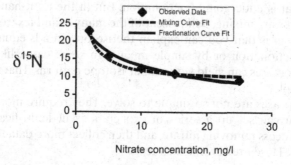

FIGURE 7.21. Nitrogen isotope values of nitrate as a function of nitrate concentration, and two curves fit to the same data. (Data are averaged from Figure 8 in Mariotti et al. 1988.)

trification would leave residual nitrate at low concentration and high $\delta^{15}N$. This seemed logical, but the calculated fractionation factor was small, about 5‰ instead of the usual 10–30‰ typical of denitrification.

A second scenario was that there were two sources of nitrate mixing in the groundwater, an agricultural pollution source with high concentrations and low $\delta^{15}N$, and a poorly characterized second source with low concentrations and high $\delta^{15}N$. A source with such high $\delta^{15}N$ was not typical, except for some instances where manures were involved.

So, was it scenario 1 or scenario 2? The answer was that it was hard to tell, and that at first, it was a genuine puzzle. How puzzling? Consider this, for example. Curve-fitting routines gave equally strong linear correlations for both scenarios, and it was really not possible to use statistics to distinguish between the two explanations: that is, r^2 values were the same, 0.97 (Figure 7.22). The left-hand plot in Figure 7.22 represents the closed system or Rayleigh conditions discussed in Sections 7.1 and 7.2, with a single source

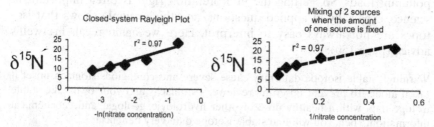

FIGURE 7.22. Derivation of the two curves shown in Figure 7.21. Data trends are explained equally well by the two opposite approaches, which involve fractionation (left; see also Section 7.2 for more details on closed system fractionation) or mixing (right; mixing analyzed with the Keeling plot approach of Sections 3.5 and 5.7).

of nitrate that is undergoing fractionation. But in the right-hand plot, two nitrate sources are mixing, reminiscent of the many examples from Chapter 5. The problem is that you can explain the isotope trends equally well by Rayleigh fractionation or by simple mixing. In some cases, like this one, these two processes can yield very similar isotope patterns. That is the point of this example.

Well, such cases are not so simple to solve. They require more data. For instance, you can set up a series of expectations if denitrification is the important process removing nitrate, and then collect more data to test these expectations. Here are several such tests.

1. Denitrification usually occurs in low-oxygen aquifers, so one could examine the water quality data for low oxygen content expected if denitrification were most important.
2. If denitrification were removing nitrate, then there should be less nitrate with higher $\delta^{15}N$ values as water moved through the aquifer. Did this removal occur along downstream transects?
3. A third possible test would be to measure the oxygen isotopes in the nitrate: do they also rise at low nitrate concentrations, a result expected for denitrification?

Getting answers to these three questions would help solve the puzzle of fractionation versus mixing.

The authors of the original study reasoned that indeed denitrification was the agent at work, so that even as agricultural activities were adding nitrate to groundwater, microbial processes were busy stripping out this nitrate in decomposition reactions. This view of nitrate dynamics might lead you to think groundwaters have a certain capacity to remove pollutant nitrate loads, a quite different view than you might have if only mixing were occurring. If it were only mixing, you might get worried that pollution levels would be getting unacceptable. Managers and citizens may care about the distinction between such fractionation versus mixing interpretations when these interpretations lead to different perspectives on capacities to handle pollutant loads. So, getting the interpretation right is often important in science, especially in applied questions, and this example shows that isotopes are not always easy to interpret. Here we again recall Fretwell's advice (page 150):

Warning! Stable isotope data may cause severe and contagious stomach upset if taken alone. To prevent upsetting reviewer's stomachs and your own, take stable isotope data with a healthy dose of other hydrologic, geologic, and geochemical information. Then, you will find stable isotope data very beneficial.

Moral: Fractionation is not always easy to interpret, but it can work for you if you also work to gain a broader understanding of your system.

7.5 Cracking the Closed Systems

You might think that the world would be full of sky-high isotope values if fractionation were busy taking out all the light isotopes, leaving the world full of higher, heavier values. But it isn't so. So, although we can learn about ideal behavior from the closed system fractionation models, they usually overestimate what happens in the real world where heavy isotope extremes appear only rarely. Other things intervene, and here we make a fanciful exploration of two such examples.

Take tuna. There they are, out there in the equatorial Pacific, feeding from an oceanic conveyor belt that slowly grows tuna food. Those of you familiar with the climate and current systems may know something of this: deep water comes up near the equator in the central Pacific, then moves on the surface both northwards and southwards. This deep water is rich in nutrients, and when it reaches the surface lit by the sun, plankton blooms result, starting the food web that soon leads to tuna. The algal use of nitrogen nutrients in this upwelling food web might lead to very high $\delta^{15}N$ values if fractionation were the only agent at work, but as it turns out, fractionation is only part of this story.

As water moves north and deep-water nitrate is depleted (Figure 7.23), nitrogen fixation of N_2 gas begins to be an important alternate N source (Figure 7.24). N fixation by cyanobacteria happens in the open sea, although admittedly we are just beginning to find out the details (Montoya et al.

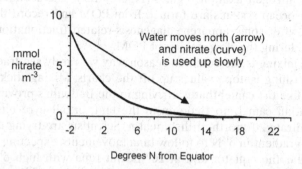

FIGURE 7.23. Diagram of nitrate dynamics near the equator in the Pacific Ocean. Deep water comes to the surface near 2° S (−2° N), then moves northward. This water is rich in nitrate. As water moves north, the nitrate is used up, fueling plankton blooms and eventually a food web that supports tuna production in the region.

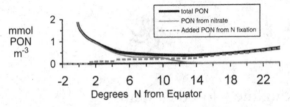

FIGURE 7.24. Particulate organic nitrogen (PON) forms from nitrate in upwelled equatorial waters, and from nitrogen fixation in waters farther to the north. Some PON sinks out and is removed from the system, and some is used in the local food webs.

2004). So let's take a spin across the Pacific, using an eye-in-the-sky isotope satellite. We see that the amount of N in particulate organic matter (POM) has a changing source, nitrate near the equator and N fixation in the north (Figure 7.24). These sources have different isotope values that put a mixing switch into the isotope baseline of the food production system.

Looking near the equator where there is little N fixation, we expect loss of light N from the nitrate pool, leaving remaining nitrate isotope-heavy, nitrate that is moving to the north. But when we start to run out of this nitrate that has increasingly high $\delta^{15}N$ values, we start to run into other things, in this case, small amounts of new N provided mostly by N fixation north of the equator (Figure 7.25). The low isotope values in the north eventually outweigh the higher nitrate-related values because, well, because there is hardly any nitrate left. So, as you run out of nitrate, you tip over into another production system, a slowly spinning ocean gyre system without regular upwelling and with a nitrogen budget based largely on N fixation. In the middle, values rise as nitrate is still important, but then go down again to the north as nitrate levels approach zero and the system tips over into a nitrogen fixation regime (Figure 7.25). The sediments below these upper ocean systems are formed from POM and record this spatial ^{15}N pattern, after some apparent diagenesis-related fractionation and ^{15}N enrichment during the slow sinking of POM.

To recap, mixing is the common reason why we rarely see fractionation in nature pushing isotope values up off the charts. Mixing sticks its nose in the tent, like the camel that is arriving in the Bedouin's proverbial tent. Pretty soon, all camel, no Bedouin in the tent, or, all gyre nitrogen, no upwelling nitrate 20° north of the equator. Scientists are trying to use this geographic gradient in $\delta^{15}N$ to follow tuna movements, expecting tuna with low $\delta^{15}N$ near the equator and at 20° N, but tuna with high $\delta^{15}N$ in the middle. In the end, the $\delta^{15}N$ map for tuna should match the $\delta^{15}N$ map known for sediments if the tuna are fairly resident and don't move too much.

Now onwards and downwards, to look at a second example of closed systems that are not truly closed. Here we go to the bottom of the sea, to the mud, with Tracey the Tuna as our guide. Tracey has this to say about the sulfur cycle down there in the mud. "Those mud bacteria make their living converting one kind of sulfur to another, sulfate to sulfide. They mine the oxygen out of those sulfate molecules in an old-time anaerobic respiration scheme, excreting the waste toxic sulfides that will turn metals (and skin) black. Part of their operation down there in the sulfate mines is isotopes, of course. Extract the lights faster, leave the heavier residue, you know the drill: light sulfide products, heavy sulfate substrates that are left over. Fractionation Frank teaches us all about it in plankton school, pre-larvae class, you know. Too bad you humans have such a poor educational system and don't learn this stuff till graduate school. But then that's the way of the world, isn't it? Look, to make it easy for you all, I made a special chart (Figure 7.26). Sulfate diffuses down from the water at the top of the mud, and heavy isotope action ensues as the sulfur microbes go after it."

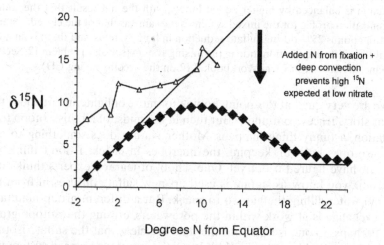

FIGURE 7.25. Isotope dynamics expected for N dynamics depicted in Figures 7.23 and 7.24. Closed diamonds = predicted PON; triangles = sedimentary organic nitrogen $\delta^{15}N$ from tops of cores in the region; diagonal line = predicted PON from nitrate only; horizontal line = predicted PON from gyre with nitrogen fixation. Closed system expectations for upwelled nitrate lead to the prediction of ever-higher $\delta^{15}N$ values in PON (diagonal line), but this does not occur. Instead, as nitrate levels approach zero, other N sources become important, in this case N from N-fixation and nitrate recycling in mid-ocean gyres located north of the equator. The isotope values in the northern gyres eventually reflect those of the more abundant N-fixation sources. Sediment values are those measured by Francois and Altabet (1994), and are higher than $PO^{15}N$ values in the upper ocean presumably because of fractionation while organic matter degrades and settles to the bottom (Saino and Hattori 1980, 1987). The ^{15}N dynamics were calculated with an assumption of 5‰ as the $\delta^{15}N$ value of nitrate, a fractionation of $\Delta = 5$‰ during nitrate use, and a 3‰ value for the northern ecosystem PON that has strong contributions of N fixation (Dore et al. 2002; see also problem 11, Section 7.14 and the answer in I Chi Workbook 7.11 on the accompanying CD).

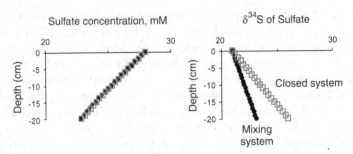

FIGURE 7.26. Sulfate concentrations (left) and isotope compositions (right) in a sediment system that is closed to new inputs (squares) or has new inputs via mixing (circles). Concentration gradients are the same in both systems, but isotope profiles differ, reflecting added sulfate in the mixing system. (The sulfate reduction rate is higher in the mixing system than in the closed system, but higher loss to sulfate reduction is balanced by higher added inputs, with the net result that the sulfate concentration profile for the mixed system is the same as that of the closed system.) Fractionation is 25‰ during sulfate reduction in both systems, and there is an additional 10‰ diffusive fractionation in the mixing system (see also problem 12, Section 7.14 and the answer in I Chi Workbook 7.12 on the accompanying CD).

We are very quiet at this point, hoping for more enlightenment about this sulfur story. Tracey continues, "But then in the midst of all this isotope fractionation, a funny thing happens. Mother Nature does something to add light isotopes back in, keeping the microbes in check. I don't think the humans have figured it out yet. One school of learned experts thinks it is diffusion, you know, as sulfate is used up, new sulfate diffuses in from the bottom water, filling the gap, so to speak. But a different group maintains that exchange is at work within the porewaters, erasing the isotope gradient. Perhaps no one is right, but the data are clear that the sulfate isotope values don't rise like they should if it were a completely closed system in a bottle." Tracey the Tuna laughs, "Something fishy is going on here!"

We look more perplexed than ever, so she also adds, "Ok, this is how it is. Even those slow-moving sediment systems are not really closed systems. Although the experts disagree on the details, they see that the system is open to some outside influence, to other processes such as stirring and diffusion, and are not completely closed off and sealed up. You humans need to understand that those closed systems don't happen that much, mostly only in people's dreams.

"There were some close-minded scientists who visited once and tried to make sense of it all, but they got the details wrong, you know. The basics and all that, sure, but they didn't understand the real action. Important point they missed: although the microbes fractionate and separate, Mother Nature restores. Keep that in mind next time you start thinking about closed systems or closed combat, something likely to drop in unexpectedly from the outside, you know?"

Leaving us to ponder her words of wisdom, Tracy swims off in a great swirl into the deep blue of the Pacific. So we ascend back to our boat, climb out, and idle back in the sun. It has been a long day on the ocean. But as the days pass, some of this starts to sink in.

The message here is that no woman is an island, and no system is truly closed. Maybe yes, in the laboratory bottle, we find the closed systems, but Mother Nature keeps probing along and generally prevents true closure in her creations: action, reaction sort of thing, really. And perhaps Tracey the Tuna was right, we humans should be learning all this in pre-L (pre-larvae) school. Maybe better late than never, though, and after all, life is not a closed book, is it?

7.6 Equilibrium Fractionation, Subtle Drama in the Cold

In this section, we consider a special class of isotope effects, equilibrium isotope effects occurring in exchange reactions. These effects are simple in many ways, yet important in the biosphere. For example, equilibrium exchange between the atmosphere and the ocean largely controls the carbon isotope value of atmospheric CO_2, with atmospheric CO_2 currently near $-8‰$ because of equilibrium exchange with $+1‰$ inorganic carbon in the surface ocean. The $-8‰$ atmospheric CO_2 value in turn strongly influences the $\delta^{13}C$ values of plants that fix CO_2, and animals that eat the plants and soils formed from the plants. Without this equilibrium exchange, all $\delta^{13}C$ values of plants, animals, and soils would be shifted upwards by about $9‰$, towards the $1‰$ value of ocean inorganic carbon. So equilibrium effects have a profound background importance in determining isotope distributions in ecological systems. Also, isotope effects in equilibrium reactions are sensitive to temperature, and, as we show at the end of this essay, they provide very interesting thermometers for studies of climate change.

This section proceeds with some philosophical ideas about equilibrium, then gives the algebra for understanding equilibrium isotope effects (EIE), and concludes with some important examples of these effects. Let's start philosophically.

Equilibrium, what does this word mean to you? It is a big word, with a lot of meanings. This word reminds us of both equality and balance, as in having a good mood that you don't want to lose, or as in standing upright in a stable position, without losing your balance and falling over: equilibrium, where diverse tendencies converge to somehow form a stabilized whole.

Chemists recognize another kind of equilibrium, where materials are exchanged, some moving in the forward direction whereas others move in the reverse direction, a kind of balance of payments. At equilibrium, there is no deficit one way or the other. The pools are not necessarily equal, no, not at all; it is just the fluxes that equate. If you think of rich nations trading with poor nations, but in a balanced way that leaves neither country richer nor poorer, then you have the equilibrium idea. The equality is in the exchange, not in the principal or wealth of the partners.

But over time, the constant interchange leads to turnover of the pools and with that turnover, a change in the character of the material in the pools. Imagine two pools, one yellow and one blue, exchanging material fluxes. The yellow pool has 200 units and exports 2 units every second to the blue pool, and the blue pool has 100 units and exports 2 units every second to the yellow pool. So, 2 = 2 immediately, and the system is balanced, right? But something is wrong, because although 2 = 2, 2 yellows do not equal 2 blues. So, it takes a while longer for the pools to exchange colors and for the colors in the fluxes to also balance. With the idea that isotopes correspond to colors, you can see that isotope equilibrium might take longer than simple chemical equilibrium that only involves masses, without considering colors. Eventually colors are no longer pure yellow and pure blue, but take on those intermediate green colors. Isotopes equilibrate like the colors of this thought experiment.

As it turns out, equilibrium also has special rules with temperature. Equilibirum occurs faster at higher temperature, but with less isotope fractionation. Equilibrium takes longer at cold temperatures, but isotope effects are larger there. In all, the equilibrium contributions to isotope drama are most important in the cold.

With this introduction, let us consider equilibrium between two molecules in a thought experiment. At the start of an exchange reaction, two different substances, A and B, have distinctly different compositions (Figure 7.27). The exchange process slowly breaks down these distinctions. But after

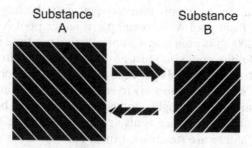

FIGURE 7.27. Two substances A and B have just been brought together and are just beginning an equilibrium exchange reaction.

prolonged exchange, there is still an important essence left on each side; homogenization does not go to completion (Figure 7.28). One aspect of the persistent difference is the isotopes, where the heavy isotope concentrates where bonds are strongest. The important points are that exchange works to homogenize the overall system, but persistent chemical differences between compounds preserve some differences, including isotope differences in a dynamic, flux-driven equilibrium.

There are other formal ways to write about equilibrium, and here we delve into the chemist's handbook for a while before turning to a simpler shorthand for ecologists. So think chemistry for a few paragraphs here.

The chemists symbolize the forward flux as a forward rate constant siphoning off material from pool A and shoving it towards pool B, with an equal rate constant for the reverse reaction. These rate constants are first-order constants, with rates a constant percentage of the pool size. (As discussed in Section 7.2, a constant percentage is the way to think about these rates, like a tax is a constant percentage on the goods you buy at a store.) If the pool sizes are equal, equal rate constants produce equal fluxes, but if one pool is much smaller as depicted in Figure 7.27 for pool B, then in an

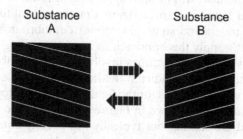

FIGURE 7.28. With passage of time, exchange promotes homogenization of isotopes in the two substances of Figure 7.27, although there are still slight isotope differences surviving this process, indicated by the milder differences in shadings of the two substances.

equilibrium situation, its export rate constant must be larger to counter-balance the flux from pool A. For example, if larger pool A has a 1% flux per time step * 200 units pool size = 2 units fluxing per time step, then pool B with only 100 units pool size must have a higher 2% flux per time step to maintain the balance, 2% flux per time step * 100 units pool size = 2 units fluxing per time step. Here you start to see that there is some math think-ing in these equilibrium thoughts, math that focuses on balanced fluxes between the pools. This balance is also called steady state because things aren't changing, and so are steady over time. The balance is steady, and the general dynamic or state of the system is at equilibrium; that is what steady state means in general, and more specifically, this means that fluxes are equal to maintain the balanced steady state.

With this focus on fluxes, now it is time to write the math for the fluxes as: Flux = (pool size) × (the rate constant), or for a steady-state system with two pools, P_1 and P_2,

$$P_1 * k_F = P_2 * k_R,$$

where P is the pool size (200 or 100 in our example), and k_F and k_R are the rate constants in the forward and reverse directions, respectively (1% and 2% in our example). So to check for equilibrium, we can substitute values: $P_1 * k_F = P_2 * k_R$ or $200 * 1\% = 2 = 100 * 2\% = 2$, and yes, we see a balance here, an equilibrium balance. Chemists rearrange this general equation into a general formula for these simple equilibria; that is, if

$$P_1 * k_F = P_2 * k_R \quad \text{then} \quad k_F/k_R = P_2/P_1.$$

So, in the end, although you and I may think of equilibrium in terms of mood or balancing in a good spot, chemists have refined this idea to a precise math formula that can be stated in words as this: in a simple equi-librium system, the ratio of the rate constants determines the ratio of pool sizes. If you keep thinking that the fluxes are equal, you can follow this train of thought, and chemists will be smiling at you, proud that you have taken a simple idea and found its expression in a mathematical formula.

Now that we understand something about equilibrium thinking at the chemical level, let's apply this approach to isotopes, writing separate equa-tions for the heavy and light isotopes involved in an equilibrium exchange where superscripts H and L indicate the heavy and light isotopes, respec-tively. The equilibrium reactions for the light and heavy isotope components of substances A and B are shown in Figure 7.29. The rate constants for the isotopes involved are similar, but typically not exactly equal for the light versus heavy isotopes.

At equilibrium, fluxes are equal in Figure 7.29 so that we can write:

$$^L A * {}^L k_F = {}^L B * {}^L k_R$$

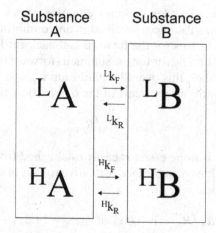

FIGURE 7.29. Equilibrium reactions for the light isotope components (^{L}A and ^{L}B) and heavy isotope components (^{H}A and ^{H}B) of the substances A and B involved in an exchange reaction. The rate constants for the reactions are given as kinetic k values, with superscripts denoting light (L) and heavy (H) isotope components, and subscripts denoting forward (F) and reverse (R) reaction directions.

for the upper reaction involving the light isotopes and

$$^{H}A * ^{H}k_{F} = ^{H}B * ^{H}k_{R}$$

for the lower reaction involving the heavy isotopes. These equalities can be rewritten as

$$^{L}k_{F}/^{L}k_{R} = ^{L}B/^{L}A \quad \text{and} \quad ^{H}k_{F}/^{H}k_{R} = ^{H}B/^{H}A.$$

Dividing the light terms by the heavy terms

$$\left(^{L}k_{F}/^{L}k_{R}\right)/\left(^{H}k_{F}/^{H}k_{R}\right) = \left(^{L}B/^{L}A\right)/\left(^{H}B/^{H}A\right)$$

then rearranging,

$$\left(^{L}k_{F}/^{H}k_{F}\right)/\left(^{L}k_{R}/^{H}k_{R}\right) = \left(^{H}A/^{L}A\right)/\left(^{H}B/^{L}B\right).$$

If α is generally the isotope fractionation factor, $\alpha = {}^{L}k/^{H}k$, and R is the isotope ratio in substance X, $R_{X} = {}^{H}X/^{L}X$, then the previous equation can be rewritten as

$$\alpha_{F}/\alpha_{R} = R_{A}/R_{B}.$$

This interesting (but complex-looking) equation states that the isotope ratio-of-ratios of two substances involved in an equilibrium gives the ratio of the fractionation factors for the forward and backward reactions. And if the overall fractionation factor for the summed (forward plus reverse) reaction is $\alpha_{EQ} = \alpha_F/\alpha_R$, then this overall equilibrium isotope effect is given by the measurable quantity at the heart of the δ definition, the ratio-of-ratios:

$$\alpha_{EQ} = R_A/R_B.$$

So, this equilibrium isotope effect is easily established from measurements of isotope compositions of substance A relative to that of substance B, using the familiar δ notation:

$$\delta_{A,B} = (R_A/R_B - 1)*1000 \quad \text{or} \quad \alpha_{EQ} = 1 + \delta_{A,B}/1000.$$

Note that if the measurements of R_B and R_A are not made relative to one another, but rather made in the more normal way, versus a common standard, a slightly different expression for α_{EQ} results. That is, if the general formula for δ is

$$\delta_X = (R_X/R_{STANDARD} - 1)*1000,$$

then

$$\delta_A = (R_A/R_{STANDARD} - 1)*1000 \quad \text{and} \quad \delta_B = (R_B/R_{STANDARD} - 1)*1000.$$

Rearranging these equations to separately solve for R_A and R_B,

$$R_A = (R_{STANDARD}/1000)*(1000 + \delta_A)$$

and

$$R_B = (R_{STANDARD}/1000)*(1000 + \delta_B)$$

so that if

$$\alpha_{EQ} = R_A/R_B$$

then

$$\alpha_{EQ} = (1000 + \delta_A)/(1000 + \delta_B).$$

Overall, these equations show how the isotope values ultimately derive from rate constants involved in equilibrium reactions. Although the math takes a while, none of it is hard and we end up with a simple expression that tells us that isotope compositions of substances involved in the equilibrium reaction readily give the equilibrium isotope effect for the overall reaction. Said another way, this result is a little more surprising, because the math shows that measurements made of the pools at equilibrium, rather than the rates at equilibrium, will give the overall isotope fractionation effect. Although the rates generate the isotope fractionation, the pools express this fractionation under equilibrium conditions, so convenient measurements of pools allow insight into rate processes. For convenience and consistency in this book, here the pool with the heavier isotope value is designated pool A, and the pool with the lighter isotope value is pool B, so that α_{EQ} values are >1.

But as promised previously, there is also a simpler algebra for all of this, one that ignores the rate constants and is more directly based on δ values, with Δ terms denoting fractionation factors involved in the exchanges. Consider two pools with isotope compositions δ_A and δ_B, with isotope fractionations Δ_A and Δ_B occurring as material leaves the pools (Figure 7.30). With this notation, we again use the idea that at steady state, fluxes are equal. The fluxes are $\delta_A - \Delta_A$ and $\delta_B - \Delta_B$ and must balance at equilibrium, so that

$$\delta_A - \Delta_A = \delta_B - \Delta_B.$$

The overall fractionation for the equilibrium exchange, $\Delta_{EQ} = \Delta_A - \Delta_B$, is:

$$\Delta_{EQ} = \Delta_A - \Delta_B = \delta_A - \delta_B.$$

You can see that this algebra is much simpler than using ratios-of-ratios of rate constants, and this simpler δ-based algebra is used in most of this book.

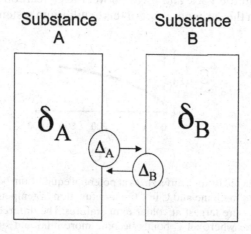

FIGURE 7.30. Isotope exchange reactions at steady-state for two substances A and B, with the reaction dynamics given using the δ and Δ notation.

However, it is more satisfying in some ways to keep the rate constants in mind rather than this simpler algebra of δ and Δ. The rate constants help us navigate thought problems such as the following. If you raise the temperature of an equilibrium exchange reaction, what happens to the isotope compositions? The answer is not all that obvious but in general, isotope substitutions in molecules have smaller effects at higher temperatures where all bonds are more energetic. Differences in isotope rate constants get smaller at higher temperatures, and there is less overall isotope fractionation in hotter equilibrium systems. Conversely, larger fractionations apply at cooler temperatures. The changes in isotope compositions of equilibrium mixtures turn out to be a nice way to estimate temperatures for ores and minerals in geological settings, and temperature dependencies are also important for isotopes in many current global biogeochemical cycles, for example, in the global exchange of atmospheric carbon dioxide with CO_2 dissolved in the sea.

The equilibrium calculations alerted early geochemist pioneers to the fact that isotope reactions should respond to temperature (Figure 7.31), with smaller effects at high temperatures where bonds are increasingly unstable anyway. This realization led to many, many geological applications where isotopes are used to establish temperatures in ancient ores and rocks. Textbooks oriented towards geologists outline many of these equilibrium applications in detail (reviewed by Hoefs 2004 and Faure and Mensing 2004).

This approach has been used also in the ocean where it is biological specimens such as bivalves and foraminifera that provide the temperature readings, based on shell carbonate formed in equilibrium with seawater (Figure 7.32). The equilibrium occurs because of carbonate chemistry, when CO_2 hydrates to form bicarbonate and then carbonate forms from the bicarbonate. These reactions are reversible, so that equilibrium develops as oxygen atoms shuttle back and forth between CO_2, carbonate, and water. Shells form from the carbonate part of this equilibrium system, and isotopes

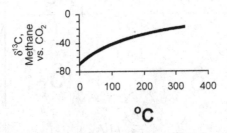

FIGURE 7.31. Predicted isotope differences at potential equilibrium between two types of carbon (C), C in methane and C in CO_2, as a function of temperature. Note that isotope differences are largest at colder temperatures. The differences decrease at higher temperature where all C bonds become more fluid and similar. At higher temperatures bonds become more energetic, and small neutron-related isotope differences become less important and less pronounced. (Data from Urey 1947).

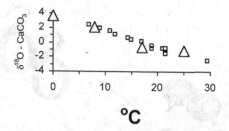

FIGURE 7.32. Oxygen isotope variations in marine carbonates as a function of temperature. This temperature variation is due to an equilibrium isotope effect between oxygen in water and oxygen in carbonates. Smaller fractionations (indicated by lower $\delta^{18}O$ values in this case) occur at higher temperatures where bond differences become smaller for the heavy versus light oxygen isotopes. Triangles = theoretical calculated values (McCrea 1950); squares = field data. Field data are for marine molluscs (mostly bivalves; Epstein et al. 1953). Paleontologists have used this relationship to estimate temperatures of ancient seas from oxygen isotopes in seashells.

record the temperature of the equilibrium, the temperature at which the shells were formed. The oxygen isotopes in the carbonate provide the temperature readings, and Mildred Cohn was an early pioneer in these equilibrium studies, developing the CO_2–H_2O exchange technique for oxygen isotope analysis (Cohn and Urey 1938).

Contemplation of these equilibrium temperature effects for the carbonate–H_2O system also led to development of an isotope paleothermometer used in estimating past global temperatures (Urey 1948). The initial thinking that oxygen isotopes function as paleothermometers for marine life has been updated (Chappell and Shackleton 1986; Shackleton 1987), so that $\delta^{18}O$ is not strictly a temperature recorder, but also reflects changes in the $\delta^{18}O$ of seawater that occur on geological time scales and in some local estuarine environments. Today the $\delta^{18}O$ record in carbonates (and in ice cores) provides an important record of past climate change, indicating cooler and warmer times in the history of our planet.

Conclusion

The idea of equilibrium is a simple one, and gives rise to precise and elegant ways of understanding isotope variations in terms of the balance between various rate constants. These rate constants change with temperature, so that isotope distributions can also indicate temperature in some simple systems. Although most natural systems are much more complex than simple chemical systems that reach equilibrium, some equilibrium reactions such as the exchange of CO_2 with lakes and oceans are important at local and even global scales. Fractionation in these equilibrium reactions adds isotope color and contrast for ecologists to appreciate and use.

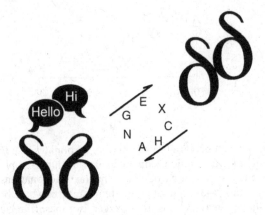

7.7 A Supply/Demand Model for Open System Fractionation

You may have noticed that although we encountered open systems in the first section of this chapter, we have otherwise ignored open systems—until now, that is. This section is the first of three sections that focus on fractionation in open systems. Open systems are actually the most common settings for fractionation. Box-and-arrow diagrams familiar to ecologists are open system diagrams. Arrows show inputs and outputs to pools or reservoirs represented by the boxes. This section considers open systems more carefully, with several aims: (1) understanding the isotope algebra of open systems, (2) applying this algebra to a well-studied isotope photosynthesis example, and (3) deriving a general supply/demand model for isotope fractionation in open systems.

Open System Algebra for One-Box Models

Open systems have both inputs and outputs, and at steady state where concentrations are not changing, the inputs equal the outputs. The isotope corollary is that at steady state, the input isotope flux equals the output isotope flux. Let's consider this more closely. In the open system diagram of Figure 7.33 (top panel), a mass flux P_{IN} enters a central pool P_1 then exits to pool P_2. The lower panel of Figure 7.33 shows the isotope action, with fractionation occurring at the exit of P_1. The isotope composition of this flux (or "isoflux") is δ_{P2}. With this orientation, now consider the steady-state assumption that inputs equal outputs. This requires that for the upper diagram, $P_2 = P_{IN}$, and for the lower diagram, that $\delta_{P2} = \delta_{IN}$. The equation for isotope fractionation in the lower diagram is

$$\delta_{P1} - \Delta_1 = \delta_{P2},$$

FIGURE 7.33. Diagram of an open system with
only one output.

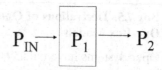

$$\delta_{IN} \rightarrow \delta_{P1} \,\text{\textcircled{Δ_1}} \rightarrow \delta_{P2}$$

so that

$$\delta_{P1} = \delta_{P2} + \Delta_1 = \delta_{IN} + \Delta_1.$$

This simple-seeming expression indicates that the isotope composition of
the intermediate pool δ_{P1} is governed by the input, but also by the fractionation during exit flux. Scientists concerned with gases in the atmosphere
such as methane use this fractionation equation to balance their isotope
budgets and to estimate the source contributions represented by δ_{IN} (Tyler
1986; Snover et al. 2000).

But these equations become much more interesting when considering a
box model with two exit fluxes instead of just one exit flux (Figure 7.34).
The solution equations for this mass balance diagram (Box 7.5) yield the
fundamental open system isotope equations, so that this derivation is a generally important result.

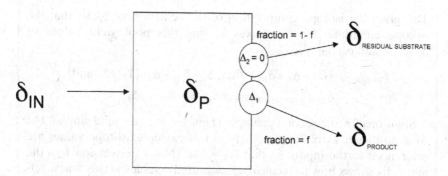

FIGURE 7.34. Diagram of an open system with substrate entering a reactor box
where product is formed with fractionation and unused substrate exits without
further fractionation. See Box 7.5 for details.

Box 7.5. Derivations of Open System Fractionation Equations for One-Box Models

Open systems have balanced inputs and outputs, and isotope dynamics in these systems are easy to predict from simple algebra given here for two generic examples. A first example considers an open system with only one product formed, with export of that product and residual unused substrate (Figure 7.34). The second example is similar, but the substrate completely reacts to form two exported products (Figure 7.35) instead of one product and an exported residual substrate.

For the first example (Figure 7.34), the problem is to calculate the isotope values of a central pool, the unused substrate and the product exiting the system. Specifically, what are the values of δ_P, $\delta_{\text{RESIDUAL SUBSTRATE}}$, and δ_{PRODUCT} when only one product is forming with fractionation (Δ_1) and residual substrate exits the system without further fractionation ($\Delta_2 = 0$)?

This is a steady-state problem, with inputs equal to outputs. The steady state also requires that "isotopes in" must equal "isotopes out," that the isotope fluxes in and out, the "isofluxes," are balanced. These isofluxes are given simply by subtracting Δ fractionations from δ values. For example, the δ value of the flux to product from the central pool is:

$$\delta_{\text{PRODUCT}} = \delta_P - \Delta_1,$$

where δ_P is the δ value of the central pool. With this isoflux algebra one can write the steady-state equations for balanced isoflux inputs and outputs:

$$\delta_{\text{IN}} = f(\delta_P - \Delta_1) + (1 - f)(\delta_P - \Delta_2),$$

and when $\Delta_2 = 0$,

$$\delta_P = \delta_{\text{IN}} + f\Delta_1.$$

This gives the isotope composition of the central pool δ_P, so that the isotope equations for the fluxes leaving this pool yield values of δ_{PRODUCT} and $\delta_{\text{RESIDUAL SUBSTRATE}}$:

$$\delta_{\text{PRODUCT}} = \delta_P - \Delta_1 = \delta_{\text{IN}} + f\Delta_1 - \Delta_1 = \delta_{\text{IN}} - \Delta_1 * (1 - f) \quad \text{and}$$
$$\delta_{\text{RESIDUAL SUBSTRATE}} = \delta_P - 0 = \delta_{\text{IN}} + f\Delta_1.$$

Solutions for the second example (Figure 7.35) are quite similar, but Δ_2 is no longer zero. For simplicity in this example, isotope values are referenced to the inputs, so that $\delta_{\text{IN}} = 0‰$. (Note: Conversion 1 in the appendix shows how to recalculate measured δ values versus a new reference such as δ_{IN}, so that this assumption of $\delta_{\text{IN}} = 0‰$ is easily realized via recalculation of measured data versus the input isotope value).

Mass balance again applies so that

$$\delta_{IN} = 0 = f(\delta_P - \Delta_1) + (1-f)(\delta_P - \Delta_2)$$

which rearranges to

$$\delta_P = f(\Delta_1 - \Delta_2) + \Delta_2.$$

Once the isotope composition of the central pool δ_P is known, the isoflux equations for material leaving this pool yield values of $\delta_{PRODUCT\ 1}$ and $\delta_{PRODUCT\ 2}$:

$$\delta_{PRODUCT\ 1} = \delta_P - \Delta_1 = f*(\Delta_1 - \Delta_2) + \Delta_2 - \Delta_1 \quad \text{or}$$
$$\delta_{PRODUCT\ 1} = \Delta_1*(f-1) + \Delta_2*(1-f)$$

and

$$\delta_{PRODUCT\ 2} = \delta_P - \Delta_2 = f*(\Delta_1 - \Delta_2).$$

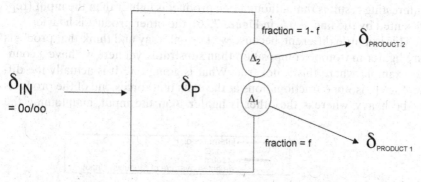

FIGURE 7.35. Diagram of an open system with substrate entering a reactor box where two products are both formed with fractionation. See Box 7.5 for details.

There are other general aspects of these seemingly simple equations. The equation

$$\delta_{SOURCE} - \Delta = \delta_{PRODUCT}$$

can be used to understand what happens when a mixture is split apart with fractionation. Interestingly, results are very nearly just the reverse of mass balance equations that apply when two substances are mixed together. The difference is that during fractionation, one of the splits has a fractionation adjustment in it:

$$\text{Mixing: } \delta_{SAMPLE} = f*\delta_1 + (1-f)*\delta_2$$

$$\text{Fractionation: } \delta_{SAMPLE} = f*(\delta_1 - \Delta) + (1-f)*\delta_2.$$

The uniting principle in these equations is mass balance, the accounting idea that inputs must equal outputs for overall amounts and also for component isotopes. Chapter 5 considered many variations of mixing equations, useful when thinking about mixing in real-world examples. The fractionation equations given in Box 7.5 also enable a nuanced and useful approach to understanding fractionation in real world examples.

The isotope modeling for open systems thus far is simple, but as you might imagine, more complex isotope results emerge quickly when there are multiple reactions leading from one box. These types of multiproduct reactions are common in the complex metabolic systems typical of many organisms and ecosystems '(Shearer et al. 1974; Hayes 2001). In fact, much more complex results arise from even slightly more complex models, as the following example shows. Consider the case where two products are formed, each with a fractionation involved (Figure 7.35). Here we again find equations and solutions based on the mass balance idea that inputs equal outputs, but this time the problem concerns two products (Figure 7.36, Box 7.5). The interesting result is that although one product is lighter than the input (represented by the line at 0‰ in Figure 7.36), the other product is heavier.

This is a little different, because we normally say and think that products are lighter in isotope composition than substrates, yet here we have a counterexample where this is not true. What is going on? It is actually the difference in isotope fractionations at the exits that forces one of the products to be heavy, whereas the other is lighter than the input, maintaining the

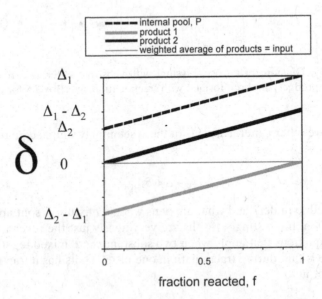

FIGURE 7.36. Graphical results for open system diagrammed in Figure 7.35, fractionation factor $\Delta_1 > \Delta_2$.

balance in the system. That last phrase is the key: mass balance requires that the two exports balance to match the input, so that if there are two separate products, one will have to be heavy to balance the other that is lighter. This makes sense, but the example serves as a caution that simple rules such as "products are lighter" do not always apply. The fundamental truth remains, however, that mass balance does apply, so that if you follow the logic of balances, you won't go wrong.

But in this example, you might ask, which of the two products carries the light isotope signal, and which one carries the remaining heavy isotope signal? The answer is that the product with the biggest isotope fractionation wins, and carries the lighter isotope signal. In this case, $\Delta_1 > \Delta_2$, so that product 1 is lighter and product 2 heavier. You can explore the relative influence of the two fractionation factors in the I Chi Worksheet for this section (see I Chi workbook 7.7 in the Chapter 7 folder on the accompanying CD). One of the interesting things to do is to switch the magnitudes of the fractionations; when $\Delta_2 > \Delta_1$ you will find that product 1 becomes heavy and product 2 is isotopically lighter than the input baseline value (Figure 7.37).

These last graphs (Figures 7.36 and 7.37) illustrate an important lesson about fractionation, that even seemingly minor changes in the ways reactions occur will produce large and dramatic changes in the observed isotope dynamics. This may seem discouraging, but it really is profiled here for two reasons. Fractionation effects can become more complex when two or more

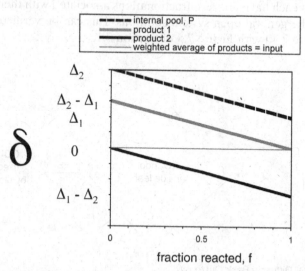

FIGURE 7.37. Graphical results for open system diagrammed in Figure 7.35, but the magnitude of the fractionation factor is reversed from $\Delta_1 > \Delta_2$ (results shown in Figure 7.36) to $\Delta_2 > \Delta_1$ (results shown here).

processes operate to make products from the same pool. Also, you have to pay strict attention to how you connect these box-and-arrow diagrams, for those connections determine how the mass balances work out in these open systems. Although I Chi is elegant, it will quickly yield complex results so that it is good to keep models simple and assumptions reasonable.

A technical note for this section on the algebra of open systems is that in some cases when working with samples whose δ values are very different than 0‰ (e.g., hydrogen isotope samples and enriched samples), the exact equations for fractionation in open systems are needed. The exact fractionation equations for open systems are given in Section 4.6 and in the appendix at the end of this book.

Open Systems and Photosynthesis

Perhaps the most widely known example of open system isotope dynamics concerns carbon isotope fractionation in photosynthesis (O'Leary 1988; Figures 7.38 to 7.40 and Box 7.6). CO_2 acquisition by a leaf starts when CO_2 diffuses across stomata into internal leaf air spaces (Figure 7.38). The leaf-internal CO_2 concentrations are reduced due to fixation by photosynthetic enzymes. CO_2 can also leak back out to the atmosphere, so that in all there are two exit possibilities for CO_2 inside the leaf, fixation to leaf sugar or back-diffusion to air (Figure 7.38). In this box model, f is considered the fraction of the incoming CO_2 flux that reacts to form leaf sugars via photosynthesis. Results from this simple, one-box model with one input and two outputs that each have nonzero fractionations associated with them (Figure 7.39) give the general open system solutions. This can be verified by comparing Figure 7.40 with Figure 7.5 of Section 7.1.

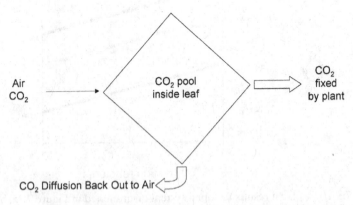

FIGURE 7.38. Initial diagram for carbon isotope fractionation during C_3 photosynthesis. CO_2 diffuses into the plant stomata and can be fixed by the plant or diffuse back out. See Box 7.6 for details.

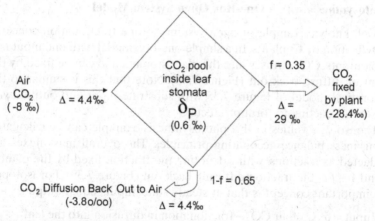

FIGURE 7.39. Carbon isotope fractionation during C_3 photosynthesis, continued from Figure 7.38. Mass balance helps budget the isotope values, starting with fractional accounting. The fraction fixed by the plant is f, and $1 - f$ is the fraction diffusing back out. Fractionations associated with these fluxes are 29‰ and 4.4‰, respectively (O'Leary 1988). With these values, mass balance algebra for this open system allows calculation the –28.4‰ isotope value of CO_2 being fixed. See Box 7.6 for further details.

These open system models for photosynthesis have proven useful in plant physiological studies of water use efficiency or WUE (Ehleringer et al. 1993 and references therein). Plants regulate the CO_2 influx and efflux dynamics primarily with their leaf stomata, with larger stomata aperture and conductance allowing more CO_2 inside the leaf for enzyme fixation and rapid plant growth. However, there is a downside to having stomata open wide, because plants will lose more water to the atmosphere. In all, there is a trade-off between carbon gain and water loss that plants seem to optimize by controlling stomatal aperture. In water-stressed conditions, leaf stomata

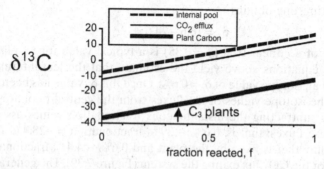

FIGURE 7.40. Results for the photosynthesis example, δ versus f with f = fraction CO_2 reacted = 0.35, a typical value for C_3 plants. See Box 7.6 for further details.

Box 7.6. Calculating Carbon Isotope Fractionation During Photosynthesis with a One-Box Open System Model

A well-known example of open system isotope fractionation concerns CO_2 fixation by C_3 plants. In a simple one-box model with one input and two outputs, CO_2 diffuses into the plant stomata and can be fixed by the plant or diffuse back out (Figure 7.38). Note that this is similar to the theoretical model of Figure 7.35 with substrate input and output, with substrate reacting to form product.

The isotope values in this photosynthesis example can be calculated using mass balance accounting principles. The overall mass fluxes are budgeted as fractions, with f denoting the fraction fixed by the plant is f, and $1 - f$ is the fraction diffusing back out (Figure 7.39). For isotopes, the important concept is that at steady state, inputs equal outputs.

Input = $\delta^{13}C$ of air CO_2 – fractionation in diffusion into the leaf
$$= -8 - 4.4 = -12.4‰$$

Output = $f*(\delta$ of carbon fixed$) + (1-f)*(\delta$ of carbon diffusing back out$)$

or   Output = $f*(\delta_P - 29) + (1-f)*(\delta_P - 4.4)$.

Combining the input and output equations by the constraint inputs equal outputs,

$$-12.4 = f*(\delta_p - 29) + (1-f)*(\delta_p - 4.4) \quad \text{or} \quad \delta_P = 24.6*f - 8.$$

The parameter f can be estimated from measurements of leaf-internal CO_2 concentrations (c_i), with f inversely related to c_i. This inverse relationship arises as follows. When carbon fixation demand for CO_2 is high in leaves (f is high), the leaf-internal CO_2 concentrations will be drawn towards zero (c_i is low). And vice versa, when demand is low (f is low), leaf-internal CO_2 pools will equilibrate with atmospheric pools and approach atmospheric CO_2 concentrations (c_i is high). Mathematically, the relationship between f and c_i is

$$f = 1 - c_i/c_a,$$

where c_a is the atmospheric CO_2 concentration. Substituting for f and rearranging, one obtains

$$\delta_P = 16.6 - 24.6*c_i/c_a.$$

A value of c_i/c_a near 0.65 ($f = 0.35$) is a typical value for C_3 plants, so that the equations above yield the estimate that the leaf-internal CO_2 pool has an isotope value of $\delta_P = 0.6‰$. Once this δ_P value has been established, the isotope values of the fluxes from the central pool are calculated by subtracting the instantaneous fractionation values associated with those fluxes; that is, 0.6‰ – 29‰ fractionation = –28.4‰ for the carbon flux that is fixed by the plant and 0.6‰ – 4.4‰ fractionation = –3.8‰ for the CO_2 flux exiting the system (Figure 7.39). The general solutions for isotope values of fixed plant carbon and unused CO_2 thus depend on f, the fractional split to form product (Figure 7.40). Note that this result for photosynthesis is simply a parameterized form of the theoretical model in Figure 7.35 for one-box open systems.

may be largely closed to retain water, and in this case most CO_2 entering the leaf will be fixed by plant enzymes, with little fluxing back out. This strong plant demand relative to supply results in low concentrations of leaf-internal CO_2. In terms of the isotope model, these water-stressed plants have a large value of f (Figure 7.39) and observed $\delta^{13}C$ values will increase (Figure 7.40). Overall, such plants retain water well, and have a high water use efficiency (WUE) that is indexed by their high $\delta^{13}C$ values. Lower $\delta^{13}C$ values often indicate ample water supplies, lower WUE, and a larger stomatal opening that allows more CO_2 entry and fixation (Ehleringer et al. 1993). Low light conditions and low N supply can also modulate the photosynthetic carbon fixation dynamics and lead to lower plant $\delta^{13}C$ values, presumably by lowering plant C demand or f (Figure 7.40).

Open Systems and a General Supply/Demand Model for Fractionation

In reality, the one-box photosynthesis model with its one-input/two-output steady-state assumptions is quite an interesting general model that gives strong predictions about how maximal fractionations can be readily reduced towards zero in many ecological reactions (Figure 7.41). When most material is routed or split towards the fixation or product formation reaction (f approaches 1 in Figure 7.41), the fractionation associated with product formation will approach zero as all incoming material is used regardless of isotope composition (Figure 7.41, middle). But what generally controls f, the routing or split towards product formation? This is the demand of the reaction, and the inputs are the supply. For example, optimal conditions for plant growth include abundant light, nitrogen availability, and water. Plants will grow rapidly in these conditions and have an increased demand for CO_2 during photosynthesis. A general result from this model is that fractionation should respond strongly to variations in the supply/demand ratio, with lower demand corresponding to larger fractionations and lower plant $\delta^{13}C$ values (Figure 7.41, bottom). Stated another way, variations in either supply or demand can affect the supply/demand ratio and observed fractionations will be smallest when the overall ratio is small. This occurs with small supply or strong demand, with a small value for the numerator or a large value for the denominator in the supply/demand ratio. Large fractionations result when the supply/demand ratio is large, with large supply or low demand.

These supply/demand relationships are a regular feature of ecological fractionations observed in many settings, including bacteria using sulfate in laboratory cultures (Canfield 2001) and phytoplankton photosynthesizing in the sea (Goericke et al. 1994; Popp et al. 1998). Recent decreases in carbon demand and productivity by polar oceanic phytoplankton have been inferred from increased carbon isotope fractionations in marine organisms

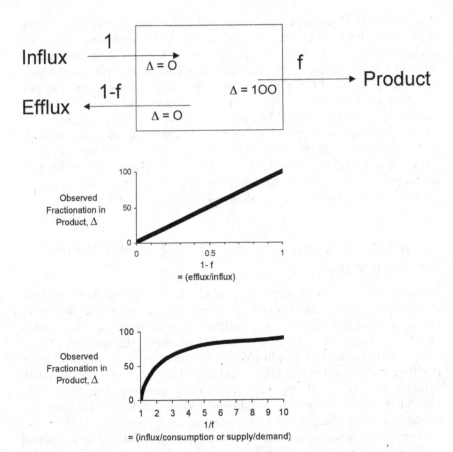

Figure 7.41. Generalized open system model for one box that has one entry flux and two exit fluxes (top panel). The starting material lies outside the box and is an assumed reference material with $\delta = 0‰$ that does not change in this reaction. Material fluxes in and out, with a fraction f of the influx forming product, and the remaining fraction $1 - f$ fluxing out as efflux. Fractionations (Δ) associated with influx to the box and efflux from the box are assumed to be 0‰, but there is a strong fractionation (maximum = 100 units) associated with the consumption step where material is removed to form product via bond formation. In this system, the observed or net isotope fractionation during product formation can approach zero when little material exits the system via diffusion or advection (middle panel), or when supply from the entry flux is much less than demand represented by the flux to form product (bottom panel). Maximal fractionations are expected when most entering material leaves the system unconsumed (middle panel) and when entry supply greatly exceeds consumption demand (bottom panel). Note that the bottom two panels have different x-axes both related to f, the fraction of material that is forming product (top panel). The middle panel shows the net fractionation Δ values observed in the product as a function of $1 - f$, and the bottom panel shows the net fractionation Δ values as a function of $1/f$. The possible range in f is 0 to 1.

(Schell 2000) although some aspects of this inference have been challenged (Cullen et al. 2001) and rebutted (Schell 2001).

These model results also help explain two observations commonly made in the stable isotope literature. The first common observation is that for chains of linked reactions, the isotope fractionation that is expressed is associated with the rate-limiting step. In terms of these open system models, this idea translates as this: the reaction having the least flux governs the fractionation. Stated again in different words, the less noticeable minor flux controls the overall fractionation. Model results are consistent with this general idea, with large fractionations in product associated with minor fluxes to product and a large efflux, that is, when f approaches 0 and $1 - f$ approaches a value of 1 in the middle panel of Figure 7.41. But what happens when the reaction splits into two equal parts ($f = 0.5$) so that there is no minor reaction? In this case, the fractionation is intermediate (Figure 7.41, middle). In reality, the rule that the slow reaction sets the fractionation is true only when reactions differ strongly in their kinetics, so that one reaction is much, much slower than the other. Many uptake reactions that largely determine the isotope values of plants and bacteria are poised in these more intermediate terms, without an overwhelmingly slower flux, and are perhaps better understood in terms of supply and demand. These uptake reactions include such processes as C_3 photosynthesis (Figures 7.39 and 7.40), sulfate reduction (Rees 1973), and nitrate incorporation (Neeboda et al. 2004).

A second common observation in the stable isotope literature is that slower reactions often yield larger fractionations. This can be explained by the balance between two reactions, typically a diffusion or advection reaction that has a small associated isotope fractionation (see Box 7.2 in Section 7.1), and a product formation reaction where bonds are created and fractionations may be consequently large. In this supply/demand model with the two competing reactions of efflux and product formation (Figure 7.41, top panel), slower reaction means less product formation and less overall demand (Figure 7.41, bottom panel), so larger fractionations are expected. Generally, where slower reaction leads to larger fractionations, there are likely dual controls of the overall reaction. These controls can be thought of as influx and fixation or alternatively, as supply and demand (Figure 7.41). A moral here is that fractionation is not constant, but can vary from near-zero values to maximal values, depending on supply/demand considerations.

In conclusion, the open system modeling described here links fractionation with biochemical and ecological processing of materials. This means that information about fractionation is process-level information, with variations in fractionation often expressing fundamental aspects of supply/demand relationships. Although effects of fractionation seem complex, and are often split between product and substrates, this complexity can yield valuable insight into how processes are split and balanced in nature. This is perhaps the good news about fractionation. The bad news is that fractionation can upset simple mixing models if it is not carefully measured and accounted for.

7.8 Open System Fractionation and Evolution of the Earth's Sulfur Cycle

Here we look at real-life examples where fractionation rather than mixing is the key to solving ecological puzzles. We focus on sulfur dynamics that evolved through the long geological history of our planet. One major theme is that supply/demand dynamics control expression of isotope fractionation at both long-term geological and shorter-term ecological scales. This essay also profiles actual laboratory experiments that show how to measure isotope fractionation. But let's start at the beginning.

Early sulfur isotope studies concerned isotopes in meteorites, asking whether the primordial sulfur in meteorites and in our planet Earth was uniform, or was it already somehow varied, showing the effects of mixing from different stellar sources? That was a beginning question for investigations about the Earth's sulfur cycle. Early studies showed that meteorites were indeed very uniform in their $^{34}S/^{32}S$ isotopic compositions (McNamara and Thode 1950). Meteorites were so uniform that the sulfur standard was soon chosen as sulfur from one of the meteorites, the Canyon Diablo meteorite that hit our planet about 50,000 years ago, leaving a big crater you can still visit in northern Arizona.

But working with those meteorites led to an interesting perspective. And that perspective was this: because the Earth inherited a uniform isotope composition from the time of its origin, then isotope deviations that would emerge over geological time were likely due to fractionation, and especially fractionation in biology and ecology. In essence, if we inherited a level playing field of isotope distributions as did the meteorites, emerging biology would create isotope lows and highs that would be easy to identify. Viewed in this way, the flatline baseline from meteorite studies was an advantage, a good point for starting the planetary biology clock: where would life tick in, with its characteristic separation of the isotopes via fractionation?

The short answer is that biology has created a record of increasing fractionation, from low to medium to current high levels of fractionation (Figure 7.42). What would cause the increase? Most experts think it was tied to increasing levels of one of the sulfur compounds, sulfate that is the most oxidized (rich in oxygen) of the many types of inorganic S compounds. Three of these compounds are listed in Table 7.3, where you can see that sulfate is oxygen-rich with four oxygen atoms. The two other S compounds, elemental sulfur and hydrogen sulfide, completely lack oxygen. The early Earth was poor in oxygen gas and rich in these reduced compounds that lack oxygen, rich in elemental sulfur (the yellow brimstone "rock that burns") and hydrogen sulfide (Table 7.3). The evolution of the biosphere is

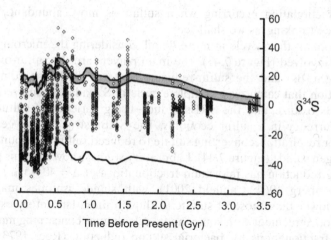

FIGURE 7.42. Sulfur isotope variations in sulfides become larger as earth evolved over time, with the geological time line running from right to left in this graph. The increased isotope variations reflect increasing importance of bacterial sulfate reduction from 3.5 billion years ago, and buildup of oxygen in the atmosphere in more recent times. The shaded area between the two top lines represents isotope compositions of sulfate, and the single bottom line represents a hypothetical maximum fractionation for sulfides with isotope values offset 55‰ lower than sulfate isotope values. (From Canfield 1998; used with the permission of the author and Nature Publishing Group. Copyright 1998.)

the story of the gradual transition from reduced to oxidized conditions, from a biosphere where highly oxidized compounds such as sulfate were rare to the current times when they are abundant. The isotope fractionation is thought to follow this transition in abundance of sulfate, with more bio-

TABLE 7.3. Three Important Sulfur Compounds.

SO_4^{2-}, sulfate	The most oxygen-rich and most "oxidized" sulfur compound. Anaerobic bacteria, the SRBs, or sulfate-reducing bacteria, use the four oxygens in this sulfur compound for respiration. Sulfate levels were likely very low on the early earth, but increased as oxygen levels increased. Sulfuric acid is H_2SO_4.
S^o, elemental sulfur	The yellow flammable mineral we know as brimstone (brimstone = "burning rock"). Volcanoes have been a rich source of S^o on earth since our planet was formed.
H_2S, hydrogen sulfide gas	The least oxidized and most "reduced" of all the sulfur compounds. This gas is poisonous to most life because sulfide combines with iron in mitochondria and blocks the enzymes there that are responsible for respiration and breathing. Hydrogen sulfide is a deadly compound on a par with cyanide, and only specially adapted organisms, especially some anaerobic sulfur bacteria, can tolerate this gas.

logical fractionation occurring when sulfate is more abundant. This is supply-side thinking, as we shall see.

Let's look at the S cycle in more detail, considering the microbial communities involved (Figure 7.43). The main players are the sulfur-oxidizing bacteria (SOBs) and the sulfur-reducing bacteria (SRBs), a community combination that catalyzes a complete cycle of S oxidation and reduction. The SOBs oxidize, and the SRBs reduce, using varieties of sulfur as an energy currency in a sulfur economy. The SRBs strip the oxygen from sulfate for respiration, converting sulfate to reduced sulfur compounds such as hydrogen sulfide (Figure 7.44). Laboratory studies show that this process of sulfate reduction has maximum fractionations of $\Delta = 40\text{--}45\text{\textperthousand}$ (Kaplan and Rittenberg 1964; Canfield 2001), with sulfide isotopes maximally 40–45‰ lower than those of sulfate. Although these laboratory estimates are in good agreement with some model calculations concerning maximum possible fractionations for bacterial sulfate reduction (Rees 1973), there are recent reports suggesting that in some field situations microbial sulfate reduction may take place with larger fractionations of 72–77‰ (Wortmann et al. 2001; Rudnicki et al. 2001). These larger fractionation values are near a value of 74‰ predicted from quantum mechanical considerations for maximum possible fractionation between sulfate and sulfide at 25 degrees C (Tudge and Thode 1950).

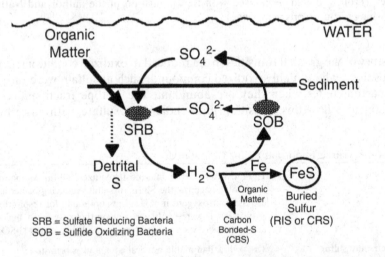

FIGURE 7.43. Diagram of an aquatic sulfuretum in which microbial partners oxidize and reduce sulfur in a complete cycle. Sulfur-oxidizing bacteria (SOBs) oxidize sulfides to sulfate with light or oxygen. Sulfate-reducing bacteria (SRBs) reverse the process, reducing sulfate to sulfides. The microbial flora influence the sulfur compositions of sediments, especially adding sulfides via reaction with iron and organic matter. Detrital sulfur from plants would be the sole source of sedimentary sulfur if SRBs were not generating sulfides. RIS = reduced inorganic sulfur, CRS = chromium reducible sulfur.

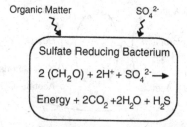

FIGURE 7.44. The process of bacterial sulfate reduction requires both sulfate and an energy source, labile carbon. Sulfate reduction is actually a kind of respiration. In the absence of free oxygen (O_2), bacteria use the 4 oxygens in sulfate instead. A byproduct of this anaerobic respiration is sulfide that can accumulate in sediments.

Investigating ancient ecology, geologists found evidence for these large fractionations of 30–75‰ in recent rocks of the last 500 million years, but rarely in the ancient ones older than 2.3 billion years (Figure 7.42). Did that mean that microbial sulfate reduction was absent on the early Earth, and that the biological sulfur cycle is only a recent invention? This would be at odds with recent molecular clock evidence that indicates sulfur-using microbes were already present 3.4 billion years ago (Shen et al. 2001). So, what was wrong with the isotopes?

Here it is time to put on your isotope thinking cap, and contemplate the possible early Earth, which by all indications was poor in sulfate (Anbar and Knoll 2002; Knoll 2003). The laboratory studies of maximum fractionations typically feed the SRBs large amounts of sulfate to saturate sulfate metabolism, but what if sulfate levels were very low? You might think of this as sulfate starvation, so that in a supply/demand model of fractionation, supply would be very low, and all sulfate available might be used regardless of isotope content. This is a no fractionation scenario. Laboratory results with sulfate-reducing bacteria support this idea (Habicht et al. 2002). Also studies of ^{33}S anomalies in ancient sulfur point to low sulfate levels in the earlier history of the Earth (Farquhar et al. 2000, 2002). The evidence for low sulfate supplies suppressing the expression of fractionation seems convincing from multiple lines of evidence.

But alternative thinking now enters the scene, with some scientists wondering whether this interpretation is really right. Could there be other forces at work that would affect this simple sulfate story? In particular, what about the SOBs, the sulfur-oxidizing bacteria that would have been active to complete the cycle, oxidizing sulfides and sulfur to regenerate sulfate? The SOBs include the spectacular photosynthetic bacteria, colored microbes that harvest light in anaerobic reducing conditions and oxidize hydrogen sulfide and sulfur to sulfate (Figure 7.45). The photosynthetic bacteria may have been part of the earliest microbial communities on the

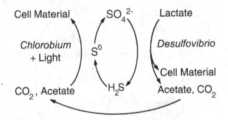

FIGURE 7.45. A laboratory sulfuretum involving *Desulfovibrio vulgaris*, a sulfate-reducing bacterium, and *Chlorobium phaeobacteroides*, a photosynthetic bacterium that oxidizes elemental sulfur (S°) and sulfide. S° is an important intermediate in sulfur cycling in this sulfuretum, and will accumulate inside the *Chlorobium*. (Data from Fry et al. 1988.)

planet, contributing to layered mats called stromatolites ("layered rocks") where, fueled by sunlight, the complete sulfur cycle may have turned and turned again, a near endless cycle of repeated oxidations and reductions (Figure 7.45). What fractionations did the SOBs contribute to the isotope pot, and what were the interactions between SOBs and SRBs?

Scientists asking such questions began studying the whole microbial sulfur cycle in the laboratory by adding both SRBs and SOBs to the same flask, creating a miniature ecosystem that completely turns the sulfur cycle through both oxidation and reduction steps, the "sulfuretum" ecosystem (Figure 7.45). And almost immediately a peculiar result emerged. Although the SRBs reduce sulfate with the usual or "normal" fractionation in which light isotopes react faster (Figure 7.46, first 30 hours), the SOBs oxidize sulfide with a small reversed or "inverse" isotope effect (Figure 7.46, hours 40–52), in which the product of oxidation, elemental sulfur, is enriched in the heavy isotope rather than depleted. In the inverse reaction, the heavy isotope flavor was reacting faster than the light isotope flavor, in apparent violation of the laws of kinetic isotope effects. What was happening?

Isotope detective work showed that at the bottom of the sulfide oxidation reactions was a small EIE rather than a KIE (here Mr. Polychaete is testing you to see if you remember what KIE and EIE stand for; see Box 7.3 for a reminder if you don't know). Fast equilibrium in the reaction

$$H + HS^- = H_2S$$

favored concentration of the heavy isotope in the H_2S gas that was the substrate the SOBs were actually using. H_2S is a dissolved gas and freely moves across membranes, so it made sense that SOBs would use this H_2S gas quickly. Using the isotope-heavy H_2S accounted for this unusual inverse KIE that at its core was really an EIE. Studies to complete the sulfur cycle showed that oxidation of elemental sulfur by the photosynthetic SOBs also

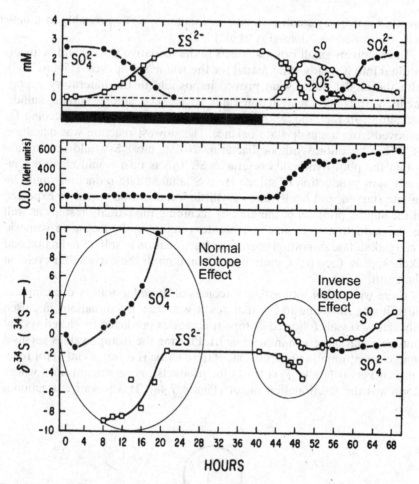

FIGURE 7.46. Sulfur cycling in a laboratory sulfuretum with both sulfate-reducing bacteria and sulfide-oxidizing photosynthetic bacteria. Top panel: the dark bar indicates dark conditions where only sulfate reduction was happening, then lights were turned on and photosynthetic bacteria rapidly oxidized sulfide to elemental sulfur ($S°$) and finally to sulfate, completing the S cycle. Thiosulfate ($S_2O_3^{2-}$) appeared transiently as part of the sulfur cycling near the end of the experiment. Middle panel: accumulation of $S°$ inside the photosynthetic bacteria increased the opacity or optical density of the culture after lights were turned on. Bottom panel, large circle: during bacterial reduction of sulfate, isotopes changed in accordance with a normal kinetic isotope effect, with light S ($\delta^{34}S < 0‰$) accumulating in the sulfide pool (ΣS^{2-}) and heavy S ($\delta^{34}S > 0‰$) accumulated in the residual sulfate (SO_4^{2-}) pool. Bottom panel, smaller circle: starting at 40 hours when lights were turned on, sulfide was rapidly oxidized to $S°$, and isotopes changed in an apparently reversed or "inverse" manner, with light S accumulating in the residual sulfide pool and heavy S accumulating in the $S°$ product. This apparent inverse isotope effect was actually the result of a fast equilibrium isotope effect between two sulfide pools, with $S°$ formed from the heavier of the two sulfide pools. See text for further explanation. (From Fry et al. 1988. Used with permission, American Society of Microbiology.)

involved only very small isotope fractionations, <2‰, probably in kineti-
cally controlled reactions (Fry et al. 1988).

The relatively small isotope effects in the oxidative sides could be incorporated into a steady-state model for the whole sulfur cycle (Figure 7.47),
with the model doing what proved impossible in the laboratory experiments, getting all the reaction rates to balance among sulfate, sulfide,
and sulfur. In fact, the laboratory experiments suggested an outcome far
removed from a steady-state balance. The slowest reaction was oxidation
of S° by the photosynthetic bacteria, so that most S would accumulate
within the photosynthetic bacteria as S°. This is turn would mean a relatively slow production of sulfate from S°, that sulfate reduction would be
sulfate starved, and finally that very little fractionation would be expected
in the sulfide pool. So in the end, by including the oxidation steps as well
as the reduction steps in the laboratory sulfur cycle, the same scenario
emerged: sulfate starvation prevented fractionation in sulfate reduction and
likely kept isotope fractionation at a minimum in the early sulfur cycle on
the earth.

There were other interesting outcomes of these laboratory experiments.
First, it was satisfying to see that yes, it was true, fractionations in closed
laboratory vessels followed isotope trajectories predicted by closed system
isotope equations given in Section 7.1. Plotting the isotope values versus a
simple transformation of f, the extent of reaction (i.e., versus $\ln(f)$ for reactants or versus $(f * \ln(f))/(f - 1)$ for products), gave straight lines whose
slope was the fractionation factor (Figure 7.48). This was nice; fractiona-

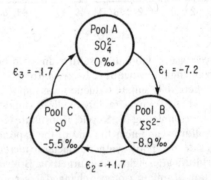

FIGURE 7.47. Summary steady-state model for isotope distributions in a model laboratory sulfuretum (Fry et al. 1988). Laboratory experiments were never at steady
state or equilibrium where all reaction rates would be equal and inputs equal
outputs by mass and by isotopes for each of the three pools. Fractionation factors
are ε values associated with closed system equations. ε values are permil fractionation factors similar to Δ values, but with opposite sign (in these examples,
this approximation holds: ε = −Δ). (Used with permission, American Society of
Microbiology.)

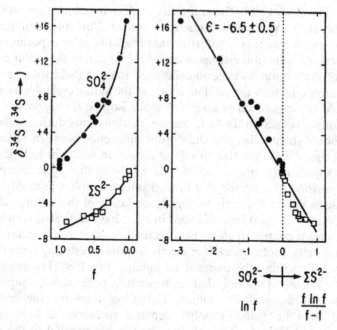

FIGURE 7.48. Calculating fractionation factors from laboratory culture data. Left panel: sulfate reducing bacteria consume sulfate and produce sulfides with a normal kinetic isotope effect. Light isotopes react faster and concentrate in the product sulfide, leaving heavy isotopes in the residual sulfate. Right panel: the fractionation factor $\varepsilon = -6.5‰$ ($\Delta = 6.5‰$) is calculated as the slope of the straight line using the closed system equations of Mariotti et al. (1981), where data are $(\ln(f), \delta^{34}S)$ for the residual sulfate and $(f * \ln(f)/(f-1), \delta^{34}S)$ for the product sulfide, with f = fraction of unreacted substrate. (From Fry et al. 1988. Used with permission, American Society of Microbiology.)

tions could be determined in a straightforward manner in time-course laboratory experiments! (Technical Note: the fractionations depicted in Figure 7.48 are ε values, approximately equal to $-\Delta$; see Box 2.1 and Mariotti et al. 1981 for derivations involving ε values, and Technical Supplement 7B on the accompanying CD for the parallel derivations involving Δ values.)

A second point concerned the reliability of these laboratory fractionation factors. You might wonder whether fractionations for sulfate reduction and sulfide oxidation would be the same in a mixed-species sulfuretum as they were when determined independently, in separated cultures of sulfate-reducing bacteria and sulfide-oxidizing bacteria. Here the concern is whether we can extrapolate fractionation results from simple single-species experiments to more complex, mixed-species ecosystems. And the answer to this question proved to be yes, that within experimental errors, the fractionations were the same in mixed cultures as they were in single-species cultures (Fry et al. 1988). The simplicity of determining fractionations has made laboratory experiments a mainstay of the sulfur isotope thinking and literature, and it was very comforting that the laboratory experiments provided good fractionation estimates useful for modeling the sulfur cycle.

Overall, the laboratory experiments suggested that there were only small fractionation effects in the oxidative side of the sulfur cycle, fractionations that would not cancel out or counterbalance isotope differences produced during sulfate reduction. In fact, inverse fractionations during sulfide oxidation should slightly increase the isotope difference between sulfates and sulfides (Figure 7.47) relative to sulfate reduction acting in isolation. With this investigation of the oxidative side of the sulfur cycle complete, it seemed justified to conclude that limited sulfate supplies probably limited fractionation during the early biological evolution of the sulfur cycle.

But recent ecological investigations in lakes have indicated another possible important control of sulfur isotope fractionation in low sulfate conditions. That extra control is carbon supply to sulfate-reducing bacteria, with carbon supply setting the demand for sulfate (Fry 1989; Fry et al. 1995). In a more general model that incorporates both sulfate supply and carbon-fueled demand for sulfate, increasing fractionations might be expected when the "sulfate supply/C demand" ratio increases (Figure 7.49; Goldhaber and Kaplan 1975). This model is quite parallel to the general open system model for supply/demand fractionation derived in the previous section (see Figure 7.41, bottom panel).

An hypothesis is that during the planetary oxidation events occurring two to three billion years ago, carbon availability for sulfate-reducing bacteria decreased when sulfate reducers were forced out of increasingly oxidized

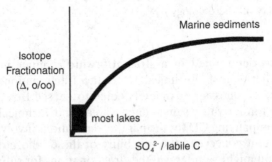

FIGURE 7.49. A conceptual supply demand model of sulfur isotope fractionation by bacteria in sediments. Both sulfate and carbon are important controls of bacterial sulfate reduction, with sulfate representing the oxygen supply and carbon representing the oxygen demand. Fractionation during sulfate reduction to form sulfides is maximal when sulfate supplies are high and carbon supplies low. Lakes generally have low concentrations of sulfate versus marine systems, <100 versus 28,000 mmol m^{-3}, respectively. The smaller sulfur isotope fractionation in lake versus marine sediments is likely due to the much lower sulfate levels in lakes, but may be due also in part to higher supplies of easily used labile carbon.

surface waters and into anaerobic muds, away from fresh sources of photosynthetic carbon. This decrease in C availability would be partly responsible for increased fractionations observed after that time, via an increase in the ratio of sulfate supply/C demand. Perhaps the next generation of sulfur scientists will find a way to test this idea by examining molecular and atomic isotope markers diagnostic of the quality and quantity of organic matter used by the sulfate reducers during this transition to a more oxidized biosphere. These ideas should be considered along with other recent ideas about biological controls of the isotope differences between sulfates and sulfides in the geological record (e.g., Jorgensen 1990; Canfield and Thamdrup 1994; Sorensen and Canfield 2004). In all, the sulfur isotope studies test ideas about biogeochemical transitions in the early history of the earth, ideas that are also being tested with carbon, oxygen, and iron isotope studies (see Rouxel et al. 2005 and references therein).

Conclusion

The early isotope geochemists viewed fractionation as a simple yes/no assay when they began studying the evolution of the sulfur cycle on earth. When did isotope peaks and valleys show up that would indicate biological fractionation of the isotopes, and how long did the preceding isotope flatline last? Just asking this simple question proved very helpful in understanding the history of sulfur cycling on this planet. However, low sulfate levels in the earliest part of the record meant that the isotope tools were very blunt and relatively uninformative during this early period that unfortunately was actually the time of most interest. Laboratory studies proved robust for probing the S cycle, and an excellent way to determine the magnitude of fractionation factors in straightforward time course experiments. But some recent ecological studies in lakes suggest that it is not just the sulfate supply that controls expressed fractionation, rather it is the balance between supply and demand that is important. Future study can test the idea that sulfur isotope fractionation in low-sulfate systems is a good indicator of carbon supply, as predicted from supply/demand models.

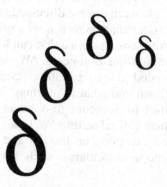

7.9 Open System Legacies

Ecologists may not realize it, but geochemists were the pioneers in showing how isotopes circulate in the biosphere. We owe much to these pioneers, and this section honors four of them: Harmon Craig, Sam Epstein, Wally Broecker, and John Hayes. Their isotope-filled careers are inspiring, and here are three short stories that highlight some of their isotope fun. The stories are set in an outdoor park that has isotope exhibits we visit during a morning stroll.

We arrive at the park entrance on a high hill and look down on a sea of grass. The sign in front of us points to "Kansas grass" on the left and "Wyoming grass" on the right. The grass is beautiful, waving in the breeze. We sit down to enjoy the good weather and scenery, and spot a plaque in the ground with this inscription: "Treasure Your Exceptions." In our guide-book, we read that one of the first studies dealing with carbon isotopes appeared in 1953, and was a survey of some 300 samples, including plants, rocks, diamonds, and fish. That early isotope biogeochemist, Harmon Craig, did not miss much. In his list of samples, sample 125 was "Kansas grass" and had an unusual, oddball isotope number, the only real oddball number in a list of 28 plant samples that included many trees, shrubs, and even "Wyoming grass." The speculative explanation given for this odd result later proved incorrect, one of the few mistakes made by that early pioneer who went on to a long and distinguished career. Not too many scientists paid attention to sample 125, until much later, 16 years later when the oddball number was discovered to be typical of a whole class of plants, C_4 plants such as corn. These plants evolved fairly recently on earth to deal better with low CO_2 conditions, and studying these C_4 plants became big business for many biologists.

This motto, "Treasure Your Exceptions," has been a guiding principle for generations of scientists, for it means that something is wrong, so there is something more you can learn. (But first, of course, you need to make sure that it is really an exception, and not just something you did wrong.) We honor Harmon Craig for honest reporting in his work that through the years showed us isotope exceptions as well as the isotope rules. He was one of the pioneers who first observed the open system photosynthetic frac-tionations in terrestrial plants that were discussed in Section 7.7.

The guidebook also states that there is now a prize offered for anyone who can develop a remote sensing device that can look out and distinguish isotopes in "Kansas grass" from isotopes in "Wyoming grass." A kind of isotope binoculars is needed to see the isotope action in nature, so that we can see what is exceptional and what is "normal." The two fields of grass exist as a test plot, waiting for someone to discover a tool that reveals the beauty of isotopes in their natural setting. We look out, and decide we like this display of treasured exception on the left and valued normal on the right. Inventing the isotope binoculars sounds hard to us, but following

Charles Hapgood's advice given in Section 1.2 of the Introduction, we are confident some complete amateur will eventually figure out such a difficult problem.

After a short rest, we walk down the hill and soon find ourselves in a field of ripening tomatoes, red and luscious in the morning sun. We meet a gardener who encourages us to eat a few tomatoes as we go, and we do just that. Farther along, we come to another plaque, "Tomatotopes, 1961." We stand eating, and read in our guidebook:

One of the earliest reports on isotopes in open biological systems focused on tomatoes, then on tomatoes plus other plants. In a survey of plant chemicals, it turned out that the carbon isotope difference between plant tissue and extracted lipids was not constant, but inverse, approaching zero at high lipid contents (Figure 7.50). The data fit an open system model based on a split in the metabolic pathways, with lighter isotopes diverted to lipids. As more and more material was diverted to lipids, the isotope values of lipids rose towards that of the input, as might be expected when all input is converted to lipid.

We leave thinking that it is interesting that the lipids that contribute to the luscious taste of tomatoes also have an isotope story out here in the open systems of nature.

We also realize that early on, Nature was experimenting with scientists to see if they could understand the isotope language, a language that is not entirely easy to decipher even when out in the open. Sam Epstein was one of those early codebreakers who began reading the rich secret language of isotopes. We munch contentedly on our last ripe tomato and marvel at how good isotopes can taste. Then we move on.

The morning is almost at an end, and we move towards the exit of the park where there is a rock garden, a kind of isotope Stonehenge. We walk up and cannot find a trace of a plaque, just two elegant scribbles "WSB" and "JMH". One of our geologist friends looks closely at some of the rocks, and lets out a low whistle. "This is really, really old, from the beginnings of the earth. In fact," she says, turning slowly, "there is a time line of the earth here, from the beginning to the present. These rocks represent the geolog-

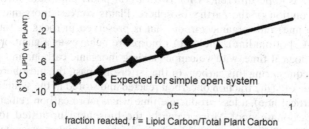

FIGURE 7.50. Carbon isotope differences between plants and lipids from those plants. (Data from Park and Epstein 1961.) Data are consistent with open system fractionation of about 9‰ during lipid synthesis.

ical history of our planet." She taps a few rock and sniffs, pulling out a pocket rock tester. "Aha," she laughs, "just as I thought. This record is not just any rock record, it is the record of organic materials through time, the biosphere coupled with geology through chemistry. Wallace S. Broecker, WSB, and John M. Hayes, JMH, famed biogeochemists could have been here indeed. And yes, they used isotopes to plumb the evolution of the carbon cycle through those many eons."

We nod, impressed, and finally find a graph on the backside of one rock, no explanation, but then we do not need so many clues any more. We see open system isotopes and an arrow (Figure 7.51). We think about this and try to puzzle it out, given our budding knowledge of fractionation in open systems developed in Sections 7.1 and 7.7. This proves too hard, but fortunately one person in our group consults our guidebook, and we learn that yes, Broecker and Hayes are honored here, for showing how an open system isotope diagram could help decipher the evolution of the biosphere.

The history of planetary oxygen dynamics is depicted here, but rests on a background concept. The background concept takes a little explanation, as follows. As carbon cycles up through the biosphere from the earth's interior, part of this carbon is split off by photosynthesis to form plant biomass and eventually organic carbon preserved in soils and rock. The remaining carbon largely forms carbonate minerals. The carbon isotopes in ancient rocks help us estimate this fractional split f between organic matter and car-

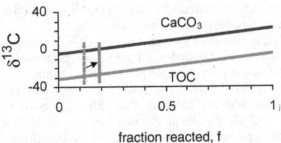

FIGURE 7.51. Isotope differences for two carbon pools important in the biogeochemical evolution of the earth's biosphere. Plants convert inorganic carbonate carbon (top line) into organic carbon that is preserved in rocks as total organic carbon (TOC, bottom line). The earth's biosphere behaves as a giant open system through geological time, with volcanoes adding inorganic carbon, and plants and sediments sequestering this carbon. In the far distant past, the biosphere operated at a low value of f (the fraction of carbon reacted and stored in the sediments, about 0.12, left vertical line), a less productive time when more carbon remained in the inorganic carbonate pool. More recently, the biosphere upshifted to a higher value of f (about 0.19, right vertical line), consistent with higher oxygenation of the biosphere. See text for further details. (From: Schopf, J. William; *Earth's Earliest Biosphere*. Copyright 1983. Princeton University Press. Reprinted by permission of Princeton University Press.)

bonates. This isotope-based estimate of f (Figure 7.51) shows organic carbon burial and is interestingly linked to oxygen accumulation in the atmosphere. More organic carbon buried means higher values of f (Figure 7.51) and less oxygen consumed, so that more oxygen accumulates in the biosphere. The end of this inference chain is that times of increased $\delta^{13}C$ values in ancient rocks were also times of increased oxygen accumulation. Yes, now we see it, a diagram of the earth's biospheric history, with increasing $\delta^{13}C$ as an indicator of planetary oxygenation, set down in an open system isotope diagram. Perhaps intended in this rock garden with its hidden cryptograph is that there are enormous mysteries on this earth and you can solve some of them with imagination, luck, and isotopes.

We also see that it can take more than one scientist to solve these mysteries of ancient planetary ecologies, and our geologist friend tells us that although honoring Broecker and Hayes is appropriate, names of many other geologists could have been added here as well. We suspect that is the intent of the organizers to add in names later, making this an Isotope Hall of Fame.

This is the end of the excursion, and we mentally admire these pioneers for the enormous breadth of their work, from molecules to the whole biosphere. Their example gives us inspiration and context for our own efforts in isotope ecology.

7.10 Conducting Fractionation Experiments

Fractionation is readily measurable in laboratory experiments, experiments that give reliable results for incorporation into the I Chi models. Here are eight points of advice should you decide that it might be worthwhile to conduct your own isotope fractionation experiments.

1. *Maximize your fractionation.* Provide enough substrate so that your reaction is not substrate-limited. If it is substrate-starved, most substrate will be consumed and fractionations will tend towards zero, yielding an artificially low fractionation estimate. To prevent this, scientists generally use well-stirred solutions, high-substrate concentrations, and even cell-free preparations in some cases to prevent slow diffusion from limiting the amount of substrate available for reaction and fractionation.

2. *Time course approach.* Many workers historically used single-point determinations of fractionation, especially isolating products after reaction had proceeded <10% in closed vessels. Under these conditions, the simple isotope difference between initial substrate and first-formed product gives a good estimate of the fractionation (see Figure. 7.4). This time-honored approach is often adequate, but sometimes can yield problematic results if there is some artifact in the initial conditions, when you set up the experiment. The alternative approach is to make time course measurements across a longer time period and extent of reaction, then fit lines or curves to the results. This gets away from any problems with initial conditions that are now represented by one or two data points out of five to ten data points, and the curve-fitting routines for the multiple time course measurement have the added benefit of yielding errors for the fractionation estimates. Focusing most of the sampling in the 20 to 80% reaction range works gives the best results with this time course, line-fitting approach. Equilibrium experiments with these approaches typically takes several hours or days (e.g., Zhang et al. 1995).

3. *Third time charm.* Most experimental work requires repetition for success. For example, the first time you establish the basic time course of the reaction, the second time you make sure you can sample and measure well throughout the time course, and the third time you actually succeed in all measurements that involve both concentration and isotope determinations. So plan for at least three tries.

4. *Multiple viewpoints.* One of the nice aspects of the time course measurements of fractionation is that you can measure fractionation from three separate viewpoints, using reactants, instantaneous product, and accumulated product (see Figure 7.4 or Technical Supplement 7B in the Chapter 7 folder on the accompanying CD for the three equations). If you can, measure fractionation in all three pools, and look for agreement among results. You gain confidence in the overall fractionation result when all measures agree. Figure 7.48 in Section 7.8 shows an example that combines measurements for substrate and accumulated product. It is desirable to have both types of data when trying to estimate the overall fractionation (Roeske and O'Leary 1984).

5. *Puzzling fractionations.* Sometimes you get unexpected results in the fractionation experiments, when the closed system equations do not lead to straight lines (e.g., Fry et al. 1985). This will require some thinking. The closed system equations apply in simple situations, when reactions are unidirectional (forward only) and single. If your results don't follow classic closed system equations, and you are sure your experimental results are correct, examine whether the assumptions of the equations have been violated. Are multiple reactions involved, or is there a significant backward reaction that you did not suspect? One interesting example to contemplate comes from studies of carbon isotope fractionation by marine plants in seawater. Inorganic carbon is present in bicarbonate, carbonate, and aqueous CO_2, but the primary substrate for the plants may be only the CO_2. As plants

withdraw CO_2, the carbonate system will replenish this substrate. In this case, the experiment is not a closed CO_2 system, even though in the larger sense it is closed for total inorganic carbon because you are not adding or removing seawater. The message here is to think carefully about your experiments, and to look carefully for unexpected reactions when you obtain odd-seeming results.

6. *Multiple products.* Some reactions have only one substrate, but multiple products. The closed system equations can work well in these instances when the multiple products are stable and do not further react (e.g., Fry et al. 1988). But if some products are formed, as intermediates (on the way to becoming other products), it is more complex. For intermediates, one needs to think about whether it is the formation or loss terms that dominate at any particular time. Intermediates that are formed, then consumed with normal kinetic isotope effects can have very low isotope values (when first formed as products), and later can have very high isotope values (when last consumed as reactants). In these cases, special modeling (think I Chi here) is needed. Steady-state models (Fry et al. 1988; Fry 2003) may be helpful for an initial understanding of these complex cases, but it is actually rare that steady state occurs. The I Chi approaches work well in the more normal (real) cases where reaction rates are not balanced completely, that is, for cases in which reaction intermediates sequentially have low then high isotope values, as formation then consumption reactions sequentially determine the isotope dynamics.

7. *Continuous fractionation estimates.* Sometimes experiments involve gases such as SO_2, H_2, N_2, or CO_2 that the mass spectrometer can sample directly, so that you can monitor fractionation in a continuous mode. This approach has been developed (Sharkey and Berry 1985; Evans et al. 1986), but rarely used in ecology, probably because it can require a dedicated, expensive mass spectrometer. However, continuously monitoring fractionation experiments seems a good idea for the future, with much more detailed time-course resolution possible following quick changes in experimental conditions. Depending on exactly how the experiments are set up, either closed system equations (Figure 7.4) or open system equations (Figure 7.5) will be appropriate for estimating fractionation in future continuous flow systems.

8. *Closing advice.* Keep the experimental systems simple. And save all the material that you may not think you want to analyze. There may be an important part of the mass balance in components you were not thinking about initially.

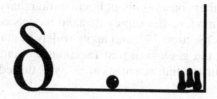

7.11 Chapter Summary

Isotope fractionation or separation is a hidden fact of life, happening continuously at the atomic level in reactions. We can blame fractionation on a quantum genie, Fractionation Frank (Box 7.1), or we can try to come to terms with the subtle effects of fractionation with logic and practice. The problem with fractionation is that you need to keep track of several things at once, especially four things involved in a mass balance: substrates, products, heavy isotopes, and light isotopes. Working with fractionation is like learning to juggle, weave, or solve crimes: there is a necessary technical skill but also an art of balance that comes with practice.

The idea of mass balance is simple, just accounting for all the sums at the end that must also add up to the starting amounts. This applies in reactions when substrates form products, and we know that as substrates disappear, products appear. Or, when lighter isotopes react more quickly to form product, the residual substrate left behind is enriched in the heavy isotopes. These statements make sense because they reflect an intuitive balance or accounting. Sections 7.1 and 7.2 elaborate on this mass balance and show that the dictates of mass balance provide a strong and elegant framework for understanding isotope fractionation, once you begin seeing all the pieces at once. This wider vision and balance is the hard part, the juggling skill that takes time and patience to learn.

Still, even for the expert there are many puzzles associated with fractionation. One puzzle is that fractionation is active during sample measurement (Section 7.3), so that you might think we could not get a true reading of isotope contents. But comparisons to standards and routine computer calculations make possible highly precise and accurate isotope determinations in spite of fractionation. A moral here is that everyday life with isotopes involves understanding fractionation and working with it, rather than ignoring it.

There are many other fractionation puzzles, for example, effects of fractionation can resemble remarkably the effects of mixing (Section 7.4), or fractionation is not fixed, but variable (Sections 7.5 to 7.7). Imagine driving a car that went slowly some days, but fast on other days. Now you see why fractionation might be a difficult subject. However, the I Chi (Isotope Power) modeling introduced earlier in the book is very helpful for working out the effects of fractionation. Two I Chi examples show fractionation at work creating isotope landscapes or patterns in the surface ocean and in bottom sediments (Section 7.5), patterns useful for following later mixing and movement of animals and compounds.

Also, there are other concepts about fractionation that apply widely, such as supply demand models. The supply demand models come from considering open systems (Section 7.7), and apply widely in many ecological settings today and in the geological past (Sections 7.8 and 7.9). Section 7.9 considers a few outstanding scientists who have turned fractionation to their advantage, providing inspiration for ecologists. For those who become truly interested in fractionation, there is advice in Section 7.10 on how to conduct your own fractionation experiments, and Technical Supplement 7A on the accompanying CD introduces how to calculate fractionation from quantum mechanical considerations. Technical Supplement 7B gives derivations of closed system equations presented in Sections 7.1 and 7.2.

The problems posed for this chapter on the accompanying CD give several ways to practice with fractionation, with many problems that challenge you to develop your own I Chi models. A summary thought for this chapter is that a little I Chi modeling, built on well-established fixed principles about fractionation, goes a long ways to helping you understand, track, and predict the dynamic effects of isotope fractionation in natural systems.

Further Reading

Section 7.1

Anderson, T.F. and M.A. Arthur. 1983. Stable isotopes of oxygen and carbon and their application to sedimentologic and paleaoenvironmental problems. In M.A. Arthur, T.F. Anderson, I.R. Kaplan, J. Veizer, and L.S. Land (eds.), *Stable Isotopes in Sedimentary Geology*. SEPM Short Course #10, Society of Economic Paleontologists and Mineralogists, Dallas, TX, pp. 1–1 to 1–151.

Benner, R., M.L. Fogel, E.K. Sprague, and R.E. Hodson. 1987. Depletion of ^{13}C in lignin and its implications for stable carbon isotope studies. *Nature* 329:708–710.

Bigeleisen, J. 1949a. The validity of the use of tracers to follow chemical reactions. *Science* 110:14–16.

Bigeleisen, J. 1949b. The relative reaction velocities of isotopic molecules. *Journal of Chemical Physics* 17:675–678.

Bigeleisen, J. 1965. Chemistry of isotopes. *Science* 147:463–471.

Bigeleisen, J. 1969. Isotope separation practice. In W. Spindel (ed.), *Isotope Effects in Chemical Processes*. Advances in Chemistry Series 89, American Chemical Society, Washington, D.C., pp. 1–24.

Bigeleisen, J. and M.G. Mayer. 1947. Calculation of equilibrium constants for isotopic exchange reactions. *Journal of Chemical Physics* 15:261–267.

Bigeleisen, J. and M. Wolfsberg. 1958. Theoretical and experimental aspects of isotope effects in chemical reactions. In I. Prigogine, *Advances in Chemical Physics*, v. 1, Wiley, New York, pp. 15–76.

Brenna, J.T. 2001. Natural intramolecular isotope measurements in physiology: Elements of the case for an effort toward high-precision position-specific isotope analysis. *Rapid Communications in Mass Spectrometry* 15:1252–1262.

Clark, M.J., B.L. Beard, and F. Albarede. 2004. Geochemistry of non-traditional stable isotopes. *Reviews in Mineralogy and Geochemistry*, vol. 55. Mineralogical Society of America anthe Geochemical Society. Washington, D.C.

Craig, H. 1953. The geochemistry of the stable carbon isotopes. *Geochimica et Cosmochimica Acta* 3:53–92.

DeNiro, M.J. and S. Epstein. 1976. You are what you eat (plus a few ‰): The carbon isotope cycle in food chains. *Geological Society of America Abstracts Program* 8:834–835.

DeNiro, M.J. and S. Epstein. 1978. Influence of diet on the distribution of carbon isotopes in animals. *Geochimica et Cosmochimica Acta* 42:495–506.

Farquhar, J., H. Bao, and M. Thiemens. 2000. Atmospheric influence of earth's earliest sulfur cycle. *Science* 289:756–758.

Farquhar, J., B.A. Wing, K.D. McKeegan, J.W. Harris, P. Cartigny, and M.H. Thiemens. 2002. Mass-independent sulfur of inclusions in diamond and sulfur recycling on early earth. *Science* 298:2369–2374.

Fernandez, I., N. Mahieu, and G. Cadisch. 2003. Carbon isotopic fractionation during decomposition of plant materials of different quality. *Global Biogeochemical Cycles* 17:1–1 to 1–11.

Fry, A. and M. Calvin. 1952. The isotope effect in the decomposition of oxalic acid. *Journal of Physical Chemistry* 56:897–901.

Fry, B. 2003. Steady state models of stable isotope distributions. *Isotopes in Environmental and Health Studies* 39:219–232.

Fry, L.M. 1962. Radium and fission product radioactivity in thermal waters. *Nature* 195: 375–376.

Hayes, J.M. 2001. Fractionation of the isotopes of carbon and hydrogen in biosynthetic processes. In J.W. Valley and D.R. Cole (eds.), *Stable Isotope Geochemistry, Reviews in Mineralogy and Geochemistry*, vol. 43. Mineralogical Society of America, Washington D.C., pp. 225–278.

Hoefs, J. 2004. *Stable Isotope Geochemistry*. Springer-Verlag, New York.

Luz, B. and E. Barkan. 2000. Assessment of oceanic productivity with the triple-isotope composition of dissolved oxygen. *Science* 288:2028–2031.

Luz, B., E. Barkan, J.L. Bender, M.H. Thiemens, and K.A. Boering. 1999. Triple-isotope composition of atmospheric oxygen as a tracer of biosphere productivity. *Nature* 400:547–550.

Mariotti, A., J.C. Germon, P. Hubert, P. Kaiser, R. Letolle, A. Tardieux, and P. Tardieux. 1981. Experimental determination of nitrogen kinetic isotope fractions: some principles; illustration for the denitrification and nitrification processes. *Plant and Soil* 62:413–430.

Martin, G.G., Y.L. Martin, N. Naulet and H.J.D. McManus. 1996. Application of ^{2}H SNIF-NMR and ^{13}C SIRA-MS analyses to maple syrup: Detection of added sugars. *Journal of Agricultural Food Chemistry* 44:3206–3213.

Martin, G.J. 1995. Inference of metabolic and environmental effects from the NMR determination of natural deuterium isotopomers. In E. Wada, T. Yoneyama, M. Minagawa, T. Ando, and B.D. Fry (eds.), *Stable Isotopes in the Biosphere*. Kyoto University Press, Japan, pp. 36–56.

McClelland, J.W., C.M. Holl, and J.P. Montoya. 2003. Relating low δ^{15}N values of zooplankton to N$_2$-fixation in the tropical North Atlantic: Insights provided by stable isotope ratios of amino acids. *Deep-Sea Research* 50:849–861.

Rayleigh, Lord. 1902. On the distillation of binary mixtures. *The London, Edinburgh and Dublin Philosophical Magazine and Journal of Science*, Series 6, 4:521–537.

Rossman, A., M. Butzenlechner, and H.-L. Schmidt. 1991. Evidence for a nonstatistical carbon isotope distribution in natural glucose. *Plant Physiology* 96:609–614.

Urey, H.C. 1939. Separation of isotopes. *Reports on Progress in Physics* 6:48–77.

Urey, H.C. 1947. The thermodynamic properties of isotopic substances. *Journal of the Chemical Society (London)*, Part 1:562–581.

Young, E.D., A. Galy, and H. Nagahara. 2002. Kinetic and equilibrium mass-dependent isotope fractionation laws in nature and their geochemical and cosmochemical significance. *Geochimica et Cosmochimica Acta* 66:1095–1104.

Zhang, J., P.D. Quay, and D.O. Wilbur. 1995. Carbon isotope fractionation during gas-water exchange and dissolution of CO$_2$. *Geochimica et Cosmochimica Acta* 59:107–114.

Section 7.2

Mariotti et al. 1981. Listed above; see Section 7.1 readings.

Section 7.4

Case, J.W. and H.R. Krouse. 1980. Variations in sulphur content and stable isotope composition of vegetation near a SO$_2$ source at Fox Creek, Alberta, Canada. *Oecologia* 44:248–257.

Fry. 2003. Listed above; see Section 7.1 readings.

Krouse, H.R. 1980. Sulphur isotopes in our environment. In P. Fritz and J.Ch. Fontes (eds.), *Handbook of Environmental Isotope Geochemistry*, vol. 1. Elsevier Scientific, Amsterdam, pp. 435–471.

Mariotti, A., A. Landreau, and B. Simon. 1988. ^{15}N isotope biogeochemistry and natural denitrification process in groundwater: Application to the chalk aquifer of northern France. *Geochimica et Cosmochimica Acta* 52:1869–1878.

Rayleigh, 1902. Listed above; see Section 7.1 readings.

Section 7.5

Altabet, M.A. 2001. Nitrogen isotopic evidence for micronutrient control of fractional NO_3^- utilization in the equatorial Pacific. *Limnology and Oceanography* 46:368–380.

Altabet, M.A. and R. Francois. 1994. Sedimentary nitrogen isotopic ratio as a recorder for surface ocean nitrate utilization. *Global Biogeochemical Cycles* 8:103–116.

Chanton, J.P., C.S. Martens, and M.B. Goldhaber. 1987. Biogeochemical cycling in an organic-rich coastal marine basin. 8. A sulfur isotopic budget balanced by differential diffusion across the sediment-water interface. *Geochimica et Cosmochimica Acta* 51:1201–1208.

Dore, J.E., J.R. Brum, L. Tupas, and D.M. Karl. 2002. Seasonal and interannual variability in sources of nitrogen supporting export in the oligotrophic subtropical North Pacific Ocean. *Limnology and Oceanography* 47:1595–1607.

Farell, J.W., T.F. Pedersen, S.E. Calvert, and B. Nielsen. 1995. Glacial-interglacial changes in nutrient utilization in the equatorial Pacific Ocean. *Nature* 377:514–517.

Galbraith E.D., M. Kienast, T.F. Pedersen, and S.E. Calvert. 2004. Glacial-interglacial modulation of the marine nitrogen cycle by high-latitude O_2 supply to the global thermocline. *Paleoceanography* 19:PA4007.

Hartmann, von M. and H. Nielsen. 1969. $\delta^{34}S$-Weste in rezenten Meeres-sedimenten und ihre Deutung am Beispiel einiger Sediment-profile aus der westlichen Ostsee. *Geologische Rundschau* 58:621–655.

Jorgensen, B.B. 1979. A theoretical model of the stable sulfur isotope distribution in marine sediments. *Geochimica et Cosmochimica Acta* 43:363–374.

Lourey, M.J., T.W. Trull, and D.M. Sigman. 2003. Sensitivity of $\delta^{15}N$ of nitrate, suspended and deep sinking particulate nitrogen to seasonal nitrate depletion in the Southern Ocean. *Global Biogeochemical Cycles* 17:7–1 to 7–18.

Montoya, J.P., C.M. Holl, J.P. Zehr, A. Hansen, T.A. Villareal, and D.G. Capone. 2004. High rates of N_2 fixation by unicellular diazotrophs in the oligotrophic Pacific Ocean. *Nature* 430:1027–1031.

Saino, T. and A. Hattori. 1980. ^{15}N natural abundance in oceanic suspended particulate matter. *Nature* 283:752–754.

Saino, T. and A. Hattori. 1987. Geographical variation of the water column distribution of suspended particulate organic nitrogen and its ^{15}N natural abundance in the Pacific and its marginal seas. *Deep-Sea Research* 34:807–827.

Section 7.6

Chappell, J., and N.J. Shackleton. 1986. Oxygen isotopes and sea level. *Nature* 324:137–140.

Cohn, M. and H.C. Urey. 1938. Oxygen exchange reactions of organic compounds and water. *Journal of the American Chemical Society* 60: 679–682.

Epstein, S., R. Buchsbaum, H.A. Lowenstam, and H.C. Urey. 1953. Revised carbonate-water isotopic temperature scale. *Bulletin of the Geological Society of America* 64:1315–1326.

Faure, G. and T.M. Mensing. 2004. *Isotopes: Principles and Applications*. John Wiley and Sons, New York.

Hoefs. 2004. Listed above; see Section 7.1 readings.

McCrea, J.M. 1950. On the isotopic chemistry of carbonates and a paleotemperature scale. *Journal of Chemical Physics* 18:849–857.

Shackleton, N.J. 1987. Oxygen, isotopes, ice volume and sea-level. *Quaternary Science Reviews* 6:183–190.

Urey, H.C. 1947. The thermodynamic properties of isotopic substances. *Journal of the Chemical Society (London)*, Part 1:562–581.

Urey, H. 1948. Oxygen isotopes in nature and in the laboratory. *Science* 108:489–496.

Section 7.7

Canfield, D.E. 2001. Isotope fractionation by natural populations of sulfate-reducing bacteria. *Geochimica et Cosmochimica Acta* 65:1117–1124.

Cullen, J.T., Y. Rosenthal, and P.G. Falkowski. 2001. The effect of anthropogenic CO_2 on the carbon isotope composition of marine phytoplankton. *Limnology and Oceanography* 46:996–998.

Ehleringer, J.R., A.E. Hall, and G.D. Farquhar. 1993. *Stable Isotopes and Plant Carbon-Water Relations.* Physiological Ecology Series of Monographs, Texts and Treatises. Academic, San Diego, CA.

Fry, 2003. Listed above; see Section 7.1 readings.

Goericke, R., J.P. Montoya, and B. Fry. 1994. Physiology of isotope fractionation in algae and cyanobacteria. In K. Lajtha and R. Michener (eds.), *Stable Isotopes in Ecology.* Blackwell Scientific, Oxford, UK, pp. 187–221.

Hayes. 2001. Listed above; see Section 7.1 readings.

Neeboda, J.A., D.M. Sigman, and P.J. Harrison. 2004. The mechanism of isotope fractionation during algal nitrate assimilation as illuminated by the $^{15}N/^{14}N$ of intracellular nitrate. *Journal of Phycology* 40:517–522.

O'Leary, M.H. 1988. Carbon isotopes in photosynthesis. *BioScience* 38:328–336.

Popp, B.N, E.A. Laws, R.R. Bidigare, J.E. Dore, K.L. Hanson, and S.G. Wakeham. 1998. Effect of phytoplankton cell geometry on carbon isotopic fractionation. *Geochimica et Cosmochimica Acta* 62:69–77.

Rees, C.E. 1973. A steady-state model for sulphur isotope fractionation in bacterial reduction processes. *Geochimica et Cosmochimica Acta* 37:1141–1162.

Schell, D.M. 2000. Declining carrying capacity in the Bering Sea: Isotopic evidence from whale baleen. *Limnology and Oceanography* 45:459–462.

Schell, D.M. 2001. Carbon isotope ratio variations in Bering Sea biota: The role of anthropogenic carbon. *Limnology and Oceanography* 46:999–1000

Shearer, G., J. Duffy, K.H. Kohl, and B. Commoner. 1974. A steady-state model of isotopic fractionation accompanyg nitrogen transformations in soil. *Soil Science Society of America, Journal* 38:315–322.

Snover, A.K., P.D. Quay, and W.M. Hao. 2000. The D/H content of methane emitted from biomass burning. *Global Biogeochemical Cycles* 14:11–24.

Tyler, S.C. 1986. Stable carbon isotope ratios in atmospheric methane and some of its sources. *Journal of Geophysical Research* 91:13232–13238.

Section 7.8

Anbar, A.D. and A.H. Knoll. 2002. Proterozoic ocean chemistry and evolution: A bioinorganic bridge? *Science* 297:1137–1142.

Canfield, D.E. 1998. A new model for Proterozoic ocean chemistry. *Nature* 396:450–452.

Canfield, D.E. 2001. Isotope fractionation by natural populations of sulfate-reducing bacteria. *Geochimica et Cosmochimica Acta* 65:1117–1124.

Canfield, D.E. and R. Raiswell. 1999. The evolution of the sulfur cycle. *American Journal of Science* 299:697–723.

Canfield, D.E. and B. Thamdrup. 1994. The production of ^{34}S-depleted sulfide during bacterial disproportionation of elemental sulfur. *Science* 266:2973–1975.

Farquhar, J., H. Bao, and M. Thiemens. 2000. Atmospheric influence of earth's earliest sulfur cycle. *Science* 289:756–758.

Farquhar, J., B.A. Wing, K.D. McKeegan, J.W. Harris, P. Cartigny, and M.H. Thiemens. 2002. Mass-independent sulfur of inclusions in diamond and sulfur recycling on early earth. *Science* 298:2369–2374.

Fry, B. 1989. Sulfate fertilization and changes in sulfur stable isotopic compositions of lake sediments. In J. Ehleringer and P. Rundel (eds.), *Stable Isotopes in Ecological Research*. Springer-Verlag, New York, pp. 445–453.

Fry, B., H. Gest, and J.M. Hayes. 1988. $^{34}S/^{32}S$ fractionation in sulfur cycles catalyzed by anaerobic bacteria. *Applied and Environmental Microbiology* 54:250–256.

Fry, B., A. Giblin, M. Dornblaser, and B. Peterson. 1995. Stable sulfur isotopic compositions of chromium reducible sulfur in lake sediments. In M. Schoonen and M.A. Vairavamurthy (eds.), *Geochemical Transformations of Sedimentary Sulfur*. American Chemical Society Symposium Series. #612, Washington, D.C., pp. 397–410.

Goldhaber, M.B. and I.R. Kaplan. 1975. Controls and consequences of sulfate reduction rates in recent marine sediments. *Soci Science* 119:42–55.

Habicht, K.S., M. Gade, B. Thamdrup, P. Berg, and D.E. Canfield. 2002. Calibration of sulfate levels in the Archean Ocean. *Science* 298:2372–2374.

Hallberg, R.O. and L.E. Bagander. 1985. Fractionation of stable sulfur isotopes in a closed sulfuretum. In D.E. Caldwell, J.A. Brierley, and C.L. Brierley (eds.), *Planetary Ecology*. Van Nostrand Reinhold, New York, pp. 285–296.

Harrison, A.G. and H.G. Thode. 1958. Mechanism of the bacterial reduction of sulfate from isotope fractionation studies. *Transactions of the Faraday Society* 53:84–92.

Jorgensen, B.B. 1977. The sulfur cycle of a coastal marine sediment (Limfjorden, Denmark). *Limnology and Oceanography* 22:814–832.

Jorgensen, B.B. 1990. A thiosulfate shunt in the sulfur cycle of marine sediments. *Science* 249: 152–154.

Kaplan, I.R. and S.C. Rittenberg. 1964. Microbiological fractionation of sulphur isotopes. *Journal of General Microbiology* 34:195–212.

Knoll, A. 2003. *Life on a Young Planet: The First Three Billion Years of Evolution on Earth*. Princeton University Press, Princeton, NJ.

Mariotti et al. 1981. Listed above; see Section 7.1 readings.

McNamara, J. and H.G. Thode. 1950. Comparison of the isotopic constitution of terrestrial and meteoritic sulfur. *Physical Review* 78:307–308.

Rees, C.E. 1973. A steady-state model for sulphur isotope fractionation in bacterial reduction processes. *Geochimica et Cosmochimica Acta* 37:1141–1162.

Rouxel, O.J., A. Bekker, and K.J. Edwards. 2005. Iron isotope constraints on the Archean and Paleoproterozoic ocean redox state. *Science* 307:1088–1091.

Rudnicki, M.D., H. Elderfield, and B. Spiro. 2001. Fractionation of sulfur isotopes during bacterial sulfate reduction in deep ocean sediments at elevated temperatures. *Geochimica et Cosmochimica Acta* 65:777–789.

Schidlowski, M., J.M. Hayes, and I.R. Kaplan. 1983. Isotopic inferences of ancient biochemistries: carbon, sulfur, hydrogen and nitrogen. In J.W. Schopf (ed.), *Earths Earliest Biosphere, Its Origin and Evolution*. Princeton University Press, Princeton, NJ, pp. 149–186.

Shen, Y., R. Buick, and D.E. Canfield. 2001. Isotopic evidence for microbial sulphate reduction in the early Archaean era. *Nature* 410:77–81.

Sorensen, K.B. and D.E. Canfield. 2004. Annual fluctuations in sulfur isotope fractionation in the water column of a euxinic marine basin. *Geochimica et Cosmochimica Acta* 68:503–515.

Thode, H.G., J. Monster, and H.B. Dunford. 1961. Sulphur isotope geochemistry. *Geochimica et Cosmochimica Acta* 25:150–174.

Tudge, A.P. and H.G. Thode. 1950. Thermodynamic properties of isotopic compounds of sulphur. *Canadian Journal of Research* B28:567–578.

Wortmann, U.G., S.M. Bernasconi, and M.E. Boettcher. 2001. Hypersulfidic deep biosphere indicates extreme sulfur isotope fractionation during single-step microbial sulfate reduction. *Geology* 29:647–649.

Section 7.9

Broecker, W.S. 1970. A boundary condition on the evolution of atmospheric oxygen. *Journal of Geophysical Research* 75:3553–3557.

Craig, H. 1953. The geochemistry of the stable carbon isotopes. *Geochimica et Cosmochimica Acta* 3:53–92.

Hayes, J.M. 1983. Geochemical evidence bearing on the origin of aerobiosis, a speculative hypothesis. In J.W. Schopf (ed.), *Earth's Earliest Biosphere, Its Origin and Evolution.* Princeton University Press, Princeton, NJ, pp. 291–301.

Park, R. and S. Epstein. 1961. Metabolic fractionation of C^{13} and C^{12} in plants. *Plant Physiology* 36:133–138.

Schopf, J.W. (ed.). 1983. *Earth's Earliest Biosphere, Its Origin and Evolution.* Princeton University Press, Princeton, NJ.

Section 7.10

Evans, J.R., D.T. Sharkey, J.A. Berry, and G.D. Farquhar. 1986. Carbon isotope discrimination measured concurrently with gas exchange to investigate CO_2 diffusion in leaves of higher plants. *Australian Journal of Plant Physiology* 13:281–292.

Fry. 2003. Listed above; see Section 7.8 readings.

Fry et al. 1988. Listed above; see Section 7.1 readings.

Fry, B., H. Gest, and J.M. Hayes. 1985. Isotope effects associated with the anaerobic oxidation of sulfite and thiosulfate by the photosynthetic bacterium, *Chromatium vinosum. FEMS Microbiology Letters* 27:227–232.

Fry, B., W. Ruf, H. Gest, and J.M. Hayes. 1988. Sulfur isotope effects associated with the non-biological oxidation of sulfide by O_2 in aqueous solution. *Chemical Geology* 73:205–210.

Hallberg, R.O. and L.E. Bagander. 1985. Fractionation of stable sulfur isotopes in a closed sulfuretum. In D.E. Caldwell, J.A. Brierley, and C.L. Brierley (eds.), *Planetary Ecology.* Van Nostrand Reinhold, New York, pp. 285–296.

Harrison and Thode. 1958. Listed above; see Section 7.8 readings.

Kaplan and Rittenberg. 1964. Listed above; see Section 7.8 readings.

Mariotti et al. 1981. Listed above; see Section 7.1 readings.

Rayleigh, 1902. Listed above; see Section 7.1 readings.

Roeske, C.A. and M.H. O'Leary. 1984. Carbon isotope effects on the enzyme-catalyzed carboxylation of ribulose bisphosphate. *Biochemistry* 23:6275–6284.

Sharkey, T.D. and J.A. Berry. 1985. Carbon isotope fractionation of algae as influenced by an inducible CO2 concentrating mechanism. In W.J. Lucas and J.A. Berry (eds.), *Inorganic Carbon Uptake by Aquatic Organisms.* American Society of Plant Physiologists, Rockville MD. pp. 389–401.

Zhang et al. 1995. Listed above; see Section 7.1 readings.

8
Scanning the Future

Overview

This chapter briefly considers the future of stable isotope ecology. The future may hold advances in technology, a more routine use of multiple chemical markers in ecological investigations, and more of that essential scientific ingredient, imagination.

8.1. *The Isotope Scanner.* We are currently limited by technology in the way we detect isotopes. We need a better isotope scanner for the future.

8.2. *Mangrove Maude.* This is the story of a real ecologist who began using isotopes, but only as part of a broader forest study that included many other types of chemical indicators. The isotopes were useful, but other indicators were actually better.

8.3. *The Beginner's Advantage—Imagine!* Science is an accumulation of facts but also a creative act with imaginative leaps. Quotes from the book *The Great Whale of Kansas* may encourage the beginning scientist that the world is open and still fresh for investigation.

Main points to learn. This final book chapter opens out with a view towards more and different kinds of tracers. In science, as in life, you find answers more rapidly by gathering different types of clues. A challenge for the future is to balance use of the isotopes with other approaches in ecology, developing real-time ways to assess, model, and understand what nature is doing each day on this planet.

8.1 The Isotope Scanner

We are waiting for the isotope scanner. If you listen to the pioneers who first worked with isotopes, they had to build their own mass spectrometers from bits and pieces in the basements of university buildings devoted to physics and chemistry. The next generation of isotope scientists hassled with vacuum lines, stopcocks, and hundreds of hours of manual sample preparation work to get their science done. Today, the isotopes arrive from auto-mated computerized systems, and the challenge is integrating and inter-preting the voluminous data, not in generating the data one slow bit at a time. What about tomorrow?

The isotopes have subtle chemical signals, with the heavy isotopes vibrat-ing in their chemical bondage a little differently than the light isotopes. So imagine a scanner, handheld of course, that is sensitive to this subtle chem-istry. Point and shoot, like today's simple cameras, and there you have your isotope number! Actually, astronomers have been doing this kind of thing for decades already, but generally not with enough precision to help us earthbound ecologists. Recent advances in laser technology enable isotope scanning of atmospheric gases, but still we cannot scan isotope values of solids, the plants, animals, and soils that ecologists study. The scanner chal-lenge awaits.

I used to canoe out in the wilderness of the 10,000 islands of Florida, wan-dering the mangrove-lined waterways, wishing for a scanner. I did have a salinity/temperature probe that I let dangle over the side, and I would glance at it now and then to see what was happening in the water, and think about how that related to the changing forest I saw all around me. I thought about how nice it would be to have an isotope scanner, one that fit in a backpack, real-time isotopes would be great. I also daydreamed about bionic uplinking, having a computer chip implanted in my left earlobe that let me uplink and downlink to the world's most powerful computer, so the data analysis and interpretation would go right along with what my senses and sensors were telling me.

Someday we will travel in reality and virtual reality at the same time, enjoying a richer life. Science will approach art much more as we find ways to expand our experience in real time. Maybe an isotope scanner will be part of it all. But for now, we don't have a scanner. In the next section, we look at what another mangrove ecologist did, using current-day technology.

8.2 Mangrove Maude

Mangrove Maude was an ecologist who loved going out in those muddy slippery mangroves, out there in the dank, sweaty, hot, and humid forests, known for their swarms of mosquitoes that defend against human intru-sion. You could find her out there in the forests, muddy to the eyebrows, and loving every minute of it. But back in the office, Ms. Maude was a natural-born skeptic, asking why and why again, especially why would isotopes help her study mangroves? Here we trace an evolving answer to that question.

Mangroves are trees, swamp trees that grow in salt water at the edges of the tropical ocean. The tide comes in to the coastal mangrove forests, covering and hiding all those many odd mangrove roots. Some roots are like big hoops you have to climb on, some are like small croquet wickets that will trip you if you don't watch out, and some are a sea of spikes ready to lance you should you fall. The tide comes in, with eels, crabs, and fish swimming through the channels and picking through the forest floor, searching for food. Then the tide goes back out, taking those predatory animals along with fallen leaves that have drifted down from the 5 to 30 m tall canopy above. Ms. Maude asks, "Where are isotopes in this forest picture?" From reading this book, you should know the answer, which is this: isotopes are everywhere, in the fish, in the water, in the leaves, in the trees. Because Ms. Maude loved plants, we began dreaming about a botanical isotope project.

In the beginning is the dream, the imagination filtering out of the pores, into the sunlight, taking form here and there in insubstantial ways. In glimmers and gleams, the thoughts come new in the morning, announcing their commanding presence after a good sleep. These mornings add up and become a rich blend that condenses in different ways, here an hypothesis, there a field trip, later a grant proposal, then a cup of coffee wondering how things will work out. In this case, the theme becomes niches of the mangrove species, the way the different species of mangroves are coexisting in forests way out there, on a small island in the far western tropical Pacific, just the right place for a budding research project that will take years to bear fruit, if at all. Plant it; then wait and see.

Well, now that we know we want to use isotopes to investigate differences among mangrove species on an island, what are the next steps? Oh, your proposal comes back with lots of feedback about why your ideas won't work, why they are unclear, and many other whys. It is at this point that you have to have some faith in yourself, to weather the storms of criticism.

One idea you hear over and over is that the isotope approach is just too narrow. The world is much bigger than isotopes, and isotopes are just a small part of a larger whole, which could be termed chemical ecology, and that might include chemistry such as pH, for instance. Better living through pH, through isotopes, through chemistry, as an old slogan goes. You look into this and find that besides isotopes, cheap analyses of many cations and trace metals are also routine and easy to obtain. So you decide on a combined approach with several kinds of chemical markers, isotopes, cations, and trace metals. Who knows which markers will describe best the niche space of the mangroves? You don't know in the beginning, but you will find out.

Off you go, you and Mangrove Maude, and one day soon, there you are, face to face with a mangrove forest way out there in the middle of the Pacific. You work at sampling bits and pieces of mangrove bark, roots, and leaves, your witches' brew for chemical ecology. Days later, you and Mangrove Maude emerge from the swamps with your packs full of tree samples. You spend the next months grinding and analyzing, filling spreadsheets with arcane bits of data, then cracking and crunching away with statistics, until it all comes out as it is. And there is a big surprise. It worked! The isotopes do separate the three mangrove species, and provide a chemical picture of their niche space (Figure 8.1). But the cations, those simple-to-measure cations, they separate the species even better (Figure 8.2). The trace metal analyses do not cleanly separate the species (data not shown), but do contribute significantly in multivariate analyses that showed chemical descriptions of niche space. Overall, the chemical approaches work surprisingly well—rejoice!

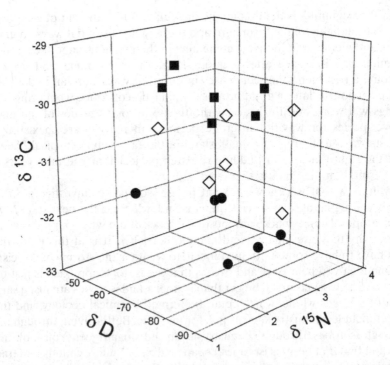

FIGURE 8.1. Isotope compositions of leaves from three mangrove species growing on the island of Kosrae, Federated States of Micronesia, western Pacific. Squares = *Sonneratia alba*; diamonds = *Rhizophora apiculata*; circles = *Brugiera gymnorhiza*.

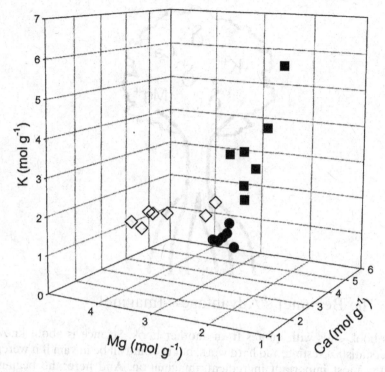

FIGURE 8.2. Cations in leaves from three mangrove species growing on the island of Kosrae, Federated States of Micronesia, western Pacific. Squares = *Sonneratia alba*; diamonds = *Rhizophora apiculata*; circles = *Brugiera gymnorhiza*.

Now Ms. Maude is puzzling away, trying to make sense of all the data, not just fixated on the isotopes. As far as the isotopes go, she is just pleased that they do a good job distinguishing the three mangrove tree species. Why do these markers separate the species? Well, answering that will take some future research, undoubtedly involving mixing and fractionation to get to the bottom of the isotope action (Newsome et al. 2007). But for now, just having chemical markers is enough to explore whether these trees are always different, or different only on this particular island.

Perhaps in the future, Mangrove Maude will use the multiple-marker, chemical ecology approach as she skips across those Pacific islands, studying the forests she loves so well. She will find some of the answers about why the mangrove species shift around in their niches from place to place, separating here, but coming together there. Someday someone will come along and ask her, "Why?" And she will know what to say.

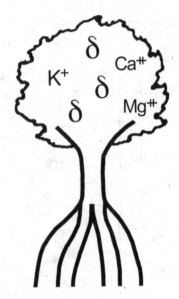

8.3 The Beginner's Advantage—Imagine!

This book ends with quotes from another book. Science is about knowledge, statistics, testing, and hard work, but it would all be in vain if it weren't
for the most important ingredient, imagination. And here, the beginner
brings new vistas unthought of, because each individual scientist gets interested in different things. Here are some thoughts to encourage you to
imagine, beginning with this quote from Albert Einstein: "Imagination is
more important than knowledge. Knowledge is limited; imagination encir-
cles the world."

The following excerpts are from *The Great Whale of Kansas* by Richard
W. Jennings (2001).

For everything that is known, there must have been a time when it wasn't. Every
sliver of knowledge that mankind has accumulated almost certainly was acquired
the hard way, preceded by a slew of goofball, wrong-headed, and false assumptions
that people believed with all their hearts but have since been cast on history's
compost heap. The earth was flat before it was round, a stationary object before it
circled the sun, the center of the universe before winding up where it is today, on
the outer edge. The history of scientific truth is a history of mistakes. . . . In the final
analysis . . . progress depends on our willingness to disregard the facts. (p. 77)

When I look at the facts, . . . I get discouraged. But when I look beyond the facts,
I see possibilities that I never saw before. (p. 133)

From where I sat, those logical assumptions felt like facts. But facts are funny
things. No matter how many you discover, there are always more you know you
should have found. . . . On any given day, the facts we know can be replaced by those
we don't. . . . Honestly, it wouldn't surprise me if after we're dead, we find out we
didn't know anything at all. (p. 137)

... very often situations aren't what they seem. If you keep digging, or simply wait, one set of facts almost always will yield to another ... it's easier to get through life if you have a good imagination. (p. 149)

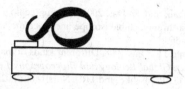

8.4 Chapter Summary

This chapter starts by thinking about the future, and what it would be like to have an isotope scanner for use in field ecology (Section 8.1). We know that we have not yet plumbed the depths of isotope information. The isotope information exists at the atom-by-atom level in nature's compounds and molecules, a level near that of DNA. This likely means there is still a very large and rich store of isotope information awaiting our routine use. We need a scanner to access this detailed isotope information, one that we can bring with us in the field for real-time isotope information.

Section 8.2 returns to the present and asks you to think creatively about science, using isotopes as one part of a tool kit for solving ecological puzzles. Experience shows that many times, chemical markers and approaches other than isotopes are more effective and cheaper for investigating natural systems (Section 8.2). It is good to keep an open mind about the facts and limitations of present-day knowledge and to proceed with imagination (Section 8.3). We who have been lifelong isotope users ("lifers") welcome you beginners who may have a big advantage for the future. That advantage is imagination that brings new thoughts and directions. Getting started with isotopes just takes a little knowledge plus some creative thinking and imagination (see Box 1.1).

Further Reading

Section 8.1

Bergamaschi, P. and G.W. Harris. 1995. Measurements of stable isotope ratios ($^{13}CH_4/^{12}CH_4$, $^{12}CH_3D/^{12}CH_4$) in landfill methane using a tunable diode laser absorption spectrometer. *Global Biogeochemical Cycles* 9:439–447.

Bowling, D.R., S.D. Sargent, B.D. Tanner, and J.R. Ehleringer. 2003. Tunable diode laser absorption spectroscopy for ecosystem-atmosphere CO_2 isotopic exchange studies. *Agricultural and Forest Meteorology* 118:1–19.

Murnick, D.E. and B.J. Peer. 1994. Laser-based analysis of carbon isotope ratios. *Science* 263: 945–947.

www.LGRinc.com. 2005. Website of a commercial company that markets lasers for isotope analysis of CO_2, H_2O and CH_4 in the field.

Section 8.2

Allen, J.A., K.C. Ewel, and J. Jack. 2001. Patterns of natural and anthropogenic disturbance of the mangroves on the Pacific Island of Kosrae. *Wetlands Ecology and Management* 9: 279–289.

Domi, N., J.M. Bouquegneau, and K. Das. 2005. Feeding ecology of five commercial shark species of the Celtic Sea through stable isotope and trace metal analysis. *Marine Environmental Research* 60:551–569.

Ewel, K.C., R.R. Twilley, and J.E. Ong. 1998. Different kinds of mangrove forests provide different goods and services. *Global Ecology and Biogeography Letters* 7:83–94.

Fisk, A.T., S.A. Tittlemier, J.L. Pranschke, and J.R.J. Norstrom. 2002. Using anthropogenic contaminants and stable isotopes to assess the feeding ecology of Greenland sharks. *Ecology* 83:2162–2172.

Fry, B. and K.C. Ewel. 2003. Using stable isotopes in mangrove fisheries research. *Isotopes in Environmental and Health Studies* 39:191–196.

Fry, B. and T.J. Smith III. 2002. Stable isotope studies of red mangroves and filter feeders from the Shark River estuary, Florida. *Bulletin of Marine Science* 70:871–890.

Kitao, M., H. Utsugi, S. Kuramoto, R. Tabuchi, K. Fujimoto, and S. Lihpai. 2003. Light-dependent photosynthetic characteristics indicated by chlorophyll fluorescence in five mangrove species native to Pohnpei Island, Micronesia. *Physiologia Plantarum* 117:376–382.

Lajtha, K. and J.D. Marshall. 1994. Sources of variation in the stable isotopic composition of plants. In K. Lajtha and R.H. Michener (eds.), *Stable Isotopes in Ecology and Environmental Science*. Blackwell, Oxford, UK, pp. 1–21.

Lin, G. and L.d.S.L. Sternberg. 1992. Differences in morphology, carbon isotope ratios, and photosynthesis between scrub and fringe mangroves in Florida, USA. *Aquatic Botany* 42: 303–313.

Newsome, S.D., C.M. del Rio, S. Bearhop and D.L. Phillips. 2007. A niche for isotopic ecology. *Frontiers in Ecology and the Environment* 5:429–436.

Okada, N. and A. Sasaki. 1995. Characteristics of sulfur uptake by mangroves: an isotopic study. *Tropics* 4:201–210.

Pinzon, Z.S., K. Ewel, and F.E. Putz. 2003. Gap formation and forest regeneration in a Micronesian mangrove forest. *Journal of Tropical Ecology* 19:143–153.

Thorrold, S., C. Latkoczy, P.K. Swart, and C.M. Jones. 2001. Natal homing in a marine fish metapopulation. *Science* 291: 297–299.

Section 8.3

Jennings, R.W. 2001. *The Great Whale of Kansas*. Houghton Mifflin, Boston.

Appendix
Important Isotope Equations and Useful Conversions

Important Equations

1. Definition of δ (Section 2.1)

$$\delta^H X = [(R_{SAMPLE}/R_{STANDARD}) - 1]*1000,$$

where X = H, C, N, O, or S, the superscript H gives the respective heavy isotope mass of that element (^{2}H, ^{13}C, ^{15}N, ^{18}O, or ^{34}S), and R is the ratio of the heavy isotope to the light isotope for the element, ^{2}H/^{1}H, ^{13}C/^{12}C, ^{15}N/^{14}N, ^{18}O/^{16}O, or ^{34}S/^{32}S.

International standards listed in Table 2.1 have these $R_{STANDARD}$ values: 0.00015576 for δ^2H (SMOW), 0.0003799 for δ^{17}O (SMOW), 0.0020052 for δ^{18}O (SMOW), 0.01118 for δ^{13}C (VPDB), 0.0036765 for δ^{15}N (AIR), 0.0078772 for δ^{33}S (VCDT), and 0.0441626 for δ^{34}S (VCDT).

2. % Heavy Isotope, Atom % or ^{H}AP (Section 2.1)

The % heavy isotope is also known as atom % of the heavy isotope or ^{H}AP

$$^H AP = 100*(\delta+1000)/[(\delta+1000+(1000/R_{STANDARD})]$$

(Conversion 4 below derives this equation from the δ definition. Also note that this equation is not strictly exact for O and S that have more than two stable isotopes, but the equation still provides an excellent approximation of ^{H}AP for ^{18}O and ^{34}S in almost all cases).

3. Fractionation (Sections 2.1, 4.6, 7.1–7.2, 7.5–7.7, Technical Supplement 2C)

$$\alpha$$

is the fractionation factor given in terms of kinetic (k) rate constants for light (L) and heavy (H) isotope-substituted molecules,

$$\alpha = {}^L k / {}^H k.$$

$$\Delta$$

is the fractionation factor in positive ‰ (permil) units,

$$\Delta = (\alpha - 1) * 1000.$$

Fractionation in All Reactions

(Sections 2.1, 4.6, 7.1, Technical Supplements 2C and 7B.)

Approximate Instantaneous Fractionation: $\delta_{\text{PRODUCT}} = \delta_{\text{SOURCE}} - \Delta.$

Exact Instantaneous Fractionation: $R_{\text{PRODUCT}} = R_{\text{SUBSTRATE}} / \alpha.$

Fractionation in a Closed System

(Sections 7.1–7.2, Technical Supplement 7B.) In a simple forward reaction with one product formed from substrate in a closed system with isotope fractionation Δ or α, there are approximate and exact isotope equations for residual substrate (RS), instantaneous product (IP), and accumulated product (AP), with isotope values expressed relative to initial substrate (INPUT). The isotope compositions are related to the fraction f of substrate converted to product, where f is the fraction reacted.

Approximate Equations

$$\delta_{RS} = \delta_{\text{INPUT}} - \Delta * \ln(1 - f)$$

$$\delta_{IP} = \delta_{\text{INPUT}} - \Delta * [1 + \ln(1 - f)]$$

$$\delta_{AP} = \delta_{\text{INPUT}} + \Delta * ((1 - f)/f) * \ln(1 - f).$$

Exact Equations

$$R_{RS} = R_{\text{INPUT}} * (1 - f)^{1/\alpha - 1}$$

$$R_{IP} = (1/\alpha) * R_{\text{INPUT}} * (1 - f)^{1/\alpha - 1}$$

$$R_{AP} = [R_{\text{INPUT}} - (1 - f) * R_{RS}]/f.$$

Fractionation in an Open System

(Sections 4.6, 7.1, 7.7.) In an open system, product is formed from substrate in a continual manner, with both product and residual substrate exiting the site of reaction.

Approximate Equations

$$\delta_{RS} = \delta_{INPUT} + \Delta * f$$

$$\delta_{PRODUCT} = \delta_{INPUT} - \Delta * (1-f)$$

Exact Equations

$$R_{RS} = R_{INPUT} * (f*\alpha + 1 - f)$$

$$R_{PRODUCT} = R_{INPUT} * (f + (1-f)/\alpha)$$

Fractionation in an Equilibrium Exchange Reaction

(Section 7.6.) Where two substances A and B are involved in an exchange reaction, and substance A becomes relatively enriched in heavy isotopes, the equilibrium fractionation factor for the overall reaction can be given exactly as α_{EQ} or approximately as Δ_{EQ}.

Approximate Equation

$$\Delta_{EQ} = \delta_A - \delta_B$$

Exact Equation

$$\alpha_{EQ} = R_A/R_B = (1000 + \delta_A)/(1000 + \delta_B)$$

4. I Chi Equations for a One-Box Open System Model (Sections 4.3, 4.6, 7.7)

I Chi equations give changes in mass and isotopes between time intervals t and $t+1$. The equations specify gains with isotope mixing, and losses with isotope fractionation.

Gains

Where m terms are masses involved in the mixing

$$m_{t+1} = m_t + m_{GAIN}$$

Approximate Equation for Isotope Mixing during Gains

$$\delta_{t+1} = (\delta_t * m_t + m_{GAIN} * \delta_{GAIN})/m_{t+1}$$

Exact Equations for Isotope Mixing During Gains

The exact equations for δ_{t+1} during mixing gains require converting δ to ^HAP, calculating mixing, then reconverting to δ. With ^HAP calculated from δ,

$$^HAP = 100 * (\delta + 1000)/[(\delta + 1000 + (1000/R_{STANDARD}))]$$

mixing is given by

$$^HAP_{t+1} = \left(\,^HAP_t * m_t + m_{GAIN} * \,^HAP_{GAIN}\right)\big/m_{t+1}$$

To convert $^HAP_{t+1}$ to δ_{t+1}, first solve for atom % for the light isotope, $^LAP = 100 - \,^HAP$, then

$$\delta_{t+1} = \left[\left(\,^HAP/\,^LAP\right)\big/R_{STANDARD} - 1\right]*1000$$

Losses

$$m_{t+1} = m_t - m_{LOSS} \quad \text{and} \quad f = \text{fraction lost from the substrate pool}$$
$$f = m_{LOSS}/m_t = 1 - m_{t+1}/m_t.$$

Approximate Equation for Isotope Fractionation during Loss

$$\delta_{t+1} = \delta_t + \Delta * f$$

Exact Equation for Isotope Fractionation During Loss

$$\delta_{t+1} = (\delta_t + 1000)*(f*\alpha + 1 - f) - 1000$$

5. Simple Two-Source Mixing (Section 5.3)

Two sources have different δ values and mix to produce a sample. The fractional contributions from the sources to the sample are $f_1 + f_2$ where

$$f_1 + f_2 = 1, \quad f_2 = 1 - f_1$$
$$f_1 = (\delta_{SAMPLE} - \delta_{SOURCE2})/(\delta_{SOURCE1} - \delta_{SOURCE2})$$

6. Two-Source Mixing with One or More Spiked Samples (Technical Supplement 6B)

For samples that have been spiked with heavy isotope to very high levels, for example, $\delta^{13}C > 4000$ or $\delta^{15}N > 12{,}000‰$, the δ notation becomes inexact for mixing (see Technical Supplement 6B) and requires conversion to atom % or ^HAP:

$$^HAP = 100 * (\delta + 1000)/[(\delta + 1000 + (1000/R_{STANDARD})]$$

The exact mixing equation for f_1, the fractional contribution of source 1, is always:

$$f_1 = \left({}^{H}AP_{SAMPLE} - {}^{H}AP_{SOURCE2} \right) / \left({}^{H}AP_{SOURCE1} - {}^{H}AP_{SOURCE2} \right)$$

7. Two-Source Mixing by Mass and Weighting (W) Factors (Section 5.4)

$$\delta_{SAMPLE} = (m_1 * W_1 * \delta_{SOURCE1} + m_2 * W_2 * \delta_{SOURCE2}) / (m_1 * W_1 + m_2 * W_2)$$

8. Calculating Source Contributions when Two Sources Have Different Concentrations (Section 5.4)

When two sources combine to form a mixture, isotopes help monitor the source contributions at two levels. At the first level, isotopes budget the fractional contributions by element, for example, $\delta^{15}N$ measurements budget elemental N contributions by the equations given above:

$$f_1 = (\delta_{SAMPLE} - \delta_{SOURCE2}) / (\delta_{SOURCE1} - \delta_{SOURCE2}) \quad \text{and} \quad f_2 = 1 - f_1$$

But when sources have different concentrations of the element being budgeted, mixing calculations extend to a second, more general, level involving total masses being mixed, as is perhaps illustrated best by example. Suppose that $\delta^{15}N$ measurements show that two sources each contribute N equally to a mixed sample ($f_1 = f_2 = 0.5$), but source 2 contains a 2 times higher N concentration. In this case, only half as much mass from source 2 is needed to match the N contribution from source 1. The mixing is 1:1 for nitrogen, but because of the different N concentrations, the mixing is 2:1 by total mass. To calculate the contributions at this more general level of total masses involved, usually weightings (W) are assigned to reflect the different concentrations, for example, with W_1 and W_2 representing different % N values for the sources. The fractional contributions to the total mass (f_{TOTAL}) from the two sources is

$$f_{TOTAL1} = f_1 * W_2 / (f_1 * W_2 + f_2 * W_1) \quad \text{and} \quad f_{TOTAL2} = 1 - f_{TOTAL1}$$

9. Blank Corrections for Contaminants Contributing to a Sample; Keeling Plots (Sections 3.5, 5.7)

Blanks contribute to observed results in many mixing situations such as laboratory analysis. The effects of a blank can be factored out using two-source mixing equations where the contaminating blank is fixed in both isotope value and amount. In this case, the mass balance mixing equations are:

$$m_{OBSERVED} = m_{TRUE} + m_{BLANK}$$

and

$$\delta_{OBSERVED} * m_{OBSERVED} = \delta_{TRUE} * m_{TRUE} + \delta_{BLANK} * m_{BLANK}$$

where the observed δ values are a mixture of true sample and contaminating blank. This can be rearranged to:

$$\delta_{OBSERVED} = \delta_{TRUE} + (\delta_{BLANK} - \delta_{TRUE}) * (m_{BLANK} / m_{TRUE})$$

This rearrangement yields a straight line when laboratory data for different-sized replicates of the same sample is plotted as (x,y) data in the form $(1/mass, \delta)$ so that

$$y - intercept = \delta_{TRUE} \quad and \quad Slope = m_{BLANK} * (\delta_{BLANK} - \delta_{TRUE})$$

The mass of the blank, m_{BLANK}, usually can be obtained from direct measurement of blanks, allowing calculation of the δ_{BLANK} from the regression line:

$$\delta_{BLANK} = \delta_{TRUE} + slope / m_{BLANK}$$

Note: it is often difficult to directly measure the δ_{BLANK} for small samples, but the approach outlined here extrapolates δ_{BLANK}, essentially by observing effects of the blank on actual samples. Finally, with known values for m_{BLANK} and δ_{BLANK}, corrections can be made to all experimental data:

$$\delta_{TRUE} = (\delta_{BLANK} * m_{BLANK} + \delta_{OBSERVED} * m_{OBSERVED}) / (m_{OBSERVED} - m_{BLANK})$$

This approach has also been used in other instances of two-source mixing where the second source is not a contaminant, but just a source fixed in both mass and isotope values, for example, in Keeling plots and for estimating background corrections in sediment cores (Sections 3.5, 5.7).

10. Calculating Trophic Level (TL) based on $\delta^{15}N$ (Problem 10, Chapter 5)

In food webs where the $\delta^{15}N$ values of plants, herbivores and higher-level consumers are measured, the average trophic level (TL) of a consumer can be calculated as

$$TL = 1 + (\delta^{15}N_{CONSUMER} - \delta^{15}N_{PLANT}) / 3$$

using plants as the basal level of the food web, or

$$TL = 2 + (\delta^{15}N_{CONSUMER} - \delta^{15}N_{HERBIVORE}) / 3$$

using herbivores as the basal second level of the food web. The value "3" in the denominator in these equations represents the permil (‰) increase in ^{15}N per trophic level, most recently estimated as 2.2‰ for invertebrates and 3.4‰ for vertebrates (see Chapter 5.5 for references). The "3" used in the denominator of these equations is thus an approximate average value. Plants represent TL 1 in these equations.

11. Calculating Trophic Level (TL) Corrections to $\delta^{13}C$ Food Web Data (Problem 10, Chapter 5)

There is a small average increase in animal $\delta^{13}C$ values with each trophic level. Before using $\delta^{13}C$ to make source assessments, this ^{13}C trophic fractionation should be factored out, and "corrected $\delta^{13}C$" values used for the mixing models:

$$\text{Corrected } \delta^{13}C = \text{Measured Animal } \delta^{13}C - 0.5*(TL-1)$$

The TL is usually estimated as above from $\delta^{15}N$ but can also be estimated from gut content data. The 0.5‰ ^{13}C enrichment factor varies from 0–2‰, with 0.5‰ representing an average value (see Chapter 5.5 for references). Plants represent TL 1 in this equation.

Useful Conversions

Conversion 1

Two samples are measured versus a common standard. What is the true isotope difference between the two samples?

$$\delta_1 = [(R_1/R_0)-1]1000$$
$$\delta_2 = [(R_2/R_0)-1]1000$$
$$\delta_{1,2} = [(R_1/R_2)-1]1000 = ?$$

Rearrange definitions to solve for R_1 and R_2:

$$R_1 = R_0 *(\delta_1 + 1000)/1000$$
$$R_2 = R_0 *(\delta_2 + 1000)/1000$$

Divide R_1 by R_2 and cancel R_0 and 1000 values:

$$R_1/R_2 = (\delta_1 + 1000)/(\delta_2 + 1000)$$

Subtract 1 and multiply by 1000:

$$\delta_{1,2} = \{[(\delta_1 + 1000)/(\delta_2 + 1000)] - 1\}*1000 \quad \text{or}$$
$$\delta_{1,2} = [(\delta_1 - \delta_2)/(\delta_2 + 1000)]*1000$$

for example, $\delta_1 = -6‰$, $\delta_2 = -16‰$, $\delta_{1,2} = 10.16‰$, not 10‰.

Note that this relationship is often given in a slightly modified form when the δ values are expressed as fractions instead of in ‰ units, for example, if $\delta_1 = -0.006$ not $-6‰$ and $\delta_2 = -0.016$ not $-16‰$ then:

$$\delta_{1,2} = (\delta_1 - \delta_2)/(\delta_2 + 1)$$

with the resulting $\delta_{1,2}$ value also expressed as a fraction, 0.01016. In this usage, permil values result when the scientist consciously multiplies by 1000, so that the fractional δ values of -0.006, -0.016, and 0.01016 become permil values of $-6‰$, $-16‰$, and 10.16‰ (*Annual Review of Plant Physiology and Plant Molecular Biology* 40:503–537; 1989)

Conversion 2

A sample is measured versus standard 1. How should you express the isotope value of this sample value versus standard 2, when the values of standards are known?

$$\delta_1 = [(R_{SAMPLE}/R_1) - 1]1000$$
$$\delta_2 = [(R_1/R_2) - 1]1000$$
$$\delta_3 = [(R_{SAMPLE}/R_2) - 1]1000 = ?$$

Rearrange definitions to solve for R_{SAMPLE} and R_2:

$$R_{SAMPLE} = R_1 *(\delta + 1000)/1000$$
$$R_2 = (1000 * R_1)/(\delta_2 + 1000)$$

Divide R_{SAMPLE} by R_2,

$$R_{SAMPLE}/R_2 = R_1 *(\delta_1 + 1000)*(\delta_2 + 1000)/(R_1 *1,000,000)$$

Cancel R_1 values, subtract 1 and multiply by 1000,

$$\delta_3 = [(\delta_1 + 1000)*(\delta_2 + 1000) - 1,000,000]/1000$$

for example, $\delta_1 = -10$, $\delta_2 = -20$, $\delta_3 = -29.8$, not $-30‰$.

Conversion 3

This conversion deals with a common laboratory problem, calibrating a new tank of laboratory gas for use as a standard for δ measurements. To perform the calibration, you use a known reference compound. These known reference compounds are available from agencies such as the National Institute of Standards (NIST) in the United States and the International Atomic Energy Agency (IAEA) in Austria. You measure the known reference compound versus the unknown tank gas, then calculate the isotope ratio (R_1) of the new tank gas. With this value, it is then possible to routinely use the new tank gas as a working standard. But to publish your results, you have to recalculate δ values of samples measured versus this tank gas in terms of the primary international standard, for example, VPDB in the case of carbon isotopes. This problem has two parts, (a) calibration of the unknown tank standard, and (b) recalculation of results versus an international standard.

a. You have an unknown lab standard gas and a reference compound whose δ value has been calibrated versus the international primary reference standard. You measure this reference compound versus the unknown tank gas. What is the isotope ratio R_1 for the tank gas?

$$\delta_1 = [(R_{\text{REFERENCE COMPOUND}}/R_1) - 1]*1000 \quad \text{and}$$
$$\delta_2 = [(R_{\text{REFERENCE COMPOUND}}/R_{\text{INTERNATIONAL REFERENCE}}) - 1]*1000$$
$$R_1 = ?$$

where $R_{\text{INTERNATIONAL REFERENCE}}$ is the known isotope ratio in an international standard reference material, for example, 0.01118 for carbon isotope measurements using the VPDB standard given in Table 2.1.

Rearrange equations for R_1 and $R_{\text{REFERENCE COMPOUND}}$:

$$R_1 = R_{\text{REFERENCE COMPOUND}} *1000/(\delta_1 + 1000)$$
$$R_{\text{REFERENCE COMPOUND}} = R_{\text{INTERNATIONAL REFERENCE}} *(\delta_2 + 1000)/1000$$

substituting for $R_{\text{REFERENCE COMPOUND}}$,

$$R_1 = R_{\text{INTERNATIONAL REFERENCE}} *(\delta_2 + 1000)/(\delta_1 + 1000)$$

For example, $\delta_1 = -10$, $\delta_2 = -20$, $R_{\text{INTERNATIONAL REFERENCE}} = 0.01118$; then R_1 = 0.011067 and the δ value for your tank standard versus the international reference material = -10.10‰. You have now calibrated your unknown tank gas (a tertiary standard) using a known reference material (a secondary standard) that was calibrated originally against an international reference material (the primary standard).

b. And once you know R_1, the ratio value of your previously unknown tank gas, how do you convert δ values measured against this tank gas to δ values referenced to an international standard? For example, you measure a second sample versus this tank gas as $\delta_3 = [(R_{\text{SAMPLE2}}/R_1) - 1]*1000$. What is δ_4, the δ value of this second sample expressed versus the international standard?

So if

$$\delta_3 = [(R_{SAMPLE2}/R_1)-1]*1000$$

$$\delta_4 = [(R_{SAMPLE2}/R_{INTERNATIONAL\ REFERENCE})-1]*1000 = ?$$

Rearrange the δ_3 equation for $R_{SAMPLE2}$:

$$R_{SAMPLE2} = R_1*(\delta_3+1000)/1000$$

and substitute this value in the equation for δ_4:

$$\delta_4 = (\delta_3+1000)*(R_1/R_{INTERNATIONAL\ REFERENCE})-1000$$

For example, if $R_{INTERNATIONAL\ REFERENCE} = 0.01118$, $R_1 = 0.011067$ as determined above, and δ_3 measures –30‰, $\delta_4 = $ –39.8‰. With these calculations, you can now use the newly calibrated tank gas and publish δ_4 values expressed relative to the primary international standard.

Conversion 4

How does one convert from δ to R_{SAMPLE}, ^HF and ^HAP (atom % of the heavy isotope) when $\delta = [(R_{SAMPLE}/R_{STANDARD}) - 1]1000$? And how does one convert back to δ from ^HAP values?

Calculate R_{SAMPLE}, the ratio of heavy-to-light isotope from rearranging the δ definition:

$$R_{SAMPLE} = [(\delta/1000)+1]*R_{STANDARD}$$

(Note: see Table 2.1 for $R_{STANDARD}$ values). To calculate ^HF and ^LF, the respective fractions of heavy and light isotope in the sample, remember that R is the ratio of heavy-to-light isotopes, or:

$$R = {}^HF/{}^LF \quad \text{and that} \quad {}^HF+{}^LF=1, \quad \text{so that} \quad R = {}^HF/(1-{}^HF)$$

Substituting for R_{SAMPLE} and rearranging, one obtains:

$$^HF = (\delta+1000)/[\delta+1000+(1000/R_{STANDARD})]$$

To calculate atom % for the heavy isotope, ^HAP

$$^HAP = 100*{}^HF \quad \text{so that}$$

$$^HAP = 100*(\delta+1000)/[(\delta+1000+(1000/R_{STANDARD})]$$

To convert from ^HAP to δ, solve for atom % for the light isotope,

$$^LAP = 100 - {}^HAP, \quad \text{then}$$

$$\delta = [({}^HAP/{}^LAP)/R_{STANDARD} - 1] * 1000$$

Conversion 5

You find ratio (R) values for a standard, but want to know the fractional abundances or F values. For example, for an oxygen standard you find that $^{17}R = 0.000402$ and $^{18}R = 0.0020052$, what are F values?

You can write the following equalities.

$$^{17}R = 0.000402 = {}^{17}F/{}^{16}F$$

$$^{18}R = 0.0020052 = {}^{18}F/{}^{16}F$$

$$^{16}F + {}^{17}F + {}^{18}F = 1$$

Simplify the equations by removing the R terms:

$$^{17}F = 0.000402 * {}^{16}F$$

$$^{18}F = 0.0020052 * {}^{16}F$$

$$^{16}F + 0.000402 * {}^{16}F + 0.0020052 * {}^{16}F = 1 \quad \text{so that}$$

$$^{16}F = 1/(1 + 0.000402 + 0.0020052)$$

Solving the last equation for ^{16}F then substituting the ^{16}F value in the two previous equations, one obtains: $^{16}F = 0.997598581$, $^{17}F = 0.000401035$, and $^{18}F = 0.002000385$.

Index